Comparative Physiology:
Water, Ions and Fluid Mechanics

Comparative Physiology: Water, Ions and Fluid Mechanics

Edited by

K. SCHMIDT-NIELSEN
James B. Duke Professor of Physiology,
Department of Zoology, Duke University

L. BOLIS
Professor and Chairman of General Physiology,
University of Messina

S. H. P. MADDRELL
Principal Scientific Officer,
Agricultural Research Council Unit of Invertebrate Chemistry and Physiology,
Department of Zoology, University of Cambridge
Fellow and College Lecturer, Gonville and Caius College, Cambridge

CAMBRIDGE UNIVERSITY PRESS
Cambridge
London   New York   Melbourne

Published by the Syndics of the Cambridge University Press
The Pitt Building, Trumpington Street, Cambridge CB2 1RP
Bentley House, 200 Euston Road, London NW1 2DB
32 East 57th Street, New York, NY 10022, USA
296 Beaconsfield Parade, Middle Park, Melbourne 3206, Australia

First published 1978

Printed in Great Britain at the
University Press, Cambridge

Library of Congress Cataloguing in Publication Data
International Conference on Comparative Physiology,

3d, Crans, Switzerland, 1976.

Comparative physiology.

Includes index.

1. Body fluids – Congresses. 2. Biological transport –
Congresses. 3. Physiology, Comparative – Congresses.
I. Schmidt-Nielsen, Knut, 1915–   II. Bolis, Liana.
III. Maddrell, S. H. P. IV. Title.
[DNLM: 1. Physiology, Comparative. 2. Ions –
Metabolism. 3. Water – Metabolism. QP31.2 C737]
QP90.5.157 1976   591.1'9212   77–7320
ISBN 0 521 21696 6

Contents

Contributors

T. ABRAHAMSSON: Department of Zoophysiology, University of Göteborg, Fack, S-40033 Göteborg, Sweden

O. S. BAMFORD: Department of Zoophysiology, University of Aarhus, DK-800 Aarhus, Denmark

T. J. BRADLEY: Department of Zoology, University of British Columbia, Vancouver, British Columbia, V6T 1W5, Canada

D. J. ELLAR: Department of Biochemistry, University of Cambridge, Tennis Court Road, Cambridge CB2 1QW, England

C. P. ELLINGTON: Department of Zoology, University of Cambridge, Downing Street, Cambridge CB2 3EJ, England

R. FÄNGE: Department of Zoophysiology, University of Göteborg, Fack, S-40033 Göteborg, Sweden

J. FISCHBARG: Eye Research Division, College of Physicians and Surgeons, Columbia University, New York, NY 10032, USA

S. E. FRANKLIN: School of Biological Sciences, University of Sydney, Sydney 2006, Australia

E. GONZÁLEZ: Centro de Biofísica y Bioquímica, Instituto Venezolano de Investigaciones Cientificas (IVIC), PO Box 1827, Caracas, Venezuela

A. E. HILL: Physiological Laboratory, University of Cambridge, Downing Street, Cambridge CB2 3EG, England

K. JOHANSEN: Department of Zoophysiology, University of Aarhus, DK-800 Aarhus, Denmark

H. D. JONES: Department of Zoology, University of Manchester, Manchester M13 9PL, England

T. KIRCHHAUSEN: Centro de Biofísica y Bioquímica, Instituto Venezolano de Investigaciones Cientificas (IVIC), PO Box 1827, Caracas, Venezuela

W. KNÜLLE: Institut für Tierphysiologie und Angewandte Zoologie, Freie Universität Berlin, Haderslebener Strasse 9, 1000 Berlin 41, West Berlin, West Germany

H. LINARES: Centro de Biofísica y Bioquímica, Instituto Venezolano de Investigaciones Cientificas (IVIC), PO Box 1827, Caracas, Venezuela

A. P. M. LOCKWOOD: Department of Oceanography, University of Southampton, Southampton SO9 5NH, England

G. LYKKEBOE: Department of Animal Physiology, University of Nairobi, PO Box 30197, Nairobi, Kenya

x   Contributors

J. MACHIN: Department of Zoology, University of Toronto, Toronto, Ontario M5S 1A1, Canada

S. H. P. MADDRELL: Department of Zoology, University of Cambridge, Downing Street, Cambridge CB2 3EJ, England

J. MAETZ: Groupe de Biologie Marine, Département de Biologie du Commissariat à l'Energie Atomique, Station Zoologique, 06230 Villefranche, France

G. M. O. MALOIY: Department of Animal Physiology, University of Nairobi, PO Box 30197, Nairobi, Kenya

S. NILSSON: Department of Zoophysiology, University of Göteborg, Fack, S-40033 Göteborg, Sweden

J. NOBLE-NESBITT: School of Biological Sciences, University of East Anglia, University Plain, Norwich NR4 7TJ, England

M. J. O'DONNELL: Department of Zoology, University of Toronto, Toronto, Ontario M5S 1A1, Canada

M. PEAKER: Agricultural Research Council Institute of Animal Physiology, Babraham, Cambridge CB2 4AT, England

T. J. PEDLEY: Department of Applied Mathematics and Theoretical Physics, Silver Street, Cambridge CB3 9EW, England

J. E. PHILLIPS: Department of Zoology, University of British Columbia, Vancouver, British Columbia, Canada

G. DE RENZIS: Groupe de Biologie Marine, Département de Biologie du Commissariat à l'Energie Atomique, Station Zoologique, 06230 Villefranche, France

D. RUDOLPH: Institut für Tierphysiologie und Angewandte Zoologie, Freie Universität Berlin, Haderslebener Strasse 9, 1000 Berlin 41, West Berlin, West Germany

R. C. SCHROTER: Physiological Flow Studies Unit, Imperial College, Prince Consort Road, London SW7 2AZ, England

E. SKADHAUGE: Institute of Medical Physiology A, The Panum Institute, Blegdamsvej 3, DK-2200 Copenhagen N, Denmark

M. A. SLEIGH: Department of Biology, University of Southampton, Southampton SO9 3TU, England

B. TEINSONGRUSMEE: College of Fisheries, Kasetsart University, Bangkok, Thailand

G. W. WHARTON: Acarology Laboratory, Ohio State University, 484 West 12th Avenue, Columbus, Ohio 43210, USA

G. WHITTEMBURY: Centro de Biofísica y Bioquímica, Instituto Venezolano de Investigaciones Cientificas (IVIC), PO Box 1827, Caracas, Venezuela

T. ZEUTHEN: Physiological Laboratory, University of Cambridge, Downing Street, Cambridge CB2 3EG, England

Preface

This book contains the papers presented at the International Conference on Comparative Physiology held at Crans-sur-Sierre in September 1976.

One of the strengths of the comparative approach to physiology is that one can find out from a range of organisms what is fundamental about solving any particular physiological problem. This is often shown most clearly when the organisms selected are those that are faced with the most extreme physiological stresses. For example, because of their small size terrestrial arthropods face daunting problems of water economy; some of them achieve the impressive feat of absorbing water vapour from the air. Chapters in the first part of this book describe the fascinating ways in which this is achieved by different organisms. We were fortunate enough to have papers from all the leading workers in this new and intriguing field. Other articles in the first section describe recent advances in the thorny problem of how water movements across epithelia are generated by solute movements.

The theme of the second section of the book is animals under the stress of living in ionically extreme or 'unbalanced' environments. Included are accounts of how bacteria, crustaceans, insects, fishes and birds cope with unusual difficulties in ionic regulation.

In the final section we have, as in our previous conferences, brought together biologists and physicists to illuminate a particular topic. This time the spotlight was turned on the role of fluid mechanics in biology. The biologists described various sorts of fluid flows occurring both inside and around the outside of organisms and covered such topics as ciliary and flagellar beating, hydrostatic skeletons and insect flight. These papers were most satisfyingly complemented by more theoretical ones on the underlying hydrodynamics and aerodynamics involved.

SIMON MADDRELL

xi

Part 1

Water transport and uptake

(A) Mechanisms of epithelial fluid transport

1. Intracellular gradients in electrical potential and ion activity in absorptive epithelia

By T. ZEUTHEN

Transport of isotonic solutions across epithelial cell layers is determined by the tissue permeabilities and the distribution of chemical and electrical potentials within the various tissue compartments, among these the intracellular compartment. However, measurements of the intracellular electrical potential by means of microelectrodes have been associated with some difficulties. In each species a spectrum of values is usually obtained, for example in the *Necturus* gallbladder values ranging from -40 mV to -80 mV have been reported (Frömter, 1972; Reuss & Finn, 1975) and in the rabbit ileum, values from -15 mV to -55 mV (Rose & Schultz, 1971; see also the review by Petersen, 1976 on the electrophysiology of glands). It is difficult to see how different cells in the same tissue could have different electrical potentials when the cells are known to be coupled electrically via low impedance pathways (Loewenstein, Nakas & Socolar, 1967). Some authors now assign the average value to the intracellular electrical potential of the unimpaled cell, whereas others argue that the lower values are caused by leaky penetrations. Furthermore, some authors systematically record lower values than others. The potentials which have been obtained, and the different criteria applied, result in different electrical potentials being assigned to the unperturbed cells. For example: in rabbit gallbladder, -45 mV was reported by Frizzell, Dugas & Schultz (1975) but -70 mV by Van Os (1975); in tortoise small intestine, -8 mV by Wright (1966) but -40 mV by Gilles-Baillien & Schoffeniels (1965); and in rabbit small intestine, -35 mV by Rose & Schultz (1971) but -10 mV by Field & Curran (quoted by Schultz & Curran, 1968).

The central problem is therefore the choice of criteria for a valid recording. These criteria are based on the assumption that the intracellular electrical potential is constant throughout the cytoplasm, and can be formulated as follows: (1) The change in recorded electrical potential on impalement should be abrupt. This abruptness of the potential step on entry has never been quantitatively described in epithelial cells, only in symmetrical cells like *Amphiuma* red cells and Ehrlich ascites tumour cells

4   T. Zeuthen

(Lassen *et al.*, 1971, 1974). (2) The intracellularly recorded electrical potential should remain constant for a given time. To this it is important to add that the potential should remain constant at the value recorded immediately (< 2 ms) after impalement of the membrane. If the electrical potential should plateau at a more negative value than that measured at penetration, this does not mean that the new value is the result of the gradual completion of a seal between the membrane and the electrode, in which case the value at the plateau should correspond to the value recorded at impalement. In any case the potentials recorded from epithelial cells are rarely constant, but show superimposed drifts towards more negative, together with both stepwise and oscillatory, potential changes (Frömter, 1972; Cremaschi, Henin & Ferroni, 1974, fig. 3; Frizzell, Dugas & Schultz, 1975, fig. 2).

One may therefore conclude either that microelectrodes are a poor tool for recording intracellular electrical potentials or that the measurements are interpreted using wrong assumptions. The purpose of this paper is to suggest that the latter applies, and to show that there exists an intracellular gradient of electrical (and chemical) potential. First, I shall summarize the results obtained in the rabbit ileum *in vivo* (Zeuthen & Monge, 1975) and in the *Necturus* gallbladder *in vitro*. These last results have been presented in abstract (Zeuthen, 1976*a,b*), and are currently in preparation. Secondly, I shall argue that microelectrode penetration is a good experimental approach in so far as no known electrode artefact could have caused the recordings of the gradients discussed below. Thirdly, I shall approach the problem another way and discuss, on a theoretical basis, mechanisms which could cause intracellular gradients.

The possibility of intracellular gradients has been discussed by Lindemann & Pring (1969), and described in frog skin and toad bladder (Chowdhury & Snell, 1965, 1966).

Summary of results

Electrodes

Single-barrelled glass microelectrodes (Frank & Becker, 1964) were used for recording the electrical potential. When ion activity was concurrently recorded, double-barrelled electrodes were employed. On barrel contained the ion-selective membrane and recorded a potential related to the chemical potential of the ion. The other barrel (the reference barrel) was filled with a salt solution and recorded the electrical potential. All measurements were made relative to a large external reference electrode. By subtraction, the chemical potential, and therefore the activity, could be obtained. The Na^+-sensitive electrodes were constructed as described by Zeuthen (1975), Cl^-- and K^+-sensitive electrodes as described by Zeuthen, Hiam & Silver

(1974). All types of electrodes had a total tip diameter of less than 1 μm and a taper of the shank from 1:10 to 1:20. Hydration, and therefore shunting through the glass walls of the electrodes recording the electrical potential were minimized by using the electrodes within 2 h of manufacture (Tasaki *et al.*, 1968). The influence of the chosen salt solution in the reference barrel on the recorded electrical potentials and ion activities was assessed by comparing the results obtained when the salt solution was 2 M KCl, 2 M Na_2SO_4, or 2 M NH_4NO_3.

Rabbit lower ileum *in vivo*
Mechanics of penetration

The absorptive cells of the rabbit ileum are about 40 μm long and 5 μm broad and packed in a columnar fashion. Because of the villous structure of the epithelium, the route of the electrode advance will be at an angle to the mucosal surface, and several cells will be penetrated when the electrode is advanced from the mucosal solution into the subserosa or in the reverse direction (Fig. 1).

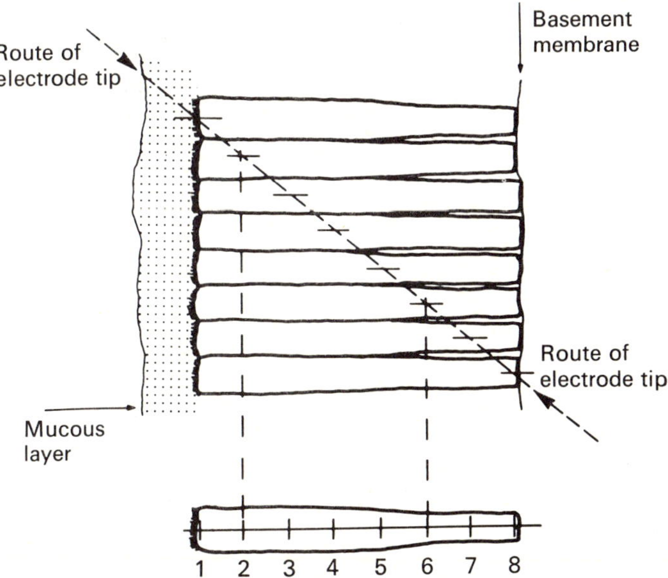

Fig. 1. The orientation of an electrode track within the epithelial cell layer of the rabbit ileum. From the number of cells impaled during an excursion, as assessed from the number of steps in the electrical profile (Fig. 2), and from the magnitude of the electrical potential of each step, the profile in an individual cell can be constructed; if the number of steps were 8, each new step would correspond to an additional depth of about ⅛ of the total cell length.

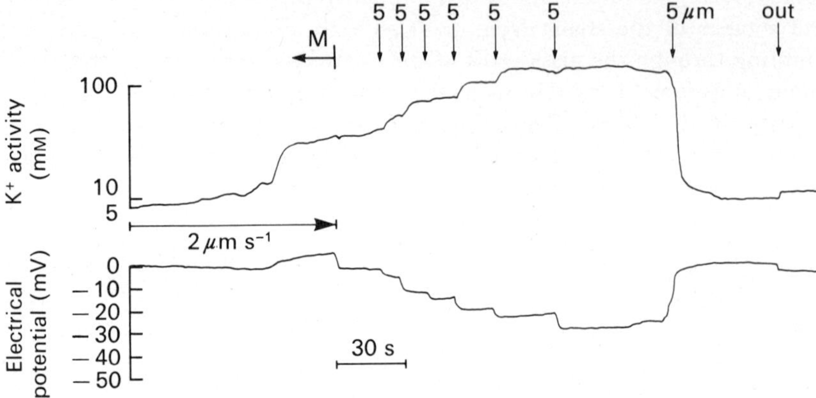

Fig. 2. Simultaneously recorded intracellular profiles of electrical potential and K$^+$ activity in the rabbit ileum. The electrode was first advanced continuously through the mucous layer (M) and then stepwise by 5 μm (arrows) through the cells towards the serosum. Prior to penetrating the mucosal membrane the electrode recorded increasing K$^+$ activity and a slightly positive electrical potential. Part of this could be an artefact caused by accumulation of the salt solution from the reference barrel in the space between the tip and the mucus and the mucosal membrane. However, the intracellularly recorded K$^+$ activities were largely independent of whether the salt solution in the reference barrel was 2 M KCl or 2 M NaCl.

Electrical profile

When electrodes were advanced from the mucosal solution (i.e. Krebs' solution), a staircase-type of profile of the electrical potential was obtained; each step being increasingly more negative (Fig. 2). When the electrode was advanced from the serosal side, a profile with decreasing negative steps was obtained. By iontophoretic staining each individual step could be identified as being recorded intracellularly from one cell; adjacent steps being recorded from adjacent cells. From the number of steps recorded the electrical profile inside each individual cell could be constructed (Fig. 3). If the number of steps in a recording was n, two consecutive electrical potentials must have been recorded from points spaced by a length equivalent to $1/(n-1)$ of the total cell length. The profile thus obtained was linear, ranging from -6.5 mV in the mucosal end to about -38.7 mV in the serosal end. With a cell height of 45 μm, this corresponds to a gradient of 0.85 mV μm^{-1}.

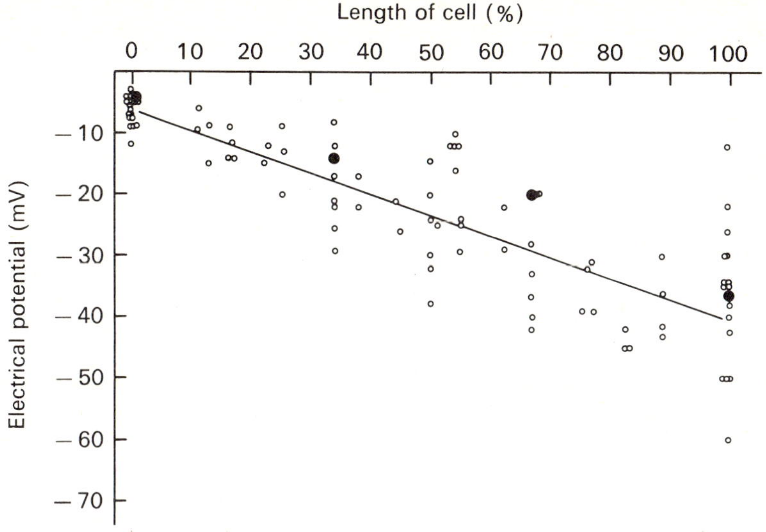

Fig. 3. The intracellular electrical potential as a function of the length of the cell. 100% corresponds to the serosal end. The profile was constructed as described in Fig. 1. The 96 points comprise 17 penetrations from 10 animals. Four of the penetrations were commenced from the serosal side, an example of which is shown by the filled circles. The linear regression time has a slope of -0.85 mV μm^{-1} (upper confidence limit, -0.95 mV μm^{-1}; lower confidence limit, -0.73 mV μm^{-1}) if the cell height was taken as 45 μm.

Ion activities

The K$^+$ activity also varied with depth below the mucosal membrane (Fig. 2). Each stepwise increase in electrical potential achieved by the advance of the electrode was associated with a stepwise increase in the K$^+$ activity. Thus the K$^+$ activity varied from 45 mM in the mucosal end to about 160 mM in the serosal end. In some preliminary experiments, the Cl$^-$ activity was found to vary from about 80 mM in the mucosal end to about 10 mM in the serosal end of the cell.

Necturus gallbladder *in vitro*
Mechanics of penetration

The mechanical conditions during penetration of the absorbtive cells of this gallbladder differ from those in the epithelial cells of rabbit ileum. The gallbladder cells are large, about 20 μm wide and about 40 μm deep, depending on the degree of stretching. Mounted in an Ussing-chamber,

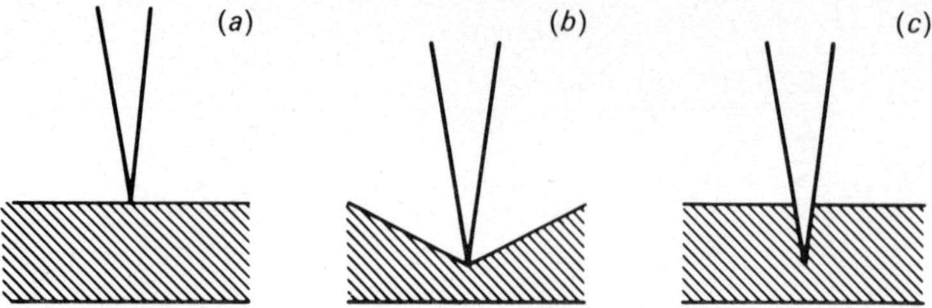

Fig. 4. A schematic presentation of the mechanical problems of penetrating into an epithelial cell layer with a microelectrode. The cell layer is represented by the two horizontal lines indicating the mucosal and the serosal membranes. (a) The electrode was advanced until it just touched the mucosal membrane, as indicated by a small change in tip potential and a 10% change in electrode impedance. The electrode was then advanced in either of two ways: (1) the piezo-drive (see text) was activated, in which case the tip of the electrode was advanced 3 μm through the undimpled mucosal membrane; (2) the advance was continued by the hand-driven micromanipulator only, in which case the mucosal membrane was indented (about 40 μm) prior to penetration (b); when penetrated the tip would be positioned deep inside the cell (c).

they form a flat layer. Thus, if the track of penetration were perpendicular to the mucosal membrane, the electrode usually remained inside one cell only. When single-barrelled electrodes were advanced from the mucosal side in steps of about 3 μm by a hand-driven micromanipulator, the mucosal membrane was indented prior to penetration (Fig. 4b). The amount of dimpling was independent of the tip diameter and averaged 37 μm for electrodes of impedance between 5 and 60 M ohm (measured in saline). After being penetrated, the tissue would return to its original shape, in effect moving the tip deep into the tissue (Fig. 4c). Thus a further advance of only 14 μm caused the tips of the electrodes to penetrate the basal serosal membrane, again independently of the impedance of the electrode. The dimpling of the mucosal membrane could be avoided by attaching the electrode to a piezo-electrical crystal which was mounted on the hand-driven micromanipulator. When the electrode, after being advanced through the mucosal solution, just touched the mucosal membrane, the piezo-electrical crystal was activated. This caused the electrode to be advanced, within less than 0.1 ms, a distance of 3 μm into the undimpled cell. From this position on a further advance of 43 μm by the hand-driven micromanipulator was needed before the tip of the electrode penetrated the basal serosal membrane.

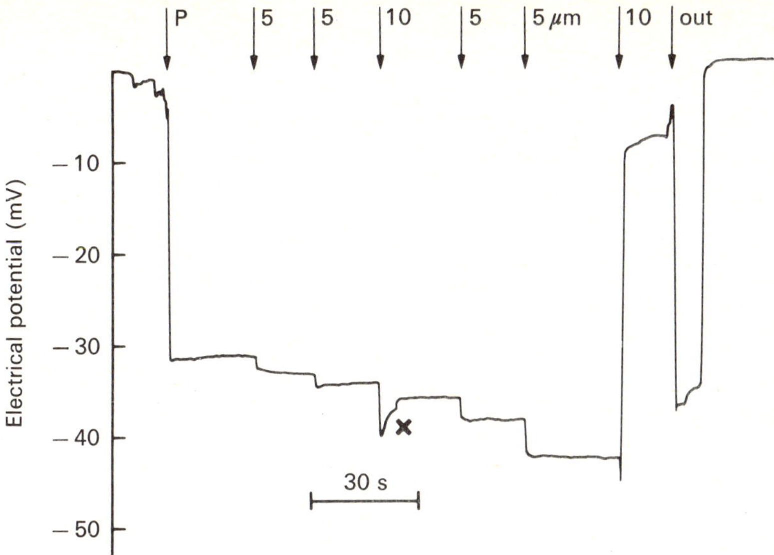

Fig. 5. A recording of the profile of the electrical potential across the epithelial cell of the *Necturus* gallbladder when the tissue was bathed in physiological saline. To avoid dimpling the mucosal membrane was penetrated (P) by advancing the electrode by a piezo-electrical micromanipulator. The advance of the electrode through the cell layer was continued by the hand-driven micromanipulator in steps of 5–10 µm. After a total advance of 40 µm, the electrode entered the subserosum. The electrode was pulled back through the cell into the mucosal solution at 'out'. The peak (x) was not typical.

Electrical profile

When the tissue was bathed in physiological saline and when the mucosal membrane was penetrated by means of the piezo-electrical crystal, then the electrical potential difference across the mucosal membrane was recorded as -29 ± 1.5 mV (22 observations, 9 animals $\pm$s.e.m.). A linear gradient of -0.6 mV μm^{-1} was recorded during the subsequent advance through the cytoplasm, and an average electrical potential difference of about -52 ± 2.5 mV was recorded across the serosal membrane (Fig. 5). When the electrode was advanced by hand only, and the mucosal membrane was dimpled, the electrode recorded during penetration first a fast linear increase in the electrical potential up to -26 ± 1.2 mV (24 observations, 4 animals) at a rate of 84 ± 15.8 V s^{-1} followed by a slower increase up to about -50 mV which lasted about 0.5 s. The first linear increase could be identified as the potential difference across the mucosal membrane, because it was similar to the one found when only this membrane was penetrated by means of the piezo-drive. On further advance of the microelectrode towards the serosal membrane, a potential of 54 ± 1.8 mV was recorded across the serosal membrane.

If the electrodes were advanced from the serosal side, the electrical potential difference across the serosal membrane was recorded as -53 ± 4.3 mV; these values were similar to those found by penetration from the mucosal side. Also, when two electrodes were placed at different depths below the mucosal membrane of the same cell they recorded different electrical potentials; measured in this way, the electrical potential gradient in the cytoplasm was calculated as -0.8 ± 0.16 mV μm^{-1}. Finally, when Na$^+$ in the bathing solutions was exchanged for K$^+$, the intracellular electrical potential became roughly constant at about -5 mV, and only towards the last quarter of the cell at the serosal end was the potential recorded as -8 mV.

Ion activities

In *Necturus* gallbladder the ion activities also depended upon the depth of recording below the mucosal membrane. For each stepwise increase in negative electrical potential obtained by advancing the electrode, a decrease in Na$^+$ or Cl$^-$ activity or increase in K$^+$ activity was recorded. The ion activities obtained with the tissue bathed in physiological saline are summarized in Fig. 6. Two groups of animals were used: animals kept at 5 °C had a K$^+$ activity which varied from about 60 mM in the mucosal end to between 107 and 150 mM in the serosal end depending on the choice of the filling solution in the reference barrel. The Cl$^-$ activity varied from about 70 mM in the mucosal end to between 33 and 40 mM in the serosal end. The electrochemical potentials of K$^+$ and Cl$^-$ were therefore approximately constant within the cytoplasm, for the chemical potential gradient, as calculated from the Nernst equation, was balanced by the gradient of electrical potential. This ranged from -28 mV in the mucosal end to -49 mV in the serosal end in these animals. However, both ions were too concentrated to be in equilibrium with the outer solutions. The Na$^+$ activity varied from about 47 mM in the mucosal end to about 21 mM in the serosal end. Animals stored at 15 °C, as compared to 5 °C, had lower intracellular Na$^+$ activities, ranging from 46 mM in the mucosal end to 13 mM in the serosal end, but higher K$^+$ activities, from 64 mM in the serosal end, and a lower Cl$^-$ activity, about 26 mM in the serosal end. The intracellular electrical potentials were less positive, being about -56 mV in the serosal end.

Thus the Na$^+$ activity was not distributed at equilibrium within the cytoplasm as both the chemical and the electrical gradients are such as to drive Na$^+$ towards the serosal side. It can be estimated that the intracellular

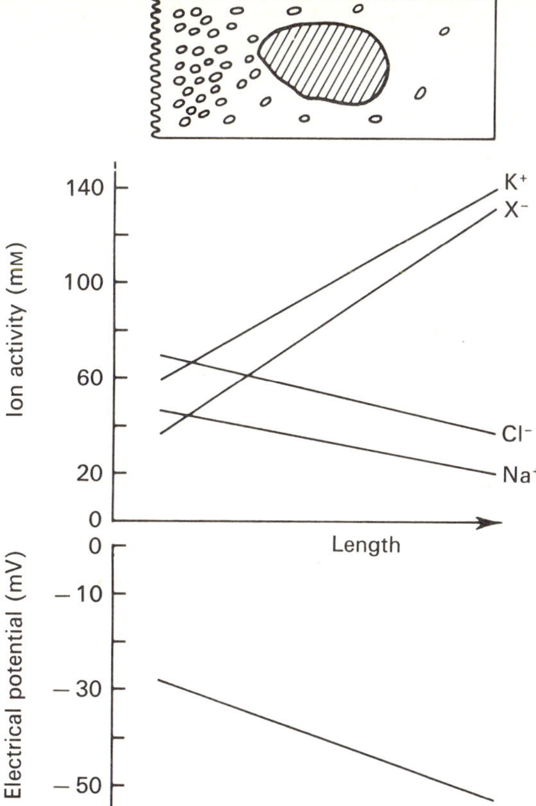

Fig. 6. Schematic presentation of the distribution of ion activities inside the epithelial cells of *Necturus* gallbladder, as recorded by ion-selective microelectrodes. The distribution of mitochondria (small ovals) and the nucleus (shaded) are indicated in the model cell above. The intracellular gradient of electrical potential is shown below. In order to fullfil the requirement of electroneutrality, there must exist a distribution of an anion (X^-) which has not been measured. The activity values are mean values obtained from animals kept at 5 and 15 °C, respectively.

flux of Na^+, calculated from the Nernst–Planck flux equation by using the measured values of the electrical and chemical potential gradients, can equal the transmural Na^+ flux (as assessed from the transmural isotonic fluid transport 5–20 μl cm^{-2} h^{-1}) only if the mobility of Na^+ in the cytoplasm was taken as less than $^1/_{10}$ of the mobility of Na^+ in saline. When ouabain (5×10^{-4} M) was added to the serosal solution, the intracellular distribution of Na^+ activity approached electrochemical equilibrium with the electrical potential gradient, within 30 min, at which time the values ranged from 80 mM in the mucosal end to about 160 mM in the serosal end.

The gradient of electrical potential was rather slowly affected by the presence of ouabain, more slowly than the gradient of Na^+ activity; in the initial 30 min period the electrical gradient and the electrical potential in the serosal end were only reduced by about 10%, and were reduced by 90% after 3 h.

When Na^+ was replaced by K^+ in the bathing solutions, the intracellular Na^+ activity was reduced to about 1 mM, and the K^+ activity (about 140 mM) and the Cl^- activity (about 90 mM) were now in electrochemical equilibrium with the bathing solutions and within the cytoplasm.

If the mucosal solution was acidified by bubbling with CO_2, the intracellular gradients of electrical and chemical potential, were abolished within 20 s.

In order to attain electroneutrality within the cell there has to exist a charge distribution (X^-) caused by an anion which was not measured (Fig. 6). Three points can be deduced about X^-: (1) Animals kept at 15 °C exhibited a steeper gradient of X^-, from 22 mM in the mucosal end to about 160 mM in the serosal end, compared with animals kept at 5 °C; here X^- varied from between 22 and 37 mM in the mucosal end to between 107 and 142 mM in the serosal end. (2) X^- become roughly constant at about 62 mM (between 50 and 70 mM) where Na^+ was exchanged for K^+ in the bathing solutions. (3) X^- was greatly reduced when the tissue was acidified with CO_2.

Specific resistivity of the cytoplasm

Current pulses were passed from an intracellular microelectrode to a large extracellular electrode. Two other microelectrodes were placed in the same cell, at a distance of 1 μm and 3–10 μm, respectively, from the tip of the current-injecting microelectrode. The difference in the evoked potentials between the two tips of the potential-recording electrodes was then a measure of the resistivity of the cytoplasm. Keeping the geometry of the electrode assembly constant and comparing the difference in evoked potentials obtained in a solution of known resistivity, a value for the resistivity could be obtained. When the tissue was bathed in physiological saline the cytoplasmic resistivity was 10 times that of saline in the (mucosal) first sixth of the cell and 5 times that of saline in the remainder. This compares well with the microanatomical findings of a high density of mucus-filled vesicles just below the mucosal membrane (see e.g. Zeuthen, 1977, fig. 1). The ratio ($\rho K^+/\rho Na^+$) between the average resistivities of the cytoplasm of cells bathed either in a Na^+-rich saline or a saline in which Na^+ was replaced by K^+, was measured as 0.61 ± 0.035 (467 observations, 6 animals, $\pm$s.e.m.). This ratio could be compared with the value that would be expected from the measured ion activities. If all ions (Na^+, K^+, Cl^- and X^-) were assumed to have equal mobilities, the ratio should be 0.88. Only if the mobilities of Na^+, and of the ions associated with X^-, were assumed to be less than one-tenth of those of K^+ and Cl^-, was the ratio calculated as 0.63–0.65. The assumption of a low mobility of Na^+ as compared with those of Cl^- and K^+ was supported by the finding that the cytoplasmic resistivity was increased in cells into which Na^+ was injected by iontophoretic or diffusional means.

Possibility of electrode artefacts

Four types of electrode artefacts may be expected with the use of glass micropipette electrodes for recording intracellular electrical potentials:

(a) incomplete sealing of the penetrated membrane around the microelectrode;

(b) lowered electrical resistance of the electrode wall due to hydration;

(c) leaking of the filling solution out of the electrode;

(d) changes in tip potentials.

It is not likely that the gradients are caused by either a leakage current or a diffusional flux along the shaft of the electrode resulting from an incomplete membrane–electrode seal. There are five reasons for this conclusion: (1) The electrical gradient was independent of the orientation of the electrode track. Thus a lower electrical potential difference across the mucosal membrane (as compared with that across the serosal membrane) was recorded whether the cell was penetrated from the mucosal side or from the serosal side. (2) Two electrodes placed at different depths below the mucosal membrane recorded simultaneously two different electrical potentials. Thus an intracellular electrical gradient exists in the penetrated cell. Now the leakage current required to produce the observed gradient of electrical potential would be more than 10^{-7} A cell^{-1}, and this current could only be supplied by the rest of the cells for a few seconds: e.g. in the *Necturus* gallbladder preparation, the apparent mucosal area of the preparation was 0.2 cm^2. With a tortuosity of 10 of the lateral serosal membranes and a membrane capacity of 1 F cm^{-2} polarized to 50 mV the total charge in the tissue would amount to 5×10^{-7} Q, equivalent to a current of 10^{-7} A lasting 5 s. Furthermore, this current would carry sufficient CA^{2+} into the cell to uncouple the cell electrically from the rest of the epithelium (Loewenstein *et al.*, 1967). As a result the impaled cell would completely depolarize much faster than is calculated above because the potential of this cell alone would be measured. Any contribution to the leakage current from an electrogenic Na^+ pump can be excluded, as most of the electrical gradient remained when Na^+ transport was stopped by adding oubain to the serosal solution. (3) Assuming that the electrical potential was constant (e.g. -60 mV) throughout the cytoplasm of the unperturbed cell, then the electrode should record this potential instantaneously on penetrating the mucosal membrane. Depending on the capacitance of the tissue and the resistance of an eventual leak, the cell would subsequently depolarize. In *Amphiuma* red cells and Ehrlich ascites tumour cells, this depolarization time was of the order of several ms (Lassen *et al.*, 1971, 1974), amply within the resolution time of the recording systems employed in this study (< 50 μs). (4) If a leak existed, one would expect the average K^+ activity to be lower and the average Cl^- activity to be higher than the values obtained from studies of intact cells. However, the average activities obtained from this study were equal to or smaller in the case of Cl^-, and equal to or larger in the case of K^+, than the values obtained by other methods: e.g. rabbit small intestine (Schultz, Fuisz & Curran, 1966; Frizzell *et al.*, 1973), and gallbladder (Diamond, 1962; Cremaschi, Smith & Wooding, 1973; Cremaschi *et al.*, 1974; Frizzell *et al.*, 1975). The average value of the Na^+ activity obtained with the electrodes (30 mM) was about half of the concentrations found by other workers cited above. Employing single-barrelled ion-selective microelectrodes in the bullfrog small intestine, Lee & Armstrong (1972) obtained an average Na^+ activity of 14 mM and a K^+ activity of 85 mM. With the double-barrelled electrodes described above (Zeuthen *et al.*, 1974), Henriques de Jesus, Ellory & Smith (1974) obtained Cl^- activities in the rabbit ileum *in vitro* between 20 mM and 80 mM. (5) The recorded Cl^- and K^+ activity changed simultaneously and momentarily with the change in the recorded electrical potentials at each stepwise advance through the cytoplasm (Fig. 2). If a leak were involved, and this leak was gradually diminished as the electrode gradually advanced into the cell, then the extrusion of excess Cl^- or the inclusion of K^+ would be a much slower process than the repolarization of the cell, as it would involve ion transport across the membrane (this argument was suggested by Dr Farmanfarmaian).

The electrodes used for recording the electrical potential were filled and used within 2 h, during which period the glass wall was considered sufficiently insulating. This could be assessed by measuring the impedance of an electrode when its tip was advanced 1–2 μm into insulating silicone grease; hereby the impendance was increased about 100-fold. As no difference in the recorded electrical potentials was found when using different salts for the internal filling solution electrode artefacts (*b*) and (*c*) could be ruled out.

In order to argue that the gradients are caused by a gradual change in tip potential one must assume the existence of an intracellular gradient of some substance capable of affecting the tip. The substance should largely disappear when Na^+ is replaced with K^+ in the external media. The substance cannot be either Na^+, K^+, Cl^- or HCO_3^- as the tip potential was tested for any influence from these ions prior to experiments. In any case the existence of such a substance would not explain the direction of the measured Na^+ gradient.

Some mechanisms which could or could not contribute to the intracellular gradients

Current

The possibility that an electrical current could cause the gradients can be ruled out. The current needed to produce the observed gradient of electrical potential would have to be as large as 10^{-2} A cm^{-2}, even if the cytoplasmic resistivity were 10 times larger than that of saline, which is about 100 ohm cm. This current is more than three orders of magnitude larger than the current that could result from the isotonic NaCl transport across the epithelium. A rate of transport of about 15 μl cm^{-2} h^{-1} of 120 mM NaCl corresponds to a Na^+ flux of about 5×10^{-10} mol cm^{-2} s^{-1} or a current of about 5×10^{-5} A cm^{-2}. Furthermore, with the tissue in an open circuit condition where no net current crosses the epithelium, an intracellular current would have to be counteracted by an extracellular current passing the leaky junctions, creating an abnormally large drop in potential here of about 5 V, if the junctions in *Necturus* gallbladder have a resistance of about 150 ohm cm^2. Finally, any mechanism based directly or indirectly on a serosally placed Na^+ pump would not explain the fact that the gradient of electrical potential was only slowly affected by the addition of ouabain to the serosal solution.

Diffusion potential

If a solution of NaCl is separated from a solution of KCl by a membrane in which the mobilities of Na^+ and K^+ are different, a diffusion potential will exist across the membrane. Thus if the Na^+ pump situated at the serosal membrane caused a K^+ influx this could create oppositely directed fluxes of Na^+ and K^+ in the cytoplasm; if the mobilities were different, a diffusion potential would develop, but as explained in the previous paragraph, the experiments with ouabain exclude this possibility.

Donnan effects

A fraction of the electrical potential difference between cell cytoplasm and the external solution is caused by intracellularly located anions which are

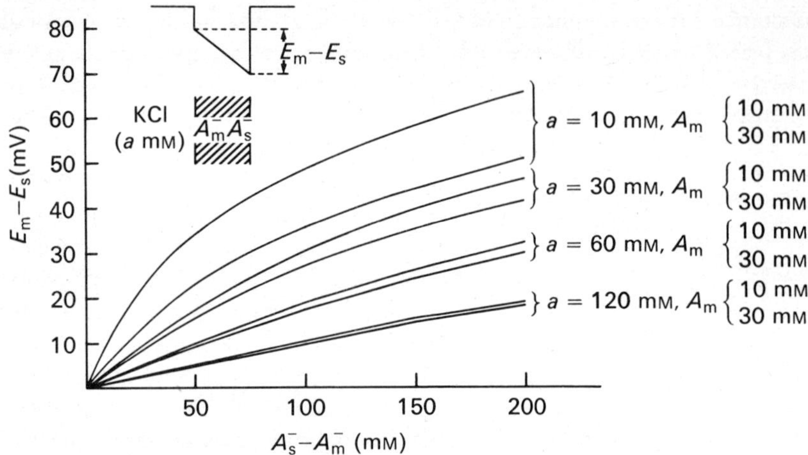

Fig. 7. The electrical potential difference $E_m - E_s$ inside a membrane in which the concentration of immobilized negative charge varied from A_m to A_s as a function of $A_s^- - A_m^-$. The membrane was surrounded by KCl of concentration a. The potential difference is plotted for different values of a and A_m.

unable to pass the cell membrane. Consider a slab, say 40 μm thick, which contains anions that are unable to leave the slab and have a low mobility within the slab. Now let the concentration of these immobilized negative charges increase linearly from A_m (mM) at one face of the membrane to a maximum of A_s at the other face. The profile of electrical potential across this membrane, when bathed in KCl at a concentration of a (mM), consists of a Donnan potential difference across the two interphases plus an electrical potential gradient within the membrane. When electrochemical equilibrium exists throughout the system and there is no binding between K^+ and Cl^- and the immobilized negative ions, the internal difference in electrical potential between the two sides of the membrane can be calculated as (Vetter, 1961, p. 53)

$$E_m - E_s = 2.3 \frac{RT}{F} \log_{10} \frac{\sqrt{(4a^2 + A_m^2)} - A_m}{\sqrt{(4a^2 + A_s^2)} - A_s} \qquad (1)$$

where 2.3 $RT/F = 59$ mV at 20 °C. All activity coefficients are taken to equal 1. $E_m - E_s$ as a function of A_s is shown in Fig. 7 at different values of a and A_m. With increasing activity a of the outer solutions, the differences in A mean less and the potential difference approaches zero. The activities inside (subscript i) and outside (subscript o) the membrane are linked via the requirement of electrochemical equilibrium which can be expressed as

$$(a_{KCl, o})^2 = a_{K^+, i} \times a_{Cl^-, i} \qquad (2)$$

(Vetter, 1961, eqn. 1.80). Thus, with an outer solution of 60 mM KCl and $A_s - A_m$ in the range 100–150 mM, the difference in electrical potential inside the slab ($E_m - E_s$) would be between 18 and 25 mV. If A_m was 10 mM, the K^+ activity would vary from about 55 mM near the site called m to about 140 mM towards the more negative site called s, and concurrently the Cl^- activity would vary from about 65 mM at m to about 25 mM at s.

In order to complete the analogy, consider a cross-section (say 20×20 μm) of the membrane discussed above, surrounded by a cell membrane. This cell membrane must then maintain the high intracellular K^+ and Cl^- activities by metabolic means, despite the outer solutions being physiological saline containing only a few mM K^+. It is far from clear at present to what degree the distribution of X^- inside the epithelial cells should be identified with an immobilized distribution of negative ions. X^- must partly be HCO_3^- (11 mM in the proximal tubule, Khuri, Bogharian & Agulian, 1974), and ATP, ADP, PO_3^- together with other metabolic intermediates. All the model shows is that the gradients of intracellular electrical potential, K^+ and Cl^- could result from a distribution of X^-, provided X^- represented an ion which was at least immobilized during transepithelial Na^+ transport.

It should be emphasized that the numbers for the activities presented here are obtained under the simplifying assumption that the activity coefficients inside as well as outside the cell equal 1.

I would like to thank Dr A. E. Hill for discussions on the contents. The study was supported by the Medical Research Council, England.

References

Chowdhury, T. K. & Snell, F. M. (1965). A microelectrode study of electrical potentials in frog skin and toad bladder. *Biochimica et biophysica acta*, **94**, 461–71.

Chowdhury, T. K. & Snell, F. M. (1966). Further observations on the intracellular electrical potential in frog skin and toad bladder. *Biochimica et biophysica acta*, **112**, 581–3.

Cremaschi, D., Henin, S. & Ferroni, A. (1974). Intracellular electric potentials in the epithelial cells of the rabbit gall-bladder. In *Bioelectrochemical Bioenergetics*, 2nd International Symposium on Bioelectrochemistry, Pont à Mousson, 1973, vol. 1, pp. 208–16.

Cremaschi, J., Smith, M. W. & Wooding, F. B. P. (1973). Temperature-dependent changes in fluid transport across goldfish gall-bladder. *Journal of Membrane Biology*, **13**, 143–64.

Diamond, J. M. (1962). The mechanism of solute transport by the gall-bladder. *Journal of Physiology*, **161**, 474–502.

Frank, K. & Becker, M. C. (1964). Microelectrodes for recording and stimulation. In *Physical Techniques in Biological Research*, vol. 5, ed. W. L. Nastuk, pp. 23–89. New York: Academic Press.

Frizzel, R. A., Dugas, M. C. & Schultz, S. G. (1975). Sodium chloride transport by rabbit gallbladder. Direct evidence for a coupled NaCl influx process. *Journal of general Physiology*, **65**, 769–95.

Frizzel, R. A., Nellans, H. N., Rose, R. C., Markscheild-Kaspi, L. & Schultz, S. G. (1973). Intracellular Cl^--concentration and influxes across the brush border of rabbit ileum. *American Journal of Physiology*, **224**, 328–37.

Frömter, E. (1972). The route of passive ion movement through the epithelium of *Necturus* gall-bladder. *Journal of Membrane Biology*, **8**, 259–301.

Gilles-Baillien, M. & Schoffeniels, E. (1965). Site of action of L-alanine and D-glucose on the potential difference across the intestine. *Archives internationales de Physiologie et de Biochimie*, **73**, 355–7.

Henriques de Jesus, C., Ellory, J. C. & Smith, M. W. (1974). Intracellular chloride activities in the mucosal epithelium of rabbit terminal ileum. *Journal of Physiology*, **244**, 31–32 P.

Khuri, R. N., Bogharian, K. K. & Agulian, S. K. (1974). Intracellular bicarbonate in single cells of *Necturus* kidney proximal tubule. *Pflügers Archiv für die gesamte Physiologie*, **249**, 295–9.

Lassen, U. V., Nielsen, A-M. T., Pope, L. & Simonsen, L. O. (1971). The membrane potential of Ehrlich ascites tumor cells. Microelectrode measurements and their critical evaluation. *Journal of Membrane Biology*, **6**, 269–88.

Lassen, U. V., Pape, L., Vestergaard-Bogind, B. & Bengtson, O. (1974). Calcium-related hyperpolarization of the *Amphiuma* red cell membrane following micropuncture. *Journal of Membrane Biology*, **18**, 125–44.

Lee, C. O. & Armstrong, W. McD. (1972). Activities of sodium and potassium in epithelial cells of small intestine. *Science, Washington*, **175**, 1261–3.

Lindemann, B. & Pring, M. (1969). A model of water absorbing epithelial cells with variable cellular volume and variable width of the lateral intercellular gaps. *Pflügers Archiv für die gesamte Physiologie*, **307**, R55–R56.

Loewenstein, W. R., Nakas, M. & Socolar, S. J. (1967). Junctional membrane uncoupling. *Journal of General Physiology*, **50**, 1865–91.

Petersen, O. H. (1976). Electrophysiology of mammalian gland cells. *Physiological Reviews*, **56**, 535–76.

Reuss, L. & Finn, A. L. (1975). Electrical properties of the cellular transepithelial pathway in *Necturus* gall-bladder. I. Circuit analysis and steady-state effects of mucosal solution ionic substitutions. *Journal of Membrane Biology*, **25**, 115–39.

Rose, R. C. & Schultz, S. G. (1971). Studies on the electrical potential profile across rabbits' ileum. *Journal of General Physiology*, **57**, 639–63.

Schultz, S. G. & Curran, P. F. (1968). Intestinal absorption of sodium chloride and water. *Handbook of Physiology*, vol. 3, *Alimentary Canal*, ed. C. F. Code, pp. 1245–75.

Schultz, S. G., Fuisz, R. E. & Curran, P. F. (1966). Amino acid and sugar transport in rabbit ileum. *Journal of General Physiology*, **49**, 849–66.

Tasaki, K., Tsukahara, Y., Ito, S., Wayner, M. J. & Yu, W. Y. (1968). A simple direct and rapid method for filling micro-electrodes. *Physiology and Behavior*, **3**, 1009–10.

Van Os, C. H. (1974). Transport parameters of isolated gall-bladder epithelium. Ph.D. thesis, University of Nijmegen, The Netherlands.

Vetter, K. J. (1961). *Elektrochemische Kinetik*. Berlin: Springer-Verlag.

Wright, E. M. (1966). The origin of the glucose-dependent increase in the potential difference across the tortoise small intestine. *Journal of Physiology*, **185**, 486–501.

Zeuthen, T. (1975). A double-barrelled Na^+-sensitive micro-electrode. *Journal of Physiology*, **254**, 8–10 P.

Zeuthen, T. (1976a). Gradients of chemical and electrical potentials in the gall-bladder. *Journal of Physiology*, **256**, 23 P.

Zeuthen, T. (1976b). Micro-electrode recording of intracellular gradients of electrical and chemical potential secretory epithelia. *Journal of Physiology*, **263**, 113–114 P.

Zeuthen, T. (1977). The vertebrate gall-bladder. The routes of ion transport. In *Transport of Ions and Water in Animals*, ed. B. Gupta, R. Moreton, I. Oschman & B. Wall. London, New York: Academic Press. (In Press.)

Zeuthen, T., Hiam, R. C. & Silver, I. A. (1974). Recording of ion activities in the brain. In *Ion Selective Microelectrodes*, ed. H. Berman & N. Herbert, pp. 145–56. London: Plenum Press.

Zeuthen, T. & Monge, C. (1975). Intra- and extracellular gradients of electrical potential and ion activities of the epithelial cells of the rabbit ileum *in vivo* recorded with microelectrodes. *Philosophical Transactions of the Royal Society*, B, **71**, 277–81.

2. Fluid transport by corneal endothelium

By JORGE FISCHBARG

The regulation of the state of corneal hydration has been known for long to be the single most important process within the ensemble of factors that determine and help maintain the physiological state of transparency of that organ. All three anatomical layers that compose the cornea play a significant role. The anterior epithelium (epithelium) constitutes an effective barrier for water and electrolytes; it can be classed as a layer of comparatively high electrical resistance, and therefore it helps to preserve the internal corneal medium. Some recent reports (Zadunaisky, personal communication; Candia, 1976) have described that, in addition, the epithelium can pump fluid out of the stroma. The extent to which this fluid movement would contribute to dehydrate the stroma seems to vary with species, and its precise role is being investigated. As for the stromal layer, the particular architecture of its layers of fibrils results in transparency, provided that the water content of it remains within prescribed limits; if left on its own, this layer will imbibe fluid, swell and become opaque. The third corneal layer, and the one around which this paper will revolve, is its posterior epithelium, better known as the endothelium. This monocellular layer presents a very low electrical resistance (Fischbarg & Lim, 1973; Lim & Fischbarg, in preparation), and is permeated by molecules as large as horseradish peroxidase (Kaye, Sibley & Hoefle, 1973), with an estimated diameter of 4.5 nm. On the other hand, the endothelium is the site of a fluid pump (Mishima & Kudo, 1967; Maurice, 1972) which counters passive leaks into the stroma by moving fluid from stroma to aqueous humour. For the case of the rabbit, where this fluid pump has been studied at length, the rate of water movement caused by the endothelium seems sufficient to account for a quasi-normal degree of stromal hydration (Mishima & Kudo, 1967; Dikstein & Maurice, 1972; Fischbarg & Lim, 1974; Hodson, 1974).

Some of the facts presently known about the corneal limiting layers (and the recent evolution of this knowledge) are summarized in Fig. 1. Undoubtedly, ionic fluxes can be studied much more easily across the high-resistance epithelium, and the identification of the ionic pumps involved at that level is a reflection of this. The evidence for ionic pumps

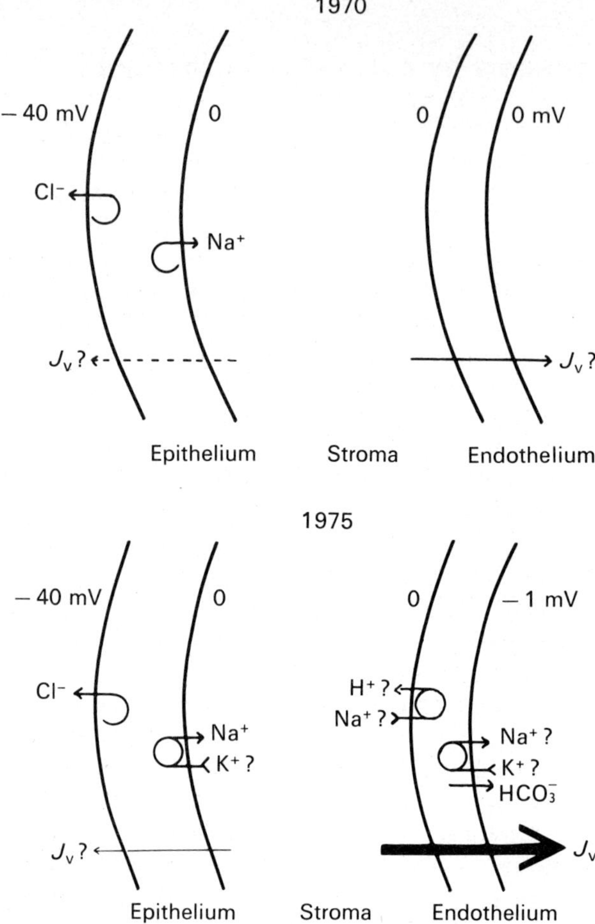

Fig. 1. Schematic diagram of the cornea, its limiting layers and the proposed or postulated transport mechanisms across them in two recent years.

at the endothelial level has so far been necessarily indirect, although the pattern of the findings was clear enough to allow us to suggest a model for them (Fischbarg & Lim, 1974). Very recently, in addition, Hodson & Miller (1976) have reported evidence for a HCO_3^- pump at the endothelial level, which is consistent with the model we have advanced.

Electrical properties of the endothelial layer

To the extent that the present communication centers on the properties of the paracellular pathway across the endothelial layer, a consideration of its electrical characteristics serves well to lay the groundwork for the ensuing analysis. As stated in previous communications (Fischbarg, 1973; Fischbarg & Lim, 1973), the values of the total resistance of the endothelium (40 to 70 ohm cm^2) experimentally measured are of the same order as the value expected if the hexagonal maze of intercellular junctions and spaces is assumed to be filled with saline. Since $R = l/\lambda c \cdot A$, using for λ and c the values of 130 cm^2 ohm^{-1} equiv.$^{-1}$ and 0.15 equiv. per 10^3 cm^3, respectively, for l the values of 1×10^{-4} cm and 11×10^{-4} cm for junctions and spaces, respectively, and taking A as the product of perimeter length (10^3 cm per cm^2 of area) times width (40×10^{-8} cm for junctions, 300×10^{-8} for spaces), R for junctions and spaces is 13 ohm cm^2 and 19 ohm cm^2, respectively, or 32 ohm cm^2 when combined. The justification for the use of the dimensions given comes from morphological work and also, in the case of the junctions, from considerations which will become apparent when the hydraulic and osmotic flows across the layer are scrutinized. The easy geometry of this layer dispells doubts which are always present in convoluted epithelia and allows accurate morphological conclusions (cf. Oh, 1963). The schemes on which this analysis rests are shown in Figs. 2 and 3.

A question that frequently arises in connection with 'leaky' epithelia such as the present one is why is it that they are actively pumping water and electrolytes across them, only to have the electrolytes leak back in considerable amounts across their low resistance? It may be added that, in view of evidence such as that given below, the question could be generalized to include water as well. The answer is complex and has not been clearly stated yet, since it might presumably require the unambiguous solution of the detailed process by which water movements are coupled to ionic translocation in these tissues. One area that appears worth investigating is whether the low resistance, rather than a hindrance, might actually be a necessary condition for efficient fluid transfer. In the light of these problems, a comparison between the properties of tight and leaky epithelia seems pertinent, and some of our work has been directed accordingly. Fig. 4 shows the behavior of the short-circuit current across our preparation. The methods involve the determination of the impedance locus and the potential difference across the layer (Lim & Fischbarg, in preparation). As can be seen there, under conditions in which fluid transport is abolished (absence of ambient HCO$_3^-$), the short-circuit current is drastically reduced, and the change is fully reversible. The actual

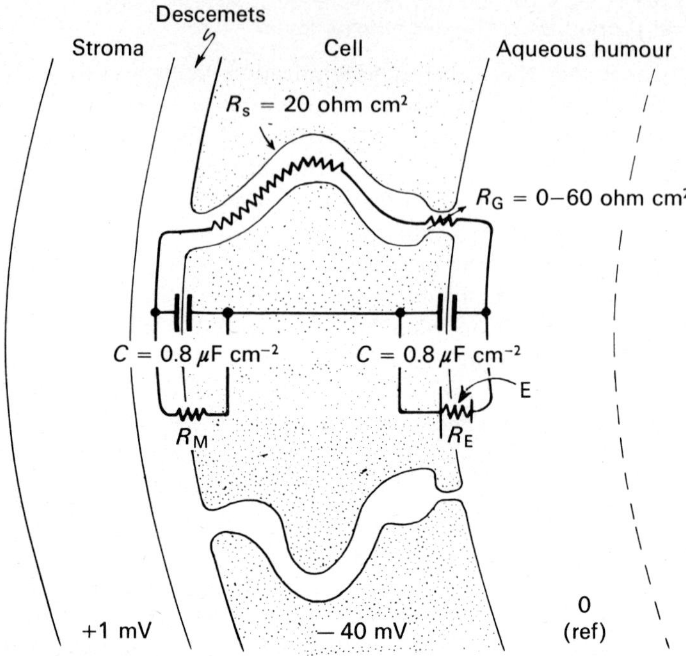

Fig. 2. Simplified electrical equivalent model superimposed on a schematic drawing of the corneal endothelium. R_S, R_G, R_M, and R_E = resistance of intercellular spaces, intercellular apical junctions, cell membranes and internal resistance of ionic pumps, respectively. The actual location of the hypothetical electrogenic pump(s) E cannot be precise; C = capacitance of cell membrane.

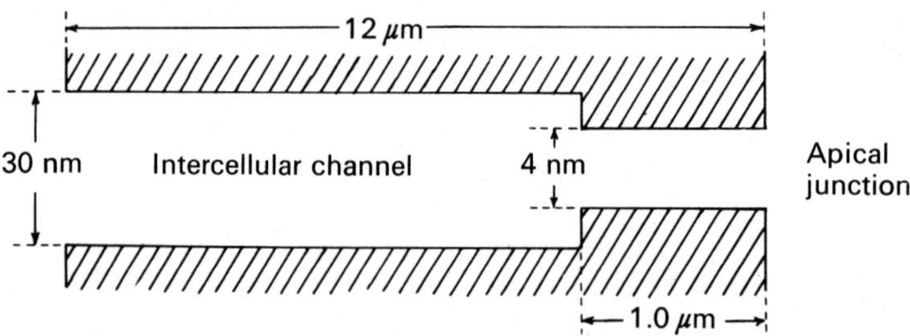

Fig. 3. Idealized drawing of the paracellular pathway across corneal endothelium. The dimensions are taken from electronmicrographs or estimated from other evidence, as in the case of the junctional width. The intercellular channels are actually convoluted, since the cell thickness is some 5–10 μm.

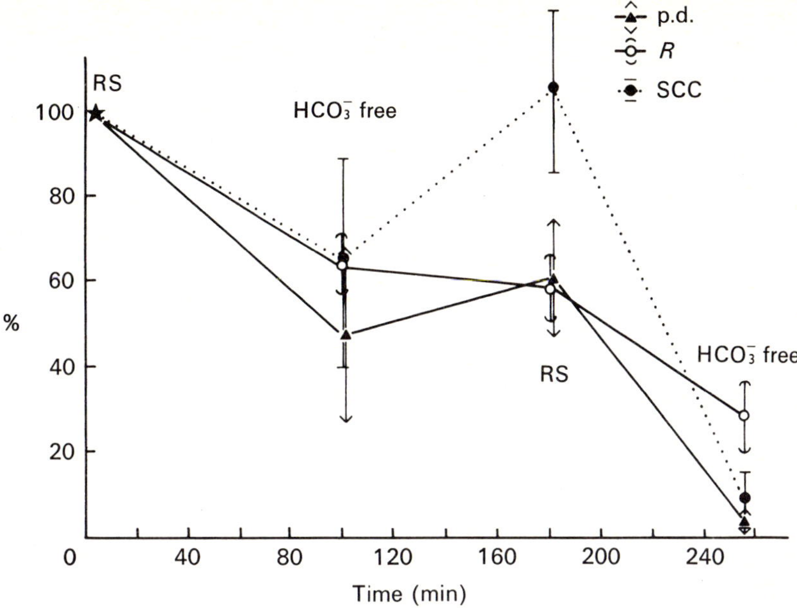

Fig. 4. Average values ($\pm$s.e.) of potential difference (p.d.), resistance (R) and short-circuit current (SCC) across corneal endothelium as a function of time and in presence of: RS = regular solution, with nutrients (adenosine, glucose, oxidized glutathione) and salts; HCO$_3^-$ free: RS with all its HCO$_3^-$ replaced by Cl$^-$. Values are expressed in percent of initial determinations for each experiment. Typical values for this series were, for the initial p.d. and R, 1.02 ± 0.03 mV and 45 ± 3 ohm cm^2, respectively, from which the initial SCC was 23 μA cm^{-2}.

correspondence of this parameter with ionic movements may be complex, as can be gathered from the model in Fig. 1, but it is nevertheless revealing that, operationally at least, this electrical behavior of a leaky layer resembles closely the classical examples of tight epithelia in presence and absence of ionic transport. There is another point that brings consistency to this picture; if the average short-circuit current value (18 μA cm^{-2}) is taken to represent the net flow of a monovalent salt (such as NaHCO$_3$) across the tissue, and if it is further assumed that salt moves together with water in an isotonic manner, the volume of fluid moved can be calculated to be 4.5 μl h^{-1} cm^{-2}. This value (4.5 μl h^{-1}) is in very good agreement with the experimental values usually determined for the rate of fluid transport; in a recent study, the average rate reported was precisely 4.5 μl h^{-1} cm^{-2} (Fischbarg & Stuart, 1975) and other reports contain similar values (cf. Maurice, 1972). This presumed isotonicity of the fluid transported will be the subject of further consideration in what follows.

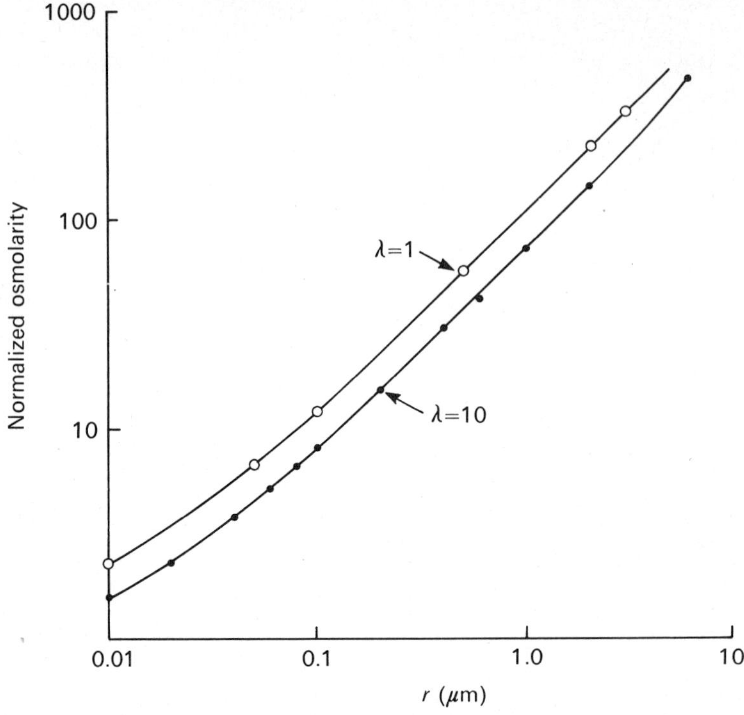

Fig. 5. The normalized osmolarity (actual osmolarity divided by the osmolarity of the ambient solution) shown as a function of the radius (r) of the idealized intercellular channels. The parameter λ, which labels the two curves, denotes the inverse of the fraction of channel (close to the apical or 'closed' end) to which the pumping sites are assumed to be restricted; thus $\lambda = 1$ means sites all along the channel, $\lambda = 10$ means sites along one-tenth of it. The values of the other parameters assumed for the model were: rate of transport $N = 8.5 \times 10^{-8}$ mOsm cm^{-2} s^{-1}; ambient osmolarity $C_0 = 0.3$ Osm; D for salt (NaCl) $= 1.54 \times 10^{-5}$ cm^2 s^{-1}; length of channels $1 = 12 \times 10^{-4}$ cm; osmotic permeability of the channel walls $P_{os} = 4.8 \times 10^{-5}$ cm s^{-1} Osm^{-1} or 26 μm s^{-1} (at 37 °C).

Solute–solvent coupling

The evidence presented so far on electrical properties, in conjunction with that described below on water permeation, suggests that this transporting system is a 'backwards' epithelium, with its intercellular junctions permeable to water, which is able to pump fluid, presumably in an isotonic fashion. To the extent that the data were not judged to be sufficient to allow the formulation of a specific model for solute–solvent coupling for this case, we decided instead to attempt to gain insight into this problem

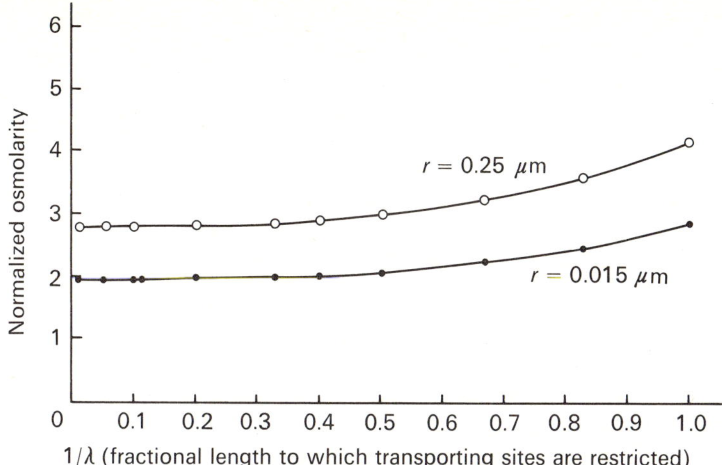

Fig. 6. The normalized osmolarity (see Fig. 5 caption) of the transported fluid is shown as a function of the distribution of transporting sites along the channels. The values of the other parameters are as in the legend to Fig. 5. Two curves are shown, one for each assumed value of channel radius (r); the values chosen (15 nm and 0.25 μm) represent limit cases for average radius of collapsed and dilated channels, respectively, according to electronmicrographs.

by evaluating the applicability of the popular 'standing-gradient' hypothesis (Diamond & Bossert, 1967) to our particular case. Rather than to adopt numerical procedures for the integration of the third-order non-linear differential equation Diamond & Bossert introduced, we preferred an analytical method (Lim & Fischbarg, 1976) which entailed a slight extension of a previously developed method (Segel, 1970). The fact that this analytical procedure yields results which are in excellent agreement with those coming from numerical methods has been detailed recently (Lim & Fischbarg, 1976). For the purposes of our initial analysis we have adopted the same boundary conditions of Diamond & Bossert's, which essentially imply that the intercellular junctions would be impermeant to both solute and solvent and that the pumping sites could be restricted to a length shorter than that of the intercellular channels. As Figs. 5, 6, 7 and 8 show, the standing-gradient model does not yield isotonic flow in our preparation under the assumptions above. Even under the most favorable limit conditions, the fluid transported remains hypertonic by a factor of two to three times with respect to the ambient osmolarity. As exemplified in those figures, either the osmotic permeability is too low, the length of the spaces too short or the diameter of them too large to account for the postulated isotonic transport across corneal endothelium. Not even a reduction in the

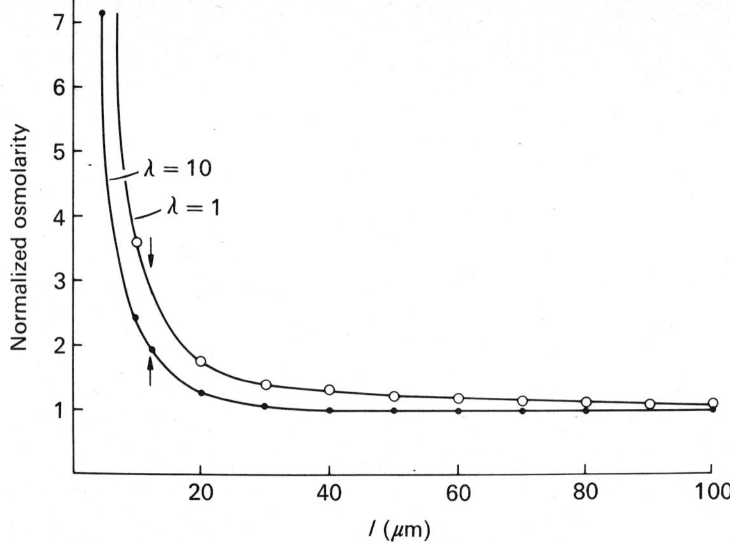

Fig. 7. The normalized osmolarity (see Fig. 5 caption) of the transported fluid as a function of the theoretical length of the channels, with the other parameters as in Fig. 5 and a radius value of 15 nm, to choose a relatively 'favorable' value for this last one. The values of λ given for the two curves are explained in caption to Fig. 5. The vertical arrows at 12 μm length show the actual measured length of the endothelial intercellular spaces.

proportion of channel length occupied by pumping sites achieves that effect.

The conclusions these results point to agree in general with those contained in two other recent reports (Hill, 1975a; Sackin & Boulpaep, 1975) in which the standing-gradient model has been examined. A question these and other similar results raise is whether changes in the original assumptions and boundary conditions in Diamond & Bossert's treatment would lead to isotonic flow. Hill (1975b) examined this matter by assuming that the intercellular junctions would be permeable to solvent, and concluded that osmosis through the junctions as the sole mechanism could not apparently explain isotonic flow. His calculations hinged on junctional osmotic permeability values derived from steady-state measurements of transepithelial osmotic permeabilities as reported in the literature. However, the possibility that those steady-state values may lead to an underestimation of the actual junctional osmotic permeability has recently become apparent. As detailed below (cf. Fischbarg, Warshavsky & Lim, 1977), for the case of the corneal endothelium the steady-state osmotic permeability of the layer under certain experimental conditions is some six times lower than its osmotic permeability immediately after a gradient

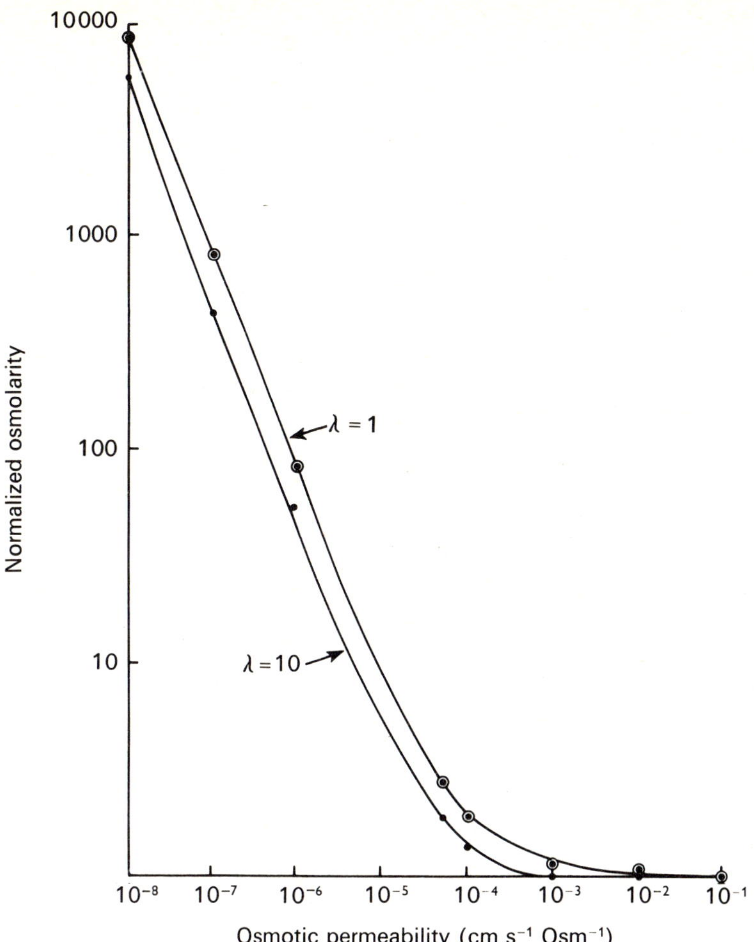

Fig. 8. The normalized osmolarity (see Fig. 5 caption) of the transported fluid is shown as a function of the assumed osmotic permeability of the channel walls. The two curves shown again represent: $\lambda = 1$, uniform distribution of pumping sites all along the channels; $\lambda = 10$, sites restricted to the apical one-tenth of the channel. Increasing λ above 10 does not measurably shift the curves 'downwards'. For comparison, the values reported elsewhere in the text for the minimum and maximum (or 'low' and 'high') osmotic conductance of the endothelial layer as a whole are 30 and 166 μm s⁻¹, or 5.4×10^{-5} and 3.0×10^{-4} cm⁴ s⁻¹ mOsm⁻¹, respectively. (From Fischbarg *et al.*, 1977).

is imposed ('zero time' P_{os}). Time-transient changes in osmotic flows have also been noted in gallbladder (Wright, Smulders & Tormey, 1972). Although in that study they were interpreted in terms of 'solute polarization' effects, the authors did point out that the actual osmotic permeability of that layer could be an order of magnitude higher than the value measured in the steady-state. For the present case, by employing the larger value for P_{os} (166 μm s^{-1} or 30×10^{-5} cm^4 s^{-1} mOsm^{-1}) reported below, in conjunction with a rate of transport of 4.5 μl h^{-1} cm^{-2} (cf. Fischbarg & Stuart, 1975; Fischbarg, Lim & Bourguet, 1977), an effective gradient of only some 4 mOsm across the intercellular junctions would explain the fluid transport observed across the corneal endothelium simply on the basis of osmosis through the junctions.

The conditions that have to be met for such a mechanism to generate isotonic flow will be explored in another communication (Fischbarg & Lim, in preparation). Initial computations have been performed with a modified 'standing-gradient' model in which the boundary condition of zero solvent velocity at the junctional end has been replaced by one that allows that velocity to assume different values as a result of water permeation through the junction. The value of that velocity for which the transported fluid becomes isotonic agrees, within the limits of experimental error, with the value expected if all the fluid transported would traverse the intercellular junctions. This adds credence to the possibility that osmosis through the junctions might constitute a viable explanation for the case of the corneal endothelium. For the case of the *Necturus* kidney proximal tubule, Sackin & Boulpaep (1975) have proposed a similar explanation. As they point out, the emerging picture resembles the 'double-membrane model' for solute-solvent coupling (Ogilvie, McIntosh & Curran, 1963) in that the concentration profiles along the channels are rather flat if it is assumed that the junctions are water-permeable; similar results have been obtained for the corneal endothelium (Fischbarg & Lim, in preparation). The conditions under which the mechanism proposed by Sackin & Boulpaep would generate isotonic flow hinge once more on the value that is assumed for the osmotic permeability of the intercellular junctions, as extrapolated from the osmotic permeability of the entire cell layer. Although so far only a possible range of values is available (cf. Whittembury *et al.*, 1959; Grandchamp & Boulpaep, 1974), this matter will probably come under closer scrutiny in the near future.

Hydraulic and osmotic water flows

The work presented in this section was motivated by the question of the possible existence of bulk water movements across the paracellular shunt of the preparation, including the intercellular junctions. All the indirect evidence pointing towards the likelihood of this possibility, such as the electrical properties reported above, some old results in which dyes as large as 1.0–2.6 nm permeated the endothelium (quoted by Maurice, 1953) and

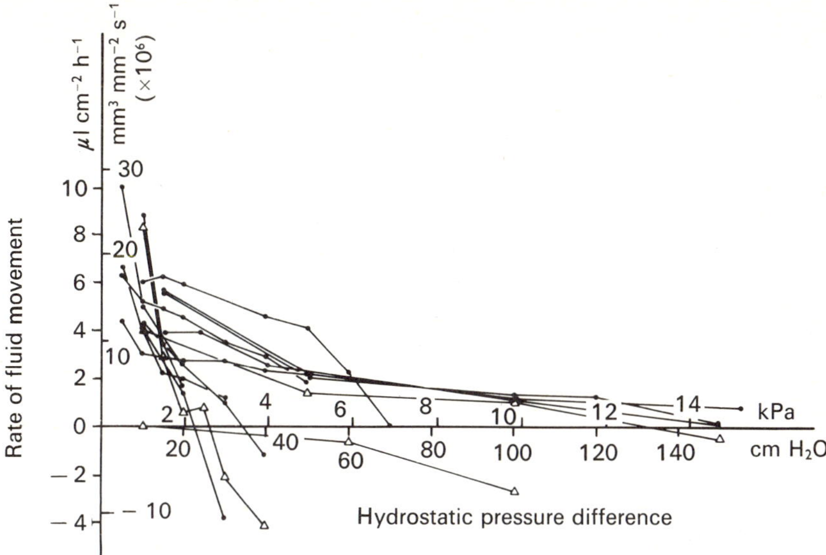

Fig. 9. The rate of fluid transport across the endothelial layer is plotted as a function of the hydrostatic pressure imposed on the inside or aqueous side with reference to that on the outside. Each curve denotes a separate experiment; the symbols represent actual rate measurements, the circles for the usual type of experiment, and the triangles for experiments in which a large fraction of stroma was dissected away. Positive rates mean movement in the physiological direction for fluid transport (stroma to aqueous humour).

measurements of very large transendothelial ionic fluxes (Maurice, 1951), has been reinforced in recent years by demonstrations that LaOH and horseradish peroxidase permeate the intercellular junctions (Leuenberger, 1973, and Kaye *et al.*, 1973, respectively). The evidence on horseradish peroxidase is especially suggestive, since its diameter can be estimated to be about 4.5 nm; channels this large would in principle be expected to be permeable to water as well, as has been argued for the case of the capillary endothelium (Karnovsky, 1967).

In order to investigate this matter, we have determined experimentally the hydraulic permeability (L_p) and the osmotic permeability (P_{os}) across the corneal endothelium. Fig. 9 shows the effect that an increased hydrostatic pressure difference has on the rate of fluid transport by this layer. As can be observed there, the rate decreases linearly with pressure for most of the range examined. The slope in Fig. 9 can be equated with a hydraulic conductance, the value of which is 83 ± 11 μm s^{-1} ($15 \pm 2 \times 10^{-5}$ cm s^{-1} Osm^{-1}) for its linear portion. A correction for the possible influence of

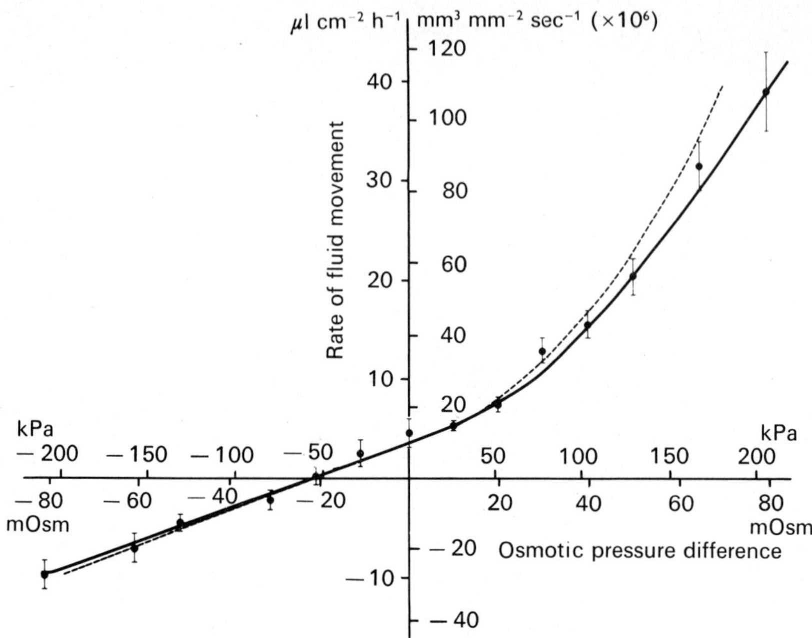

Fig. 10. Rates of fluid movement across the endothelial preparation due to a range of osmotic pressure differences. Positive osmotic pressures mean that the inside (aqueous) solution was made hypertonic by addition of sucrose; negative osmotic pressures were obtained by beginning experiments in 80 mOsm hypertonic solutions on both sides (by sucrose addition) and lowering the tonicity of the inside (the present technique does not allow easy access to the chamber whose volume is being 'clamped', the outer one in the present case). The average and s.e. are shown in each case. The solid line was fitted by eye to the data; the dashed line was calculated after correcting for possible solute polarization effects. Directions of fluid movements as in Fig. 10. The hydrostatic pressure difference was maintained at 10 cm H$_2$O. Here and elsewhere, $T = 36.80\ ^\circ$C.

unstirred layers can be introduced by assumming that the increased hydrostatic pressure difference (ΔP) results in a hydraulic leak in a direction opposite to that of fluid transport and independent of it. Using for the unstirred layers at the aqueous (δ_1) and outside (δ_2) interfaces thickness values of 350 and 850 μm, respectively, and employing the pertinent values for the other parameters (NaCl $D_1 = 1.54\times10^{-5}$ cm^2 s^{-1}, $D_2 = 1.0\times10^{-5}$ cm^2 s^{-1}, $\sigma_{\text{NaCl}} = 0.6$, $J_v = 4.5\ \mu$l h^{-1} cm^{-2}; cf. Fischbarg, Warshavsky & Lim, 1977), the contribution of the unstirred layers can be computed from the expression $C_o = C_{\text{bulk}}\ \exp\ (-J_v\ \delta^{-1}\ 2D^{-1})$ (Fischbarg, Warshavsky &

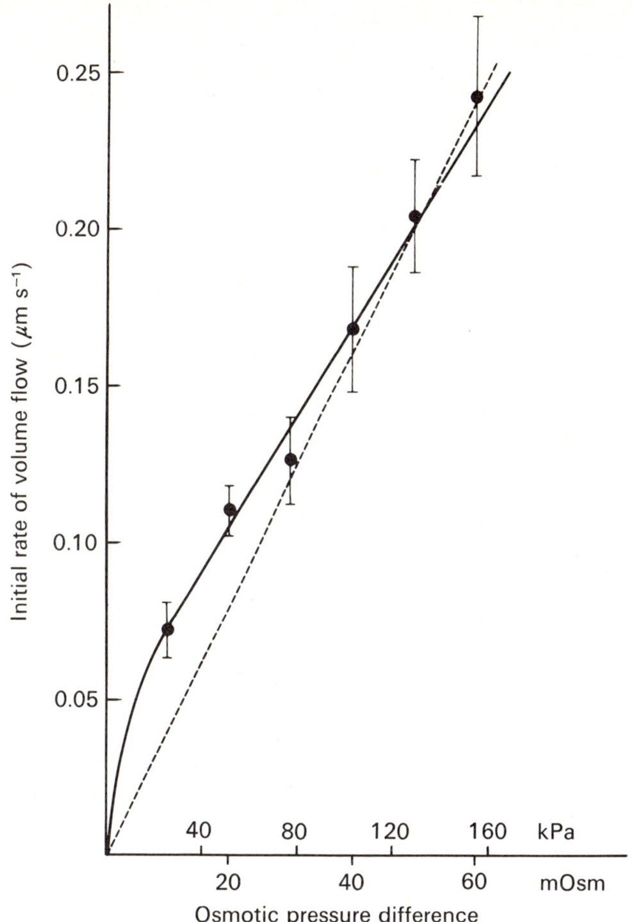

Fig. 11. The initial rates of osmotic flow obtained after the sudden imposition of an osmotic gradient across the endothelium are plotted against the value of the osmotic gradient. Volume flow was from stroma to aqueous in response to aqueous hypertonicity. Average and S.E. are shown at each gradient value examined; the solid line was fitted by eye, while the dashed line represents the chord up to the highest value.

Lim, 1977. In this fashion the corrected value for L_p increases to 114 μm s^{-1} (21×10^{-5} cm s^{-1} Osm^{-1}).

The osmotic permeability of the endothelial layer was determined with two different procedures. In one, a flow was induced across the layer by an osmotic pressure difference and the rate of volume flow was recorded by an automatic procedure (Bourguet & Jard, 1964; Fischbarg, Lim & Bourguet, 1977). After a sufficiently long time had elapsed for the rate to become stable, the steady-state rate was noted. In the other procedure, an osmotic gradient was imposed across the endothelial preparation while the stromal thickness was continuously monitored with a microscope (for a similar procedure, cf. MacRobbie & Ussing, 1961; Maurice, 1969). From the time-transient change in stromal thickness resulting from the osmotic flow out of the stroma, the endothelial P_{os} at zero time could be extrapolated (cf. Mishima & Hedbys, 1967). In both cases, osmotic gradients were obtained by the addition of sucrose. Fig. 10 summarizes the results with the first procedure, and Fig. 11 those with the second one. As can be noted in Fig. 10, the steady-state osmotic flow was not linear with respect to the driving force; similar results have been reported in other preparations (a representative example and a review can be found in Diamond, 1966). The results in Fig. 10 can be taken to show two limiting slopes, from which the two osmotic conductances for the first and third quadrants, which will also be termed 'high' and 'low' osmotic conductances, can be calculated to be 88 ± 6 μm s^{-1} and 25 ± 3 μm s^{-1}, respectively (or 16 ± 1 and 4.5 ± 0.5 cm s^{-1} Osm, respectively). If these values are corrected for the possible influence of unstirred layers, the P_{os} become 115 and 26 μm s^{-1}, respectively, for the high and low cases. As for the P_{os} values determined from the time-transient measurements, Fig. 11 shows that the behavior of the flows was reasonably linear for the range of 10 to 60 mOsm $\Delta\pi$, in contrast with the non-linearity apparent in the steady-state case (Fig. 10) for the same range. From the linear region in Fig. 11, the slope (or P_{os} value) is 166 μm s^{-1} at $t=0$. The possible influence of unstirred layers on this value can be computed from the complete solution of the diffusion–convection equation* (Pedley & Fischbarg, in preparation): they cannot apparently account for more than a 5–10% discrepancy. In a previous study in which P_{os} at $t=0$ was measured (Mishima & Hedbys, 1967), the value reported was 221 μm s^{-1}, which is precisely equal to the chord value in Fig. 11; the agreement is excellent.

The pattern that emerges from these results can be put as follows: the value for L_p, the limiting value for P_{os} at steady-state, in the first quadrant of Fig. 10, and the value for P_{os} at $t=0$ all reflect a high conductance across the endothelial layer, while the limiting value for P_{os} at steady-state in the third quadrant of Fig. 10 reflects a low conductance. The fact (noted above) that the hydraulic flow is linear with respect to its driving force while the osmotic flow at steady-state is not, taken in conjunction with the

* The solution for the semi-infinite case, with the bulk solution–unstirred layer separation as the only boundary, is good as an approximation. It is:

$$C(x,\ t) = (C_b/2)\ \{\mathrm{erfc}\,[(x-J_v t)/2\sqrt{(Dt)}] + \exp\,(J_v x/D)\ \mathrm{erfc}\,[(x+J_v t)/2\sqrt{(Dt)}]\}.$$

observation that the osmotic flow at zero time is again largely linear with its driving force, implies that the properties of the high conductance pathway can be altered as a function of time by an imposed osmotic gradient in a manner dependent on the magnitude and direction of the imposed gradients and/or flows. To the extent that the numerical sum of the low plus the high conductances (26 and 114 μm s^{-1}) comes close to the maximum conductance found for the case of P_{os} at $t = 0$ (166 μm s^{-1}), the transition observed in the steady-state P_{os} value in the first quadrant (Fig. 10) with increasing osmotic gradient could be the result of the progressive 'recruitment' of the high-conductance pathway, tending to a maximum P_{os} of the order of that at $t = 0$.

The possible location of the high-conductance pathway deserves detailed consideration. From the markedly high value of this hydraulic or osmotic conductance when compared with values reported for tight epithelia and most cell membranes (cf. House, 1974), the paracellular pathway constituted by the intercellular junctions and spaces appears likely to be the site of this high conductance. Calculations about the expected value of this conductance on the basis of assumed laminar flow across water-filled slits reinforce this explanation. The pertinent parameters are: $L_p = a^3 b/12\eta l$ l^{-1}, where: a = slit width, the unknown; b = total cell perimeter per unit area, or 10^3 cm cm^{-2} for the 1.75 cm^2 corneas of the four month old rabbits used (Oh, 1963); $\eta = H_2O$ viscosity at 37 °C, or 6×10^{-3} g cm^{-1} s^{-1}; l = length of the slits, or 1 μm from electron-micrographs; $L_p = 114$ μm s$^{-1} = 0.8 \times 10^{-11}$ cm^2 s g^{-1}. The junctional width computed is 3.9 nm. In spite of the fact that, with all the approximations involved, an exact numerical correlation would appear unlikely, the value computed is in good agreement with necessarily gross morphological estimates of 3–4 nm for the slit width and with the estimated diameter of horseradish peroxidase (4.5 nm) that permeates them. In addition to the ones above, there is yet another line of reasoning that points to similar conclusions. If, for the sake of argument and against all the evidence listed so far, the intercellular junctions are now assumed to be impermeable to water, the high P_{os} ($t = 0$) found would require a value of some 200 μm s^{-1} for the P_{os} of the cell membrane proper, assumed to be uniform all around it. Values as high as this have only been reported in red blood cells (cf. House, 1974), but are, in principle, possible. On the other hand, however, it becomes extremely difficult to account for a transition to a much lower P_{os} such as that at steady-state (26 μm s^{-1}) if such would be the case. Even an extreme collapse of the intercellular spaces that would eliminate the lateral cell membranes from consideration would decrease the total P_{os} only down to some 100 μm s^{-1}. Changes in membrane properties or inordinately high unstirred layer thicknesses (of the order of tens of centimeters) have to be invoked in addition, which makes this possibility unlikely. In contrast, if one returns to the assumption that the high-conductance pathway coincides with the intercellular junctions, the explanation for the decrease to low conductance can be more satisfactorily formulated in terms of a time-dependent change in the apparent reflection coefficient of the junctions to sucrose; the estimated size of sucrose (1.1 nm; Durbin, 1960) and the fact that it permeates the endothelium (Mishima & Trenberth, 1968) are consistent with this possibility. In a qualitative sense, at least, results such as those in Fig. 10 might arise because high rates of flow would succeed in 'sweeping' the solute away from the paracellular pathway, countering diffusion and therefore preserving the initial osmotic gradient. Undoubtedly more work is needed to validate this particular hypothesis, which in principle would seem plausible. In keeping with this general type of reasoning, the value of the low-conductance P_{os} might be, in the extreme case, the result solely of osmotic flow across the cell membranes rather than the junctions. This interpretation does not seem to pose difficulties, since P_{os} for the cell membranes in that case can be calculated to amount to 30 μm s^{-1} if assumed uniform for the series barrier of apical and lateral plus basal membranes. Although definite conclusions on this basis alone would be wanting, this P_{os} value nevertheless agrees with values reported for several other cells (cf. House, 1974), with the particular example of the 'inner membrane' of frog skin epithelium (24 μm s^{-1}, MacRobbie & Ussing, 1961) and with those of artificial bilayers (17–104 μm s^{-1} in range, with a cluster around 20–30 μm s^{-1}, cf. Huang & Thompson, 1966, and Hanai & Haydon, 1966).

Conclusion

The present paper reviews several lines of evidence which suggest that the intercellular junctions of the corneal endothelium behave as a network of slits, largely filled with ambient solution and permeable to both water and electrolytes. The supportive evidence discussed includes (*a*) the experimentally determined value of the electrical resistance, which is in the order expected from such network; (*b*) evidence from high rates of permeation by ions and non-electrolytes across that layer, and for the morphological demonstration of the passage of large markers (LaOH, horseradish peroxidase) across the intercellular junctions; (*c*) evidence presented here of a comparatively large hydraulic and osmotic permeability across the layer, numerically compatible with flow across water-filled slits. The original version of the standing-gradient model for solute–solvent coupling, when examined in detail, is shown here to predict hypertonic rather than isotonic fluid transport across this preparation. The evidence advanced relates the expected osmolarity of the transported fluid to (*a*) the radius of the channels; (*b*) their length; (*c*) the distribution of pumping sites along them; and (*d*) the osmotic permeability of the channel walls. The effects produced by modifying the original standing-gradient hypothesis so as to include intercellular junctions permeable to solvent are discussed, and the resulting possibility of osmotic flow through the junctions is evaluated in the light of existing measurements for the osmotic permeabilities across several epithelia.

The work presented here has been, for the most part, performed in collaboration with Dr J. J. Lim and Mr C. R. Warshavsky. It has been supported by USPHS Research Grants to J.F. (EY00727 and EY01080) and J.J.L. (EY01422), by Research Career Development Award EY00006 to J.F. and by a Fight for Sight, Inc. Fellowship for C.R.W.

References

Bourguet, J. & Jard, S. (1964). Un dispositif automatique de mesure et d'enregistrement du flux net d'eau à travers la peau et la vessie des amphibiens. *Biochimica et biophysica acta*, **88**, 442–4.

Candia, O. A. (1976). Fluid and Cl transport by the epithelium of the isolated frog cornea. *Federation Proceedings*, **35**, 703.

Diamond, J. M. (1966). Non-linear osmosis. *Journal of Physiology* **183**, 58–82.

Diamond, J. M. & Bossert, W. H. (1967). Standing-gradient osmotic flow. A mechanism for coupling of water and solute transport in epithelia. *Journal of General Physiology*, **50**, 2061–83.

Dikstein, S. & Maurice, D. M. (1972). The metabolic basis to the fluid pump in the cornea. *Journal of Physiology*, **221**, 29–41.

Durbin, R. P. (1960). Osmotic flow of water across permeable cellulose membranes. *Journal of General Physiology*, **44**, 315–26.

Fischbarg, J. (1973). Active and passive properties of the rabbit corneal endothelium. In Symposium on Transport and the Eye, ed. J. A. Zadunaisky. *Experimental Eye Research*, **15**, 615–38.

Fischbarg, J. & Lim, J. J. (1973). Determination of the impedance locus of rabbit corneal endothelium. *Biophysical Journal*, **13**, 595–9.

Fischbarg, J. & Lim, J. J. (1974). Role of cations, anions and carbonic anhydrase in fluid transport across rabbit corneal endothelium. *Journal of Physiology*, **241**, 647–75.

Fischbarg, J., Lim, J. J. & Bourguet, J. (1977). Adenosine stimulation of fluid transport across rabbit corneal endothelium. *Journal of Membrane Biology*. (In Press.)

Fischbarg, J. & Stuart, J. (1975). The effect of vitreous humor on fluid transport by rabbit corneal endothelium. *Investigative Ophthalmology*, **14**, 497–506.

Fischbarg, J., Warshavsky, C. R. & Lim, J. J. (1977). Pathways for hydraulically and osmotically induced water flows across epithelia. *Nature*, **266**, 71–4.

Grandchamp, A. & Boulpaep, E. L. (1974). Pressure control of sodium reabsorption and intercellular backflux across proximal kidney tubule. *Journal of Clinical Investigation*, **54**, 69–82.

Hanai, T. & Haydon, D. A. (1966). The permeability to water of bimolecular lipid membranes. *Journal of Theoretical Biology*, **11**, 370–82.

Hill, A. E. (1975*a*). Solute–solvent coupling in epithelia: a critical examination of the standing-gradient osmotic flow theory. *Proceedings of the Royal Society, B*, **190**, 99–114.

Hill, A. E. (1975*b*). Solute–solvent coupling in epithelia: contribution of the junctional pathway to fluid production. *Proceedings of the Royal Society, B*, **191**, 537–47.

Hodson, S. (1974). The regulation of corneal hydration by a salt pump requiring the presence of sodium and bicarbonate ions. *Journal of Physiology*, **236**, 271–302.

Hodson, S. & Miller, F. (1976). The bicarbonate ion pump in the endothelium which regulates the hydration of rabbit cornea. *Journal of Physiology*, **263**, 563–77.

House, C. R. (1974). *Water transport in Cells and Tissues*, p. 165, pp. 323–4. London: E. Arnold.

Huang, C. & Thompson, T. E. (1966). Properties of lipid bilayer membranes separating two aqueous phases; water permeability. *Journal of Molecular Biology*, **15**, 539–54.

Karnovsky, M. J. (1967). The ultrastructural basis of capillary permeability studied with peroxidase as a tracer. *Journal of Cell Biology*, **35**, 213–36.

Kaye, G. I., Sibley, R. C. & Hoefle, F. V. (1973). Recent studies on the nature and function of the corneal endothelial barrier. *Experimental Eye Research*, **15**, 585–613.

Leuenberger, P. M. (1973). Lanthanium hydroxide tracer studies on rat corneal endothelium. *Experimental Eye Research*, **15**, 85–91.

Lim, J. J. & Fischbarg, J. (1976). Standing-gradient osmotic flow: examination of its validity using an analytical method. *Biochimica et biophysica acta*, **443**, 339–47.

MacRobbie, E. A. C. & Ussing, H. H. (1961). Osmotic behaviour of the epithelial cells of frog skin. *Acta physiologica scandinavica*, **53**, 348–65.

Maurice, D. M. (1951). The permeability to sodium ions of the living rabbit's cornea. *Journal of Physiology*, **122**, 367–91.

Maurice, D. M. (1953). The permeability of the cornea. *Ophthalmic Literature, London*, **7**, 3–26.

Maurice, D. M. (1969). The cornea and sclera. In *The Eye*, vol. 1, ed. H. Davson, pp. 489–600. New York, London: Academic Press.

Maurice, D. M. (1972). The location of the fluid pump in the cornea. *Journal of Physiology*, **221**, 43–54.

Mishima, S. & Hedbys, B. O. (1967). The permeability of the corneal epithelium and endothelium to water. *Experimental Eye Research*, **6**, 10–32.

Mishima, S. & Kudo, T. (1967). *In vitro* incubation of rabbit cornea. *Investigative Ophthalmology*, **6**, 329–39.

Mishima, S. & Trenberth, S. M. (1968). Permeability of the corneal endothelium to nonelectrolytes. *Investigative Ophthalmology*, **7**, 34–43.

Ogilvie, J. T., McIntosh, J. R. & Curran, P. F. (1963). Volume flow in a series-membrane system. *Biochimica biophysica acta*, **66**, 441–4.

Oh, J. O. (1963). Changes with age in the corneal endothelium of normal rabbits. *Acta ophthalmologica*, **41**, 568–73.

Sackin, H. & Boulpaep, E. L. (1975). Models for coupling of salt and water transport: proximal tubular reabsorption in *Necturus* kidney. *Journal of General Physiology*, **66**, 671–733.

Segel, L. A. (1970). Standing-gradient flows driven by active solute transport. *Journal of Theoretical Biology*, **29**, 233–50.

Whittembury, G., Oken, D. E., Windhager, E. E. & Solomon, A. K. (1959). Single proximal tubules of *Necturus* kidney. IV. Dependence of H_2O movement on osmotic gradients. *American Journal of Physiology*, **197**, 1121–7.

Wright, E. M., Smulders, A. P. & Tormey, J. M. (1972). The role of the lateral intercellular spaces and solute polarization effects in the passive flow of water across the rabbit gallbladder. *Journal of Membrane Biology*, **7**, 198–219.

3. Fluid transport across *Necturus* gallbladder epithelium

By A. E. HILL

The *Necturus* gallbladder epithelium will transport fluid which is isotonic to the luminal bathing solution over a large range of osmolarities from 400 mOsm down to about 1 mOsm. At the very lowest osmolarities the secretion is hypertonic, but only by about 1 or 2 mOsmol per litre, as determined by flame photometry of the principal cations, which are Na^+ and K^+. In this condition the secretion is short lived, falling to very low values after about an hour. Electron micrographs obtained by Dr B. S. Hill show that the cells are very swollen, but that they preserve their identity and membranes. The secretion is inhibited by pretreatment with 10^{-3} M ouabain at normal plasma osmolarity; in addition the rate of fluid transfer during secretion is far in excess of that which could be produced by simple filtration through the hydraulic conductance of the preparation, driven by the small pressure difference pertaining in the experiments. It therefore seems that the fluid production is a true secretion, presumably driven by the sodium pump at the baso-lateral membranes of the epithelial cells.

The conclusions to be drawn from these experiments are the following: (i) If the cell contents are hypertonic to the secretion at the very low osmolarities then the secretion cannot be the result of osmotic equilibration across the baso-lateral membranes. (ii) If the cell contents are roughly isotonic to the secretion in the dilute condition, then two geometrical considerations come into play. The first is that standing-gradient osmotic theory as applied to the membranes bounding the interspaces requires that the separation between the membranes for a channel of this length (30 μm) should be only about a nanometre for the secretion to be 2 mOsm to a basal solution of 1 mOsm. This is not observed in electron micrographs. Secondly, the osmotic gradient over the luminal membrane could not exceed 1 mOsm, as a consequence of which the osmotic permeability of this membrane would have to be around 10^{-2} cm Osm^{-1} s^{-1} to support the observable volume flow through the system (Hill, 1975). (iii) If the luminal membrane has become a ghost, and water can easily enter the cell, then the cytoplasm is devoid of ATP and other supporting metabolites. It appears therefore that water crosses the epithelium via the junctional complex, i.e. by an extracellular route.

Recently we have been studying the entrainment of sucrose, urea, and inulin fluxes in the transepithelial water flow. These substances cross the epithelium in the secretion at a relatively high concentration. To explain the entrainment a model has been set up in which the water flow is partitioned between the interspace membranes and the junctions, and the non-electrolytes are assumed to move purely extracellularly, which is virtually certain. By assuming the sort of velocity profile which would be given by standing-gradient fluid flow into the interspaces it is possible to numerically solve the combined convection–diffusion equation for flow across the junction and down the interspace. To obtain the observable rates of non-electrolyte entrainment in the secretion, the water flow has to be adjusted until at least a very substantial fraction of it occurs across the junctions. Thus these studies confirm that the junctional route is almost certainly the major, if not the only, route of transepithelial fluid flow during secretion.

In a recent theoretical examination of fluid flow across the *Necturus* proximal tubule epithelium, Sackin & Boulpaep (1975) have also shown that standing-gradient osmosis probably contributes in a minor way to the fluid flow, and that the junctions must mediate a large fraction of the flow under conditions of isotonic transfer. How this flow across the junctions is driven will undoubtedly require a more detailed explanation during the next few years.

I should like to thank the Science Research Council of Great Britain for support and travel expenses, and the International Conference on Comparative Physiology for the invitation to attend the meeting.

References

Hill, A. E. (1975). Solute–solvent coupling in epithelia: a critical examination of the standing-gradient osmotic flow theory. *Proceedings of the Royal Society, B,* **190**, 99–114.

Sackin, H. & Boulpaep, E. L. (1975). Models for coupling of salt and water transport: proximal tubular reabsorption in *Necturus* kidney. *Journal of General Physiology,* **66**, 671–733.

4. Observations on the action of urea and other substances in opening the paracellular pathway in amphibian skins

By E. GONZÁLEZ, T. KIRCHAUSEN, H. LINARES &
G. WHITTEMBURY

Ussing & Windhager (1964) noticed that, if a frog skin is bathed on both sides with Ringer's solution, addition of urea to the solution bathing the outside to render this solution hyperosmotic resulted in a marked drop in the transepithelial resistance. Measurements of the flow of labelled sucrose (which should stay extracellularly) indicated that this decreased transepithelial resistance was the result of opening of the paracellular pathways under the action of urea. These observations have been extended to other amphibian skins and to toad bladder (DiBona & Civan, 1973; Ussing & Johansen, 1969; Wade, Revel & DiScala, 1973). Electron microscopical observations have confirmed that the phenomenon is indeed related to opening of the paracellular pathways (Ussing, 1971; Erlij & Martinez-Palomo, 1972). Most workers have centered their attention around the observation that transport of small molecules in an unexpected ('anomalous') direction may accompany the opening of the paracellular shunt pathways under given experimental conditions (Ussing, 1966; Franz & Van Bruggen, 1967; Biber & Curran, 1968). On the other hand, little attention has been paid to the observation that addition of other substances besides urea, at the same concentrations, may open the paracellular pathway to a different extent (Lindley, Hoshiko & Leb, 1964; Franz & Van Bruggen, 1967; Biber & Curran, 1968). Thus, in the kidney proximal tubule, urea has been found to be much more effective than mannitol in inducing a change both in permeability and in morphology of the paracellular pathways responsible for this increased permeability (Perez-González & Whittembury, 1974; Rawlins *et al.*, 1975).

The purpose of this paper is to review some experiments performed to study the phenomenon just described. For this purpose the effectiveness of several molecules in opening the paracellular pathway was compared. The main conclusion that stems from this work is that the paracellular shunt may be opened in the absence of osmotic gradients. Thus the opening of the paracellular pathways in epithelia is not exclusively an osmotic phenomenon. These observations relate to the paper by A. E. Hill in this volume (see Chapter 3) in that it may prove useful to study the functional relations between transepithelial active transport of sodium, and osmolar concentration of the transported fluid, under conditions in which the size of the paracellular pathway may be varied. The body of this work forms the Ph. Sc. Thesis of one of us (González, 1976). Part of this work has been published in a preliminary form (González *et al.*, 1975).

Materials and methods

Abdominal skins of *Bufo marinus* were mounted between Ussing chambers provided with electrodes to pass current and to measure the potential difference across the preparation (Ussing & Zerahn, 1951). Precautions were taken to avoid edge damage (Helman & Miller, 1971). After a suitable equilibration (control) period, several substances were added to the external bathing solution (experimental period). A first group of experiments was performed with Na_2SO_4 Ringer's solution bathing both sides of the preparation. These experiments showed that the test substances did not markedly alter the short-circuit current. Thus they could be used in subsequent experiments because they did not damage the transport properties of the skin and their action was reversible. Experiments in which sodium is being transported across the epithelium limit the interpretation of changes in the transepithelial resistance measurements to changes in the paracellular shunt pathway, since the transcellular route for ion transport which is effectively in parallel to the paracellular pathway has a resistance at least ten times lower than that of the paracellular shunt pathway (Ussing & Windhager, 1964). Consequently, we decided to increase the trans-epithelial resistance by decreasing the transcellular transport pathway. For this purpose K_2SO_4 was substituted for Na_2SO_4 in the outside bathing solution. With this procedure, values for the transepithelial resistances were of the order of 10^4 ohm cm^{-2}, to be compared with values ten times lower which were observed when the skin was transporting sodium. It is known that the paracellular pathway may be opened in the absence of trans-epithelial sodium transport (Ussing, 1966). Transepithelial resistances (R) were measured from the voltage changes induced by current pulses passed across the preparation. In this paper ratios of R are used, namely the value of R under the condition in which a given molecule had been added to the external solution ($R_{exper.}$) divided by the value of R before the molecule was added ($R_{control}$).

All bathing solutions contained in mmol dm^{-3}, $KHCO_3$, 2.4; calcium gluconate, 1; glucose, 5. In addition the solution bathing the outside of the skin contained K_2SO_4, 56, and that bathing the inside, Na_2SO_4, 56. The osmolarity of these solutions was 160 ± 1 mOsm during the control periods. The pH of the solutions was 8.0 (Whittembury, 1964).

Table 1. *Transepithelial resistance after addition of 100 mOsmol kg^{-1} of given molecules to the K_2SO_4 Ringer's solution bathing the external side of toad skin ($R_{exper.}$) expressed as a ratio of the corresponding resistance measured before ($R_{control}$) $\pm$ S.E.*

Average $R_{control}$ value was 9.61×10^3 ohm cm^{-2}. All experiments were performed at pH 8.0. Amines were added as sulfate salts. Anions as K$^+$ salts. N indicates number of skins explored

	Molecule	($R_{exper.}/R_{control}$)	N
(a)	Ethanol	1.10±0.07	16
	Ethylene glycol	0.96±0.05	24
	Glycerol	0.74±0.08	8
	Mannitol	0.62±0.05	16
	Acetone	0.98±0.03	8
	Formaldehyde	0.87±0.05	8
(b)	Urea	0.46±0.03	26
	Formamide	0.73±0.04	8
	Acetamide	0.70±0.05	8
	Biuret	0.80±0.08	8
(c)	Ethylamine	0.37±0.04	8
	Diethylamine	0.51±0.07	8
	Triethylamine	0.73±0.05	8
	Propylamine	0.54±0.03	16
	Butylamine	0.52±0.07	7
	Hydrazine	0.41±0.02	8
	Ethylenediamine	0.41±0.06	8
	Guanidine	0.47±0.09	8
(d)	Formate	0.43±0.03	8
	Acetate	0.53±0.02	24
	Butyrate	0.77±0.06	8
	Oxalate	0.36±0.05	8
	Glycine	0.61±0.06	8
	Phosphate	0.24±0.06	8
	Sulfate	0.66±0.09	8

Results and discussion

Effectiveness of molecules in lowering R

Inspection of Table 1 indicates that molecules differ in their effectiveness to lower R:

(a) Alcohols: Neither ethanol nor ethylene glycol lower R. The effectiveness of alcohols seems proportional to their number of OH groups. Not shown in the table is the observation that t-butanol and n-butanol were ineffective in lowering R. Thus an increase in the lipid solubility or in the

Table 2. *Effect of addition of urea, hydrazine and oxalate* (100 mOsmol kg^{-1}) *on transepithelial resistance in the absence* (0 *calcium*) *and in the presence of either* 1 *or* 10 mM *calcium in the solution bathing the outside of the toad skin*

Molecule	$R_{\text{exper.}}/R_{\text{control}}$ 0 mM Ca	$R_{\text{exper.}}/R_{\text{control}}$ 1 mM Ca	$R_{\text{exper.}}/R_{\text{control}}$ 10 mM Ca	N
Urea	0.32±0.05	0.46±0.03	0.47±0.02	8
Hydrazine	0.23±0.04	0.41±0.02	0.43±0.03	8
Oxalate	0.12±0.02	0.36±0.05	0.38±0.04	8

N as for Table 1; means ±S.E.

molecular weight of the probing molecule does not increase the effectiveness of the molecules to lower R. Acetone and formaldehyde were also ineffective.

(*b*) Amides: Urea was more effective than the alcohols and than any of the other amides tested.

(*c*) Amines: Amines proved to be very effective molecules in lowering R. As with alcohols, the presence of a long and heavier molecule with a higher lipid solubility diminished the ability of the amines to lower R. Thus ethyl, propyl and butylamine showed progressively decreasing R ratios. The number of free-H groups in the amino group seems important in enhancing the effectiveness of the molecule, since the R ratios decreased when H in the amino group was substituted (compare ethyl, diethyl and triethylamine). The presence of more than one amino group in a molecule did not result in an increased effectiveness (cf. ethylamine, ethylendiamine, hydrazine and guanidine).

(*d*) Anions were also very effective in lowering R. Again, increase in the length of the chain decreased the effectiveness of the molecule (cf. formate, acetate and butyrate). Anions with the lowest calcium solubility product were most effective (oxalate and phosphate). Importantly, glycine, a zwitterion, was also effective.

(*e*) Calcium ions: Bathing solutions contained 1 mmol dm^{-3} calcium gluconate. Removal of Ca^{2+} from the external bathing solution increased the ability of the molecules to induce a drop in R. This is shown in Table 2.

(*f*) Synergic effects: Urea, amines and anions were additive in their ability to induce a drop in R. Thus 100 mOsmol urea + 100 mOsmol acetate resulted in a R ratio of 0.22, not very different from that induced by 200 mOsmol of either urea (R ratio = 0.18) or acetate (R ratio = 0.15) alone. Similarly 100 mOsmol acetate+100 mOsmol propylamine resulted in an R ratio of 0.16. Thus molecules seem to act as single 'units', independently of whether they might neutralize the effect of another, as would be expected from a mixture of, for example, propylamine and acetate, where the possibility exists that they may form salts.

Role of ionic strength

The effectiveness of urea to lower R was explored at different ionic strengths under conditions in which the osmolarity of the external solution was kept constant at 160 mOsm during the initial control condition. The results are summarized in Table 3. Three solutions were used during the control period, namely K_2SO_4 Ringer's, or 1% K_2SO_4 Ringer plus 144 mOsm choline sulfate (which would give the same ionic strength as the K_2SO_4 Ringer's) or 1% K_2SO_4 Ringer's plus 144 mOsm ethylene glycol. Notice that addition of urea to the solutions of higher ionic strength (rows a, b, e, and f) produced a marked drop in R. On the other hand in solutions with reduced ionic strength 200 mOsm urea was ineffective (row d) and only 400 mOsm urea was slightly effective.

Table 3. *Effect of the ionic strength of the outside bathing solution maintained always at an osmolarity of* 160 mOsm *during the control period on the transepithelial resistance change induced by addition of* 200 *or* 400 mOsm *urea (or ethylene glycol) to that solution*

	Control solution			Added during experimental period			
	K_2SO_4 (mOsm)	E-glycol (mOsm)	Chol-SO_4 (mOsm)	E-glycol (mOsm)	Urea (mOsm)	$R_{exper.}/R_{control}$	N
a	160	—	—	—	200	0.18±0.01	8
b	16	—	144	—	200	0.19±0.04	8
c	160	—	—	200	—	0.89±0.07	8
d	16	144	—	—	200	1.00±0.04	16
e	160	—	—	—	—	0.12±0.04	12
f	16	—	144	—	400	0.12±0.03	4
g	160	—	—	400	—	0.64±0.07	16
h	16	144	—	—	400	0.60±0.11	8

N as for Table 1; means ±S.E. E-glycol = ethylene glycol; Chol-SO_4 = choline sulfate.

Table 4. *Effect of the addition of urea to the transepithelial resistance in the absence of transepithelial osmotic gradients making the outside bathing solution hyperosmotic as compared with the inside bathing solution*

The internal bathing solution was Ringer's sodium gluconate (*a*) and Na_2SO_4 (*b, c*) with the osmolarities indicated. The external bathing solution was Ringer's potassium gluconate (*a*) and K_2SO_4 (*b, c*) with the osmolarities indicated. The changes observed after addition of urea were not observed when ethylene glycol was added (instead of urea)

| | Osmolarity of bathing solutions (mOsm) | | | | | |
| | Outside solution | | | | | |
Ringer	Control solution	Urea added	Exper. solution	Inside solution	$R_{exper.}/R_{control}$	N
a Gluconate	100	100	200	200	0.87 ± 0.02	20
b Sulfate	40	200	240	240	0.54 ± 0.06	4
c Sulfate	31	259	290	310	0.51 ± 0.07	5

N as for Table 1; means $\pm$s.e.

Effect of urea in the absence of external hyperosmolarity

Results given in Table 4 show that the outside solution need not be hyperosmotic (after addition of urea) as compared to the solution bathing the inside of the skin for a drop in the value of R to be observed, provided that the external solution has a given ionic strength and provided that urea is added in sufficient amounts. Row *c* shows in addition that a significant drop in the value of R is observed even with an external solution 20 mOsm, hypo-osmotic as compared with the internal solution.

The results shown in Tables 1–4 clearly indicate that osmotic effects are not of prime importance to open the paracellular pathway.

Effect of changes in the concentration of H⁺ on transepithelial resistance

A change in the pH of the external bathing solution or of both solutions resulted in marked drop in the R ratio. This is shown in Table 5. Again the presence of an increased external osmolarity is not required to induce a crop in R (cf. Schoffeniels, 1955).

Table 5. *Effect of changes in the* H^+ *concentration on transepithelial resistance*

Skins were bathed in K_2SO_4 Ringer's solution (outside) and Na_2SO_4 Ringer's solution (inside). Results were similar whether both sides or only the external solution had an altered H^+ concentration. No effects on transepithelial resistance were observed if only the H^+ of the internal solution was changed. $N = 8$ for each experiment

H^+ mol dm^{-3}	$R_{exper.}/R_{control}$	H^+ mol dm^{-3}	$R_{exper.}/R_{control}$
7×10^{-12}	0.029 ± 0.004	1×10^{-6}	1.10 ± 0.08
1×10^{-11}	0.65 ± 0.06	1×10^{-5}	1.13 ± 0.10
1×10^{-10}	0.85 ± 0.12	1×10^{-4}	1.25 ± 0.12
1×10^{-9}	0.95 ± 0.04	1×10^{-3}	0.80 ± 0.11
1×10^{-8}	1.00 ± 0.03	4×10^{-3}	0.034 ± 0.004

Means $\pm$ S.E.

Measurements of sucrose fluxes under the conditions described above

Sucrose fluxes were measured across the skin to ascertain whether the observed changes in the R ratio were caused by changes in either the paracellular or the cellular routes for ion movement. Since sucrose is a molecule presumably confined to the extracellular space, and its flux across the skin is negligible under control conditions, a sizable increase in the transepithelial sucrose flux should reflect increased permeability of the paracellular pathways. Our measurements indicated that sucrose fluxes increased markedly in all the experimental conditions described above in which a significant drop in R was observed. Thus in all probability the observed changes in R result from opening of the paracellular pathways in toad skin.

Hypothetical model for the zonulae occludentes

Most of the changes in R described above should refer to the zonulae occludentes, since they are the place of closest apposition between cells (Farquhar & Palade, 1965). They should represent therefore the place with the highest resistance in the paracellular pathway. Measurements of dilution potentials across the toad skin as a function of changes in pH indicate that the pathways used by the ions producing the diffusion potentials have fixed negative charges at pHs between 6 and 9. These charges become neutral between pH 4 to 5 and below pH 4 the net sign of the charges becomes positive. The following hypothetical model seems compatible with the present observations. The paracellular pathway may be envisaged at this level to be lined by an array of fixed negative charges. These would arise from the predominance of anionic groups at pHs of 6 to 9 – for example, end carboxyl or phosphate groups which would be fully dissociated at these pH values – over cationic groups formed mainly by end $-NH_2$ groups, which would not be dissociated at these pH values. The net anionic charges of opposite cells would be kept together by Ca^{2+} which would form bridges crossing the intercellular junctions from one cell to another thus apposing the cells at the zonulae occludentes. Addition of urea or of amines would react (as electron-pair acceptors) with these anionic (ligand) groups thus displacing Ca^{2+}. As a consequence Ca^{2+} would be liberated from the junctions and cells would tend to separate. An addition of anions may bind Ca^{2+}, thus displacing it from its membrane location. Cells would separate under these circumstances. At a pH lower than 5 dissociation of anionic groups would be weaker. Thus the sign of the charges facing the paracellular channel would be neutral at a pH of about 5. Below pH 4 there would be a predominance of highly dissociated $-NH_3^+$ groups. Ca^{2+} would be similarly liberated under these circumstances. Recent experiments indicate that small amounts of Ca^{2+} are indeed liberated towards the outside bathing solution after addition of urea or when the pH of the external solution is lowered (González, 1976). Ca^{2+} seems to have the right size to fit into the width of a few nanometers that may exist at the level of the zonulae occludentes (Pethica, 1961; Bianchi, 1969). The possibility that hydrogen bonds intervene to keep the zonulae occludentes tight should be kept in mind. This could explain the mode of action of the polyalcohols with more than one $-OH$ group referred to before. Obviously much work remains to be performed to test the validity of this model.

Physiological significance of these observations

The fact that the permeability of the paracellular pathway is increased in the absence of osmotic gradients by either a high urea concentration or by changes in pH raises the possibility that these changes may be relevant to epithelial function. One immediate thought is the possibility that sodium transport across amphibian skins may become isosmotic when the junctions are opened. This might allow us to check present models for isosmotic transport in epithelia, as discussed in the paper by A. E. Hill in this volume.

In other transporting epithelia like the mammalian kidney tubule, it is known that the distal nephron segments have closely sealed zonulae occludentes under control circumstances (Tisher & Yarger, 1975). However, an increase in the paracellular permeability to non-electrolytes has been observed in distal nephron segments subsequent to increased intraluminal osmolarity (De Bermudez & Windhager, 1975), and even more to increased intraluminal urea concentrations (Whittembury *et al.*, 1976). It is known that the distal nephron segments are where the lowest tubular fluid pH and the highest urea concentrations are observed. Increased concentrative power of the kidney is observed in animals that receive a diet that results in high urea excretion under apropriate circumstances. It is tempting to suggest that these changes in the tubular fluid composition of the distal urine may open the paracellular pathway, although this increased permeability might not be detectable by studying lanthanum penetration. The collecting duct urine would equilibrate through this pathway and not necessarily via the transcellular pathway subsequent to the action of the countercurrent system.

It is a pleasure to thank Dr G. Chuchani for continuous advice during this work. Part of this work was supported by CONICIT and the OAS.

References

Bianchi, C. P. (1969). *Cell Calcium*. London: Butterworths & Co.

Biber, T. U. L. & Curran, P. F. (1968). Coupled solute fluxes in frog skin. *Journal of General Physiology*, **51**, 606–20.

De Bermudez, L. & Windhager, E. E. (1975). Osmotically induced changes in electrical resistance of distal tubules of rat. *American Journal of Physiology*, **229**, 1536–46.

DiBona, D. R. & Civan, M. M. (1973). Pathways for movement of ions and water across toad urinary bladder. I. Anatomic site of transepithelial shunt pathways. *Journal of Membrane Biology*, **12**, 101–28.

Erlij, D. & Martinez-Palomo, A. (1972). Opening of tight junctions in frog skin by hypertonic urea solutions. *Journal of Membrane Biology*, **9**, 229–40.

Farquhar, M. G. & Palade, G. E. (1965). Cell junctions in amphibian skin. *Journal of Cell Biology*, **26**, 263–91.

Franz, T. J. & Van Bruggen, J. T. (1967). Hyperosmolarity and the net transport of nonelectrolytes in frog skin. *Journal of General Physiology*, **50**, 933–49.

González, E. (1976). Propiedades de permeabilidad de la via paracelular de un epitelio con alta resistencia eléctrica (piel aislada de *Bufo marinus*). Ph. Sc. Thesis. Instituto Venezolano de Investigaciones Cientificas, Caracas.

González, E., Kirchhausen, T., Linares, H. & Whittembury, G. (1975). Electrical resistance and osmotic gradients in amphibian skins. In *Proceedings of the 5th International Biophysics Congress*, Abstract P-150, Copenhagen, Denmark.

Helman, S. I. & Miller, D. A. (1971). *In vitro* techniques for avoiding edge damage in studies of frog skin. *Science, Washington*, **173**, 146–8.

Lindley, B. D., Hoshiko, T. & Leb, D. E. (1964). Effects of D_2O and osmotic gradients on potential and resistance of the isolated frog skin. *Journal of General Physiology*, **47**, 774–93.

Perez-González, M. & Whittembury, G. (1974). Widening of the paracellular pathway in the kidney tubule by a transtubular osmotic gradient: passage of graded size non-electrolytes. *Pflügers Archiv für die gesamte Physiologie*, **351**, 1–12.

Pethica, B. A. (1961). The physical chemistry of cell adhesion. *Experimental Cell Research, Supplement* **8**, 123–40.

Rawlins, F. A., González, E., Perez-González, M. & Whittembury, G. (1975). Effect of transtubular osmotic gradients on the paracellular pathway in toad kidney proximal tubules: electron microscopic observations. *Pflügers Archiv für die gesamte Physiologie*, **353**, 287–302.

Schoffeniels, E. (1955). Influence du pH sur le transport actif de Na à travers la peau de grenouille. *Archives internationales de Physiologie et Biochimie*, **63**, 513–30.

Tisher, C. C. & Yarger, W. E. (1975). Lanthanum permeability of tight junctions along the collecting duct of the rat. *Kidney International*, **7**, 35–43.

Ussing, H. H. (1966). Anomalous transport of electrolytes and sucrose through the isolated frog skin induced by hypertonicity of the outside bathing solution. *Annals of the New York Academy of Sciences*, **137**, 543–55.

Ussing, H. H. (1971). Introductory remarks. *Philosophical Transactions of the Royal Society, B*, **262**, 85–90.

Ussing, H. H. & Johansen, B. (1969). Anomalous transport of sucrose and urea in toad skin. *Nephron*, **6**, 317–28.

Ussing, H. H. & Windhager, E. E. (1964). Nature of shunt path and active Na transport path through frog skin epithelium. *Acta physiologica scandanavica*, **61**, 484–504.

Ussing, H. H. & Zerahn, K. (1951). Active transport of sodium as the source of electric current in the short-circuited isolated frog skin. *Acta physiologica scandanavica*, **23**, 110–27.

Wade, J. B., Revel, J. P. & DiScala, V. A. (1973). Effect of osmotic gradients on intercellular junctions of the toad bladder. *American Journal of Physiology*, **224**, 407–15.

Whittembury, G. (1964). Electrical potential profile of the toad skin epithelium. *Journal of General Physiology*, **47**, 795–808.

Whittembury, G., Proverbio, F., González, E., Perez-González, M. & Rawlins, F. A. (1976). Molecular sieving properties of the paracellular pathway in kidney tubules. Effect of urea. In *Proceedings of the 6th International Congress of Nephrology, Florence, 1975*, pp. 118–24. Basel: Karger.

(B) Uptake of water vapour by arthropods

5. Absorption of water vapour by *Thermobia domestica* and other insects

By J. NOBLE-NESBITT

The work of Buxton (1930) and Mellanby (1932) on the larva of the mealworm, *Tenebrio molitor*, first drew attention to the possibility that insects could gain water from subsaturated atmospheres of high relative humidity (RH). Subsequently, a limited number of other insects (including the firebrat, *Thermobia domestica* – Beament, Noble-Nesbitt & Watson, 1964) and acarines have been shown to possess the same ability, with some being capable of absorbing water from RHs down to 43%, the limiting humidity for net uptake being known as the critical equilibrium humidity (CEH). In general terms it has been clearly established that these terrestrial arthropods absorb water from the vapour phase of the atmosphere into their tissues and body fluids against gradients of water activity which are always great and often very considerable indeed. Most of the known cases have been listed, and the topic reviewed, a number of times in the past two decades, notably by Beament (1954, 1964, 1965), Edney (1957), Noble-Nesbitt (1969, 1970*b*, 1973, 1977), Schmidt-Nielsen (1969), Berridge (1970), Maddrell (1971), Ramsay (1971), House (1974), Ebeling (1974) and Stobbart & Shaw (1974), and these articles represent a valuable fund of information and ideas on the subject, though some reappraisal is now required. Much of the early work was concerned with the overall process, and engendered ideas (which could be subjected to experimental investigation) regarding possible sites and mechanisms of uptake. I shall consider some recent advances in these lines of investigation. In doing so, I will concentrate on my own work, which has been mainly concerned with *Thermobia*, though I have also looked at some other species. Where appropriate, I shall relate these studies to other known cases of water vapour uptake in an attempt to draw out general principles which may be applicable to all cases, to draw attention to possible dichotomies, and so to gain a more complete perspective of the topic as a whole.

Attempts to locate site of uptake

Early attempts to locate a specific site for the uptake of water from subsaturated atmospheres came to nothing. In the absence of any clear-cut evidence indicating a specific site, it was generally assumed that uptake occurred over the whole external surface of the body, in a process that was akin to the reversal of transpiration. Such was the position six years ago, when I reported the results of a series of experiments conducted on *Thermobia* and *Tenebrio* which clearly indicated that the site of uptake was the rectum in these particular insects (Noble-Nesbitt, 1970*a,b*). Subsequent investigations (some of which are reported in Noble-Nesbitt, 1973) confirmed this conclusion for these insects. These findings aroused great interest and stimulated a number of laboratories into reinvestigating the site of uptake in other insects and other arthropods. Indeed, some of the most notable of these are represented by the authors of Chapters 5–9, and we are able to note the strides they are taking in elucidating further the process of uptake, its site and possible mechanisms. Of these, precise location of the site of uptake is a critical first step in detailed study of the process and mechanism.

It is perhaps as well to recapitulate briefly the experimental basis for the location of the site of uptake in the various species used by different investigators. Simple blocking procedures, using wax or wax–resin mixtures, have been extensively employed, and were used in my original series of experiments (Noble-Nesbitt, 1970*a,b*). The results were remarkably clear-cut, with little evidence of side effects. In summary, application of paraffin wax to different parts of the body and to the mouth so as to block it did not significantly affect uptake, once the insect had recovered from the immediate trauma of the operation, which took a few hours only. On the contrary, blockage of the anus with wax invariably prevented uptake. The rectal avenue of uptake in *Thermobia* was clearly established. Similarly, uptake was prevented in *Tenebrio* when the anus was capped with wax–resin mixture, but was restored when the capping was removed, providing this was done early enough, before secondary blockage or necrosis set in (Noble-Nesbitt, 1973). It has been suggested that blocking the anus with wax in this manner may affect the uptake process in other ways than by simple occlusion of the airway into the rectum (Okasha, 1971). That this is not the case in *Thermobia* has been shown by isolation of the abdomen tip in a separate chamber of a double-chambered array (Noble-Nesbitt, 1973, 1975). With this technique the anus remains free to open and close, but rapid uptake only occurs when the tail-end chamber is opened to an ambient atmosphere of high humidity. The humidity in the closed tail-end chamber is lowered to approximately 50 % RH, whereas the closed head-end chamber remains the same. Similar results have been obtained with *Tene-*

brio (Noble-Nesbitt, unpublished observations). Further support for the involvement of the rectum comes from the ligature experiments of Dunbar & Winston (1975) on *Tenebrio*. Posterior ligatures, presumably because they effectively isolate the main part of the body from the rectum, prevented uptake, whilst anteriorly-placed ligatures only diminished uptake. Confirmation of the effect of blocking the anus in this insect came from the work of Machin (1975). All of this experimental evidence supports the conclusion that the rectum is the site of uptake in the firebrat and the mealworm.

In contrast to this, however, as also reported elsewhere in this volume by Rudolph & Knülle (Chapter 8), are the results with ixodid ticks, which indicate an oral avenue of uptake (Knülle & Devine, 1972; McEnroe, 1973; Rudolphe & Knülle, 1974). These investigations employed both blocking and isolation procedures. Blocking experiments carried out in my own laboratory suggest that this may also be the case in argasid ticks. The site of uptake in mites is reported by Wharton (Chapter 7) to be the supra-coxal glands. These results raised the question of whether alternative sites may occur in different insect species.

I was fortunate whilst on sabbatical leave at the University of Colorado at Boulder in having a limited number of specimens of the desert cockroach, *Arenivaga investigata*, sent to me from Professor Edney's laboratory. He originally showed that this insect is also capable of absorbing water vapour from RHs down to 82.5% (Edney, 1966). A few of the specimens were successfully mounted in double-chambered arrays. They were found to take up water rapidly when the head-end only was exposed to high humidity, but not when the tail-end only was so exposed, suggesting that oral uptake may occur in this species (see Fig. 1). They were observed to extrude salivary secretions when feeding on dry food particles, thus wetting the food before ingestion. Similar behaviour was also noticed in desiccated individuals during uptake, salivary secretion apparently being extruded and then withdrawn, raising the possibility that the salivary secretion is involved in the uptake process. This system has been more extensively studied by O'Donnell and his results are reported elsewhere in this volume (Chapter 9). They support the conclusion that oral uptake occurs in *Arenivaga*.

The present conclusions concerning the site of uptake must therefore be that it can, and does, differ between species. Certainly this is the case in different arthropods as evidenced by the insects, mites and ticks. The recent results with *Arenivaga* indicate that this also applies to different species of insect. An interesting corollary of this is that not only may the site differ, but different sites may mean different mechanisms. What is clear, however, is that with a reasonably well-defined site of uptake, the process and the mechanism can be investigated with greater certainty.

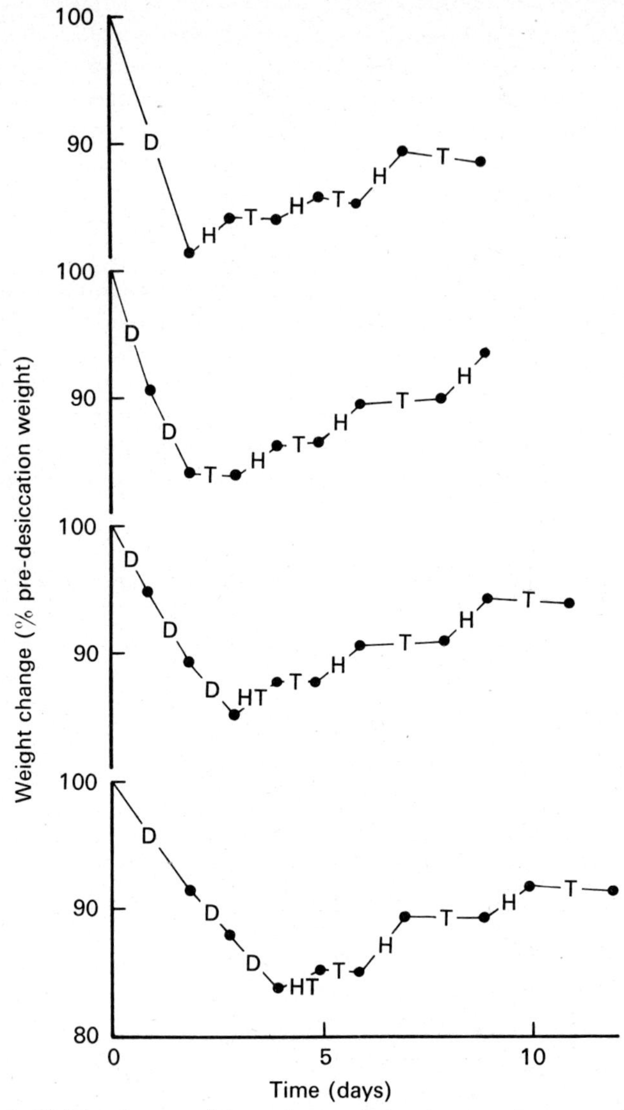

Fig. 1. Weight changes of 4 specimens of *Arenivaga* each cemented into a double-chambered array following desiccation (D) and during 3 cycles of reversible arrest of uptake in hydrating conditions (93% RH) with the tail-end only (T) and the head-end only (H) or both ends (HT) alternately exposed to the ambient humidity. Rapid uptake occurs when the head only, or head and tail, are exposed, but not when the tail only is exposed.

(*a*) Longitudinal section

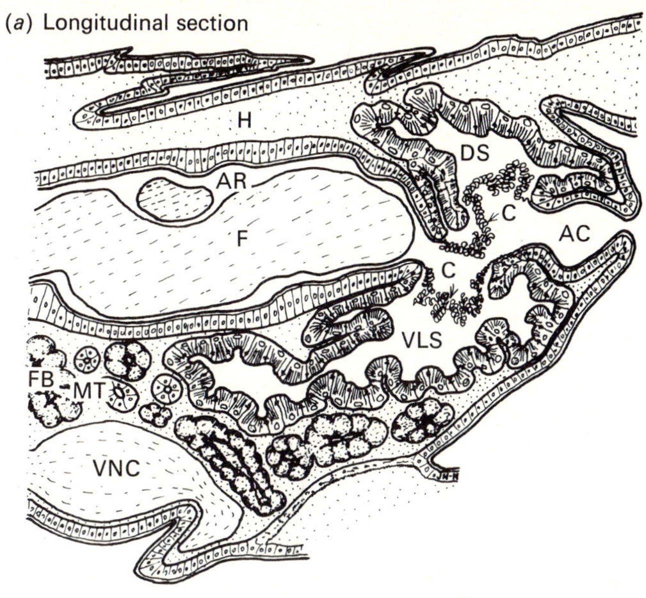

(*b*) Transverse section

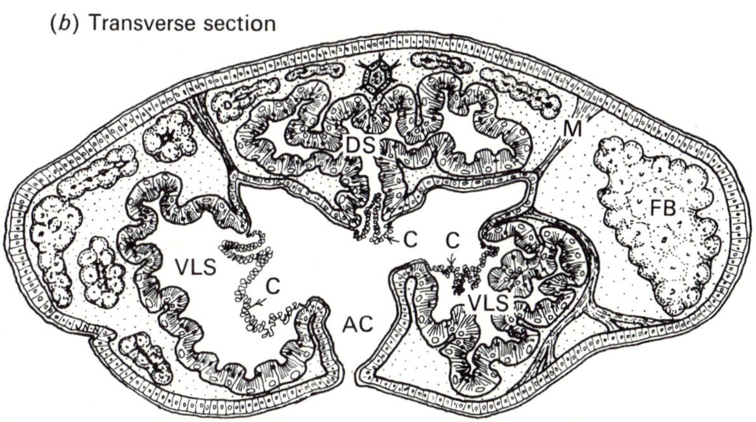

Fig. 2. Diagrams showing (*a*) vertical-longitudinal, and (*b*) transverse profiles, of the posterior sacculate region of the rectum of *Thermobia* and its relation to surrounding structures. *Key*, AC = anal canal; AR = anterior rectum; C = cuticle, lifted away from sac epithelium; DS = dorsal sac of posterior rectum; F = faecal pellet/faeces; FB = fat body; H = haemolymph/haemocoele; M = dilator muscle of posterior rectum; MT = Malpighian tubule; VLS = ventro-lateral sac of posterior rectum; VNC = ventral nerve cord.

The rectal site of *Thermobia*

The firebrat has a clearly differentiated, sacculate posterior region of its rectum (Noble-Nesbitt, 1970*a,b*, 1973; Noirot & Noirot-Timothée, 1971). The arrangement is shown in Fig. 2. The maximum volume of these sacs in a 30 mg insect is approximately 250 nl. This is the region which is thought to be the actual site of uptake in this insect. The fine structure is remarkable, with the main absorbing epithelium being a single layer of cells, thrown into folds forming the saccules. Each epithelial cell has a hypertrophied apical (luminal) plasma membrane, decorated on its cytoplasmic surface with a 15 nm wide particulate coating and thrown into pleated folds which embrace rod-like mitochondria in a very dense and regular array, orientated perpendicular to the main cell surface. Junctional complexes occur on the lateral plasma membranes, near the luminal surface. Otherwise, the lateral and basal plasma membranes are unexceptional. In fixed preparations, the overlying cuticle is usually lifted well away from the cell surface. A faintly granular material fills the subcuticular space, and may contain mucopolysaccharide (Noirot & Noirot-Timothée, 1971). This material probably penetrates between the apical folds of the cell. As we shall see later, it may play a role in uptake.

It is to this posterior sacculate region of the rectum that I have directed my investigations aimed at elucidating further the process and mechanism of uptake.

The mechanism of uptake in *Thermobia*

A number of mechanisms which have been suggested as possible means of absorbing water vapour and rendering it freely available to the internal tissues have been or are in the process of being closely examined. It is my intention to consider some of these possible mechanisms, to review what has been learned about them in recent years, and where possible to indicate the present status of each. I shall draw mainly on my own work, including current work, on *Thermobia* especially, but I shall also refer to other examples where possible, in an attempt to relate the various suggested mechanisms to the observed process of uptake.

Pressure changes

The interesting possibility that cyclical pressure changes could be involved in the mechanism of uptake was raised a few years ago in a review article by Maddrell (1971). The basic idea was that muscular compression of the rectum causing a pressure increment within the enclosed rectal air spaces would raise the RH. A sufficiently great increase in pressure would lead to condensation of water on the intima of the rectum, where it would be absorbed into the tissues. Release of pressure and replenishment of the air in the rectum from the surrounding atmosphere, perhaps during a phase of rectal musculature relaxation and expansion, would prime the system for another compression phase. Even though the possibility was an intriguing one, it had to be ruled out as a complete answer because the frequency of the cycles required for the observed maximal rate of uptake of 5 μg min^{-1} (estimated at 80 cycles s^{-1} for a rectal volume of 100 nl in *Thermobia*) was too great. Even if the rather larger estimate of 250 nl for the volume of the rectal sacs in *Thermobia* is taken, 30 cycles s^{-1} are still required, and this still seems too high. Pulsating movements of the anal valves of *Thermobia* (as distinct from wide opening of the valves) are observed during uptake periods, however, and there remained the possibility that pressure changes may make some contribution to uptake.

This possibility has been investigated by making use of the change in volume which would occur if the pressure were raised. As Maddrell pointed out, doubling of the pressure would double the RH. It would also halve the volume of the rectal air spaces. This would be the extreme situation for *Thermobia*, where CEH is below 50% RH. Intermediate pressure changes would give correspondingly smaller RH increments, and correspondingly smaller volume decrements. If we consider a 10% RH change as probably the minimum to be of any real significance in the process of uptake, then a 20% increase in pressure would be necessary, leading to a 20% decrease in volume. In terms of *Thermobia*'s rectal sacs, this change in volume would amount to approximately 50 nl, and should be measurable given a fine enough technique. To this end, I enclosed individual specimens in small volume (approximately 0.5 ml) transparent containers. A small drop of saturated potassium chloride solution was included to give 85% RH and the container sealed with a microcapillary outlet tube which itself was sealed with a paraffin (kerosene) indicator droplet. The meniscus of the oil droplet was viewed under a microscope fitted with an ocular micrometer. Using this system, volume changes as small as 0.25 nl could be measured; even smaller changes could be seen as minute movements of the meniscus. The system responded rapidly to induced volume changes in the main container in control observations. The insect could also be viewed microscopically, and movement of the anal valves was checked. During a period when uptake occurred (as checked by weight gain) and the anal valves pulsated, no measurable cyclical volume changes took place. From this, we must conclude that pressure effects play no direct, significant part in the uptake process. The pulsations of the anal valves are likely to be involved in tidal replenishment of the air in the rectal sacs (Noble-Nesbitt, 1973, 1977).

Ion-linked water transport

Active transport of inorganic ions, with water following passively, is favoured as an interpretation of water movements in many biological systems, and could easily provide an explanation for the uptake of water vapour, since concentrated solutions of pure salts will take up water from the vapour phase provided that the RH is high enough. However, even saturated solutions of some of the salts more commonly found in organisms lose water to atmospheres with RHs in the range where some insects are still capable of taking up water. Perhaps the problem can best be put into perspective by listing some of the salts together with the RH at which their saturated solutions neither gain nor lose water (the CEH) at room temperature. For potassium sulphate (K_2SO_4) it is 98% RH; for sodium sulphate (Na_2SO_4), 93% RH; for potassium chloride (KCl), 85% RH; for sodium chloride (NaCl), 75% RH; for potassium carbonate (K_2CO_3), 44% RH; for potassium nitrite (KNO_2), 43% RH and for calcium chloride ($CaCl_2$), 32% RH. Some change in these values results if the temperature changes. Those as far as NaCl could explain many examples of uptake, ranging through most of the acarines and many insects, including *Tenebrio*. For the firebrat, *Thermobia domestica* (CEH below 50% RH), the lower-range salts would be sufficient. These, however, represent extremely high ionic concentrations, and one of the problems to be solved is how such high concentrations are produced. But before considering that problem, we can ask the simple question of whether locally high concentrations of ions occur. And because we have a precise location of the site or uptake, we can direct this question specifically to particular regions of the animal.

By courtesy of the Biological Microprobe Group in Cambridge, I have been looking into this aspect in *Thermobia*, using sections of frozen-hydrated and frozen-dried material for electron-probe X-ray microanalysis. Although this work is continuing, all of the results obtained point to the absence of unphysiologically high concentrations or unusual distribution of any of the common ions, K^+, Na^+, Ca^{2+} and Cl^-, in any part of the rectal sac tissue examined. The ionic profiles obtained from cells and haemolymph indicate that normal, physiological ionic distributions are preserved throughout the examination, and suggest that any local concentration of ions would equally be preserved. From this we must conclude that local high concentrations of ions do not seem to be involved in uptake in *Thermobia*, a conclusion which is supported by the ultrastructure of the sac epithelium, which is unlike that found in other insects where water transport is thought to be ion driven.

Organic molecules

A further possibility is that organic molecules are used as part of the absorbing mechanism, either by cyclical secretion and reabsorption (when entrained water would be taken into the tissues), or by cyclical configurational changes which would alternately increase and decrease their affinity for water. The latter case would seem to require some valve-like arrangement whereby the organic material would be presented alternately to the water vapour of the air phase (during increase of affinity) and to the tissues (during decrease of affinity) for it to operate successfully. The former case is an intriguing suggestion (especially if configurational changes occur at the same time) and warrants critical investigation. Indeed, if the organic molecule is so closely associated with the epidermal cell membrane as to be effectively restricted to it, and in essence therefore is acting as a 'carrier' molecule for water, we would in fact be dealing with a case of primary transport of water, as Maddrell (1971) pointed out.

The fine-structural arrangement of the hypertrophied apical plasma membrane of *Thermobia*'s rectal sac epithelial cells, with its particulate inner coating and its close proximity to the outer mitochondrial membrane, suggests that just such a system may operate in the firebrat, though a good deal of further investigation is required before we are able to pronounce with any great degree of certainty on this possibility.

It has been suggested by Diamond (quoted in House, 1974) and Weis-Fogh (personal communication) that glycerol may be the substance involved. Glycerol will take up water from any RH and at first sight this makes it an attractive choice. The insects studied, however, cease absorption at a distinct and narrow RH range and this fact needs to be borne in mind. It seems likely that if a 'carrier' molecule is involved, it will have this same characteristic property, or at least will alter significantly over this RH range. If glycerol is involved, viscosity changes in the relevant RH range may be an important factor in limiting uptake to the higher RHs.

In the firebrat, the subcuticular mucopolysaccharide material may play an important role in the uptake mechanism, especially if it is continually being secreted and reabsorbed, or cyclically altered in configuration. It may be the cause of the presumed osmotic swelling seen in the extracellular channels of the apical region of the cells during normal fixation of desiccated individuals (Noble-Nesbitt, 1973).

Regulation

Since the site of uptake has been located with fair certainty only comparatively recently, and since the means by which uptake occurs is only now emerging in some cases, it is perhaps inevitable that little is still known regarding the important aspect of how it is regulated. We do know, however, something of what will affect uptake and this gives us some insight into how it may be regulated. I will now briefly recall some of these points.

Volume regulation

Previously desiccated *Thermobia* will rehydrate to their pre-desiccation weight over short periods. With longer periods, starvation losses become apparent, though the overall drop in weight is not *pro rata* with the loss in dry matter (resulting from respiration), but somewhat less. This has been regarded as evidence for some measure of volume regulation (Noble-Nesbitt, 1969; Okasha, 1971). In turn, this suggests that the insect has some means of monitoring volume, possibly by stretch receptors, and of reacting to it by turning uptake on or off. This could be mediated entirely through the nervous system, or in conjunction with the endocrine system. It deserves further investigation.

Phase of moulting cycle

Uptake is known to cease during the moulting cycle, yet can recommence within 3 h of ecdysis in *Thermobia* (see Fig. 3). Losses during the moulting cycle are then made up. The insect attains a weight greater than its immediate pre-ecdysis weight, giving the appearance of overcompensatory gains. Since the pre-ecdysis phase of arrest may be prolonged in some

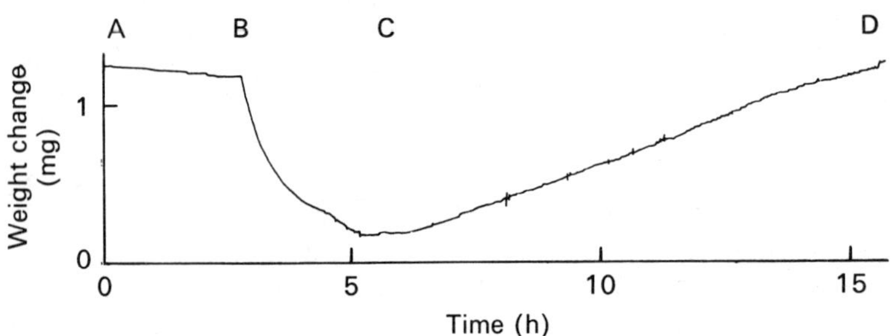

Fig. 3. Tracing from a continuous electrobalance weight record made during a moult in *Thermobia*. At A, the moult is underway with separated cuticles, and slow weight losses occur. At B, ecdysis commences and is marked by a sudden, rapid weight loss, presumably because of evaporation of moulting fluid exposed when the old cuticle splits. At C, within 3 h of ecdysis, net uptake is evident, and this continues past the pre-ecdysial weight level (D). The exuvia remained on the balance pan.

insects – more than 1 or 2 days – this is an important point to consider when results are analysed. My own long-term starvation–hydration experiments with *Tenebrio* tentatively suggest that in this insect real overcompensatory gains occur just after the moult. This may have some significance for subsequent growth when it is again allowed food.

The physiological changes during the moult in *Thermobia* are paralleled by profound changes in the fine structure of the posterior sacculate rectal epithelium, further strengthening the view that this tissue is responsible for uptake. Total regression of the mitochondria–apical plasma membrane complex occurs early in the moult and coincides with cessation of uptake. It is reformed at the end of the moult when uptake recommences (Noble-Nesbitt, 1973).

Feeding and metabolic water production

The early experiments of Buxton (1930) and Mellanby (1932) on starved mealworms indicated that metabolic water augmented the insect's water reserves, and that its production could be regulated to some extent, increasing in lower humidities. However, Mellanby was able to show that metabolic water production in starving mealworms did not account for the large weight gains found when desiccated insects were placed in high humidities. He concluded that water vapour was absorbed from the air by these starving insects.

Metabolic water will also augment the water reserves in the feeding insect, and this too may assume greater proportions when feeding occurs in drier atmospheres.

Feeding allows growth, requiring an increased water supply to maintain within tolerable limits the balances between water content, dry weight and body size. This increased supply of water comes partly from increased metabolic water production and partly from increased uptake of water vapour, in high humidities. Feeding thus stimulates uptake of water vapour. This conclusion is supported by recent studies on *Thermobia* (Okasha, 1971) and *Arenivaga* (my own observations), which show that net uptake occurs during a feeding period.

Means of regulation

The firebrat simply closes its anal aperture tightly in desiccating conditions. In hydrating conditions, the regular pulsations of the anal valves seen during continuous uptake become less regular and less frequent as net uptake falls off. In *Tenebrio* a similar mechanism may occur. In those cases where water vapour is absorbed orally, the volume of the secretions presumably increase and decrease during and after uptake. The recent work of McMullen, Sauer & Burton (1976) suggests this for ixodid ticks. This aspect is also dealt with by other contributors elsewhere in this volume (see Chapters 7–9), and I will not dwell on it here.

What is clear, however, is that the underlying regulatory mechanisms could now usefully be investigated more intensively.

General conclusions

Sites of uptake have now been located. In *Thermobia* and *Tenebrio* this is the rectum, but in *Arenivaga*, as in ixodid and probably argasid, ticks an oral site seems likely. Some possible mechanisms of uptake have been investigated. Pressure changes do not appear to play a part in *Thermobia*. Neither does inorganic-ion-linked transport, which, however, may be the basis of many of the less exacting cases of uptake in other species. We still seem to be thrown back on the possibility that *Thermobia* depends upon an organic-molecule-linked transport, possibly involving membrane-attached enzyme systems with mitochondria in close juxtaposition. In other words, to the real possibility that a primary transport of water – active transport in the strict sense – may occur. This possibility is under investigation.

Some of the results quoted for *Tenebrio* and *Arenivaga* derive from work carried out whilst I was on sabbatical leave as a Senior Fulbright-Hayes Scholar at the Department of Environmental, Population and Organismic Biology, University of Colorado, Boulder, supported by NSF grant GB14167 awarded to Professor P. W. Winston, to whom I am indebted for hospitality and facilities. The living *Arenivaga* were supplied by Professor E. B. Edney's laboratory and I am grateful to him and to Mr P. Franco, who collected and despatched them for me. The work on electron-probe X-ray microanalysis of *Thermobia* rectum was carried out in the Biological Microprobe Laboratory, Department of Zoology, University of Cambridge, supported by a grant from the Science Research Council awarded to the late Professor T. Weis-Fogh and Drs P. Echlin, B. L. Gupta, T. A. Hall and R. B. Moreton, to whom I am grateful, together with Dr D. A. Parry for facilities and Dr A. Saubermann for advice.

References

Beament, J. W. L. (1954). Water transport in insects. *Symposia of the Society for Experimental Biology*, **8**, 94–117.

Beament, J. W. L. (1964). The active transport and passive movement of water in insects. *Advances in Insect Physiology*, **2**, 67–129.

Beament, J. W. L. (1965). The active transport of water: evidence, models and mechanisms. *Symposia of the Society for Experimental Biology*, **19**, 273–98.

Beament, J. W. L., Noble-Nesbitt, J. & Watson, J. A. L. (1964). The water-proofing mechanism of arthropods. III. Cuticular permeability in the firebrat, *Thermobia domestica* (Packard). Journal of Experimental Biology, **41**, 323–30.

Berridge, M. J. (1970). Osmoregulation in terrestrial arthropods. In *Chemical Zoology*, ed. M. Florkin & B. T. Scheer, vol. 5, *Arthropoda*, Part *A*, Chapter 10, pp. 287–319. New York, London: Academic Press.

Buxton, P. A. (1930). Evaporation from the mealworm (*Tenebrio*, Coleoptera) and atmospheric humidity. *Proceedings of the Royal Society B*, **106**, 560–77.

Dunbar, B. S. & Winston, P. W. (1975). The site of active uptake of atmospheric water in the larvae of *Tenebrio molitor* (Coleoptera). *Journal of Insect Physiology*, **21**, 495–500.

Ebeling, W. (1974). Permeability of insect cuticle. In *The Physiology of Insecta*, ed. M. Rockstein, 2nd edn, vol. 6, 271–343. New York, London: Academic Press.

Edney, E. B. (1957). *The Water Relations of Terrestrial Arthropods*. London: Cambridge University Press.

Edney, E. B. (1966). Absorption of water vapour from unsaturated air by *Arenivaga* sp. (Polyphagidae, Dictyoptera). *Comparative Biochemistry and Physiology*, **19**, 387–408.

House, C. R. (1974). *Water Transport in Cells and Tissues*. London: E. Arnold.

Knülle, W. & Devine, T. L. (1972). Evidence for active and passive components of sorption of atmospheric water vapour by larvae of the tick *Dermacentor variabilis*. *Journal of Insect Physiology*, **18**, 1653–64.

McEnroe, W. D. (1973). The rectum is not the site of water vapour uptake by *Dermacentor variabilis* Say (Acarina: Ixodidae). *Acarologia*, **14**, 542–3.

McMullen, H. L., Sauer, J. R. & Burton, R. L. (1976). Possible role in uptake of water vapour by ixodid tick salivary glands. *Journal of Insect Physiology*, **22**, 1281–5.

Machin, J. (1975). Water balance in *Tenebrio molitor* L. larvae: the effect of atmospheric water absorption. *Journal of Comparative Physiology* B, **101**, 121–32.

Maddrell, S. H. P. (1971). The mechanism of insect excretory systems. In *Advances in Insect Physiology*, ed. J. W. L. Beament, J. E. Treherne & V. B. Wigglesworth, vol. 8, pp. 199–331. London, New York: Academic Press.

Mellanby, K. (1932). The effect of atmospheric humidity on the metabolism of the fasting mealworm (*Tenebrio molitor* L., Coleoptera). *Proceedings of the Royal Society* B, **111**, 376–90.

Noble-Nesbitt, J. (1969). Water balance in the firebrat, *Thermobia domestica* (Packard). Exchanges of water with the atmosphere. *Journal of Experimental Biology*, **50**, 745–69.

Noble-Nesbitt, J. (1970*a*). Water uptake from subsaturated atmospheres: its site in insects. *Nature, London*, **225**, 753–4.

Noble-Nesbitt, J. (1970*b*). Water balance in the firebrat, *Thermobia domestica* (Packard). The site of uptake of water from the atmosphere. *Journal of Experimental Biology*, **52**, 193–200.

Noble-Nesbitt, J. (1973). Rectal uptake of water in insects. In *Comparative*

Physiology, ed. L. Bolis, K. Schmidt-Nielsen & S. H. P. Maddrell, pp. 333–51. Amsterdam: North-Holland; New York: Elsevier.

Noble-Nesbitt, J. (1975). Reversible arrest of uptake of water from subsaturated atmospheres by the firebrat, *Thermobia domestica* (Packard). *Journal of Experimental Biology*, **62**, 657–69.

Noble-Nesbitt, J. (1977). Active transport of water vapour. In *Transport of Ions and Water in Animals*, ed. B. L. Gupta, R. B. Morton, J. L. Oschman & B. J. Wall. London, New York: Academic Press. (In Press.)

Noirot, Ch. & Noirot-Timothée, C. (1971). Ultrastructure du proctodeum chez le Thysanoure *Lepismodes inquilinus* Newman (= *Thermobia domestica* Packard). II. Le sac anal. *Journal of Ultrastructure Research*, **37**, 335–50.

Okasha, A. Y. K. (1971). Water relations in an insect, *Thermobia domestica*. I. Water uptake from subsaturated atmospheres as a means of volume regulation. *Journal of Experimental Biology*, **55**, 435–48.

Ramsay, J. A. (1971). Insect rectum. *Philosophical Transactions of the Royal Society* B, **262**, 251–60.

Rudolph, D. & Knülle, W. (1974). Site and mechanism of water vapour uptake from the atmosphere in ixodid ticks. *Nature, London*, **249**, 84–5.

Schmidt-Nielsen, K. (1969). The neglected interface: the biology of water as a liquid gas system. *Quarterly Reviews of Biophysics*, **2**, 283–304.

Stobbart, R. H. & Shaw, J. (1974). Salt and water balance; Excretion. In *The Physiology of Insecta*, ed. M. Rockstein, 2nd edn, vol. 5, pp. 361–446. New York, London: Academic Press.

6. Water vapour uptake by *Tenebrio*: a new approach to studying the phenomenon

By JOHN MACHIN

This paper is concerned with the site and some of the kinetic properties of water vapour absorption in larval *Tenebrio*. Water balance studies with mealworms (Buxton, 1930; Mellanby, 1932; Kalmus, 1936) were among the earliest to demonstrate that exchanges with the atmosphere were not always dictated by simple activity differences across the integument. The anomalous component of these exchanges are now attributed to active water vapour absorption.

Recent experimental evidence based on blocking or ligaturing techniques suggests that water vapour is absorbed in mealworms through the rectal complex (Noble-Nesbitt, 1970; Dunbar & Winston, 1975; Machin, 1975). It seems likely that, by keeping the anus open in high humidities, mealworms further exploit the solute concentrating abilities of the rectal complex shown by Ramsay (1964) and Grimstone, Mullinger & Ramsay (1968) to be employed in dehydrating the faeces. Although Dunbar & Winston (1975) concluded that water vapour uptake does not appear to play a very important part in water balance, the opposite is suggested by increased growth rates observed in feeding animals which had previously undergone a period of atmospheric absorption (Machin, 1975).

Some workers have questioned the value and significance of traditional techniques which mechanically interfere with the animal in order to determine the site of uptake. Okasha (1971) has argued that, no matter how carefully these experiments are performed, the experimental treatment may differentially interfere with sensory control of uptake. Although these objections have been overcome by experiments in which the mechanical interference remains unchanged while different parts of the body are alternately exposed to absorbing and non-absorbing atmospheres (Rudolph & Knülle, 1974; Noble-Nesbitt, 1975), an alternative proof of a rectal site of uptake is possible using totally unrestrained animals. The technique, devised by Ramsay (1964) uses weight adjustments of the faecal pellets in known humidities following elimination to indicate the physiological state of the rectal lumen. In my experiments, where the weights of recently fed mealworms and their freshly eliminated faeces were continuously recorded in high humidities, weight adjustments of the faeces were

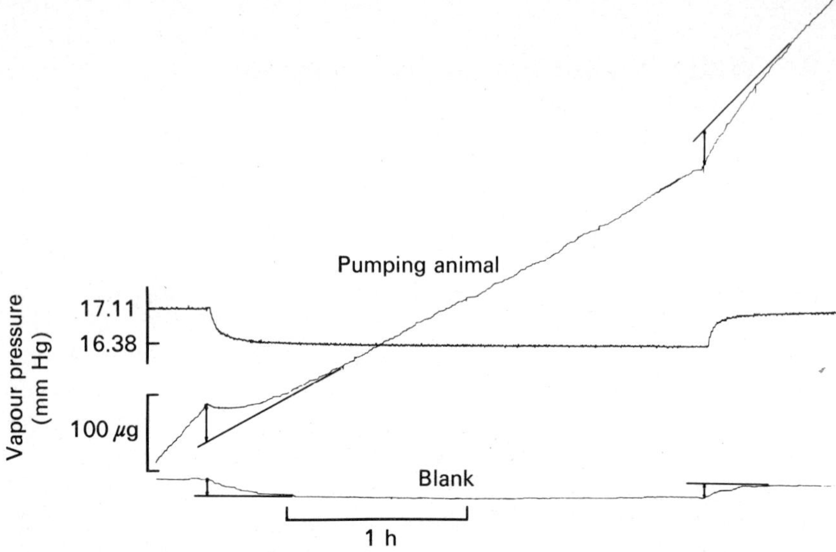

Fig. 1. Representative weight traces for pumping mealworms and empty weighing cages (blank) during two humidity changes. The vertical weight scale, 100 μg, is identical for both traces. Extrapolated lines on the weight traces indicate transitional weight changes. Humidity is shown by a simultaneous dew point hygrometer recording, calibrated in terms of vapour pressure.

identified as short-lived interruptions of the established trend. The faecal pellets (weighing an average of 33 μg) produced by intermoult animals which were not absorbing, and thus had their anuses closed, gained up to 10 μg upon elimination, showing that conditions in the rectal lumen were favourable for atmospheric absorption. Faeces from animals which were already absorbing from the atmosphere indicated no change in weight. This demonstrated that water vapour was rapidly diffusing into the rectal lumen as it was absorbed.

Previous studies of atmospheric absorption have usually involved the measurement of 'steady-state' uptake rates in constant humidity. I have developed a technique which yields valuable additional information about the absorption mechanism which exploits significant passive exchanges that follow abrupt humidity change. Fig. 1 shows some typical continuous weight recordings obtained with fasting mealworms over two humidity changes. The upper trace shows steady humidity-dependent weight gains in constant humidity. Imposed on these are significant transient weight gains or losses (indicated by the vertical arrows) depending on the direction of the humidity change. The lower trace shows a blank run with the animal

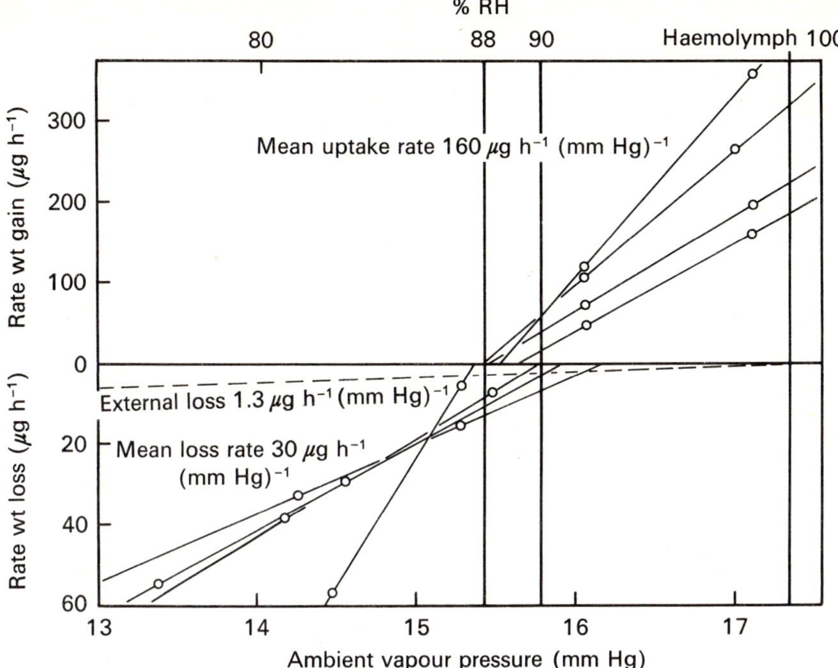

Fig. 2. Graph summarising steady-state exchanges in animals with their anuses open together with their humidity thresholds. Note that gain and loss scales are different. External losses with anuses closed are taken from Machin (1976).

cage only. This was virtually indistinguishable from the records obtained from animals with their anuses closed. Transient weight gains and losses represent water exchanges in absorbing animals because they continue to take place after humidity change in the weighing chamber is complete. In addition the amount of adjustment exceeds that caused by a surface absorption which appears to occur principally on the animal cage.

From the gain and loss rates in equilibrium conditions a picture of the steady-state parameters of the *Tenebrio* water vapour pump can be built up. It is now well established that the rate of weight gain in individual mealworms increases linearly with ambient humidity above a threshold of about 88% RH. Beament (1965) was the first to realise the potential significance of this threshold value, obtained in this study by extrapolating uptake rates to zero, as an indicator of the type of pump involved. He was impressed by the fact that the threshold, expressed in relative humidity terms was unaffected by temperature and seemed to be a constant characteristic of the animal involved. Before being absolutely sure, however, that threshold values have a physical reality and give a measure of the lowest vapour pressure generated by the pump, two conditions have to be met. These are that errors caused by integumental evaporation from the haemolymph are negligible and that uptake rates in different humidities are governed by physical and not biological variables. It can be seen in Fig. 2 that cuticular transpiration based on data from Machin (1976) is indeed small in relation to the gains and losses from absorbing animals. Confirmation that gains actually do involve the exchange of water with a compartment of low vapour pressure comes from the fortunate chance that some absorbing mealworms keep their anuses open for some time after they are exposed to subthreshold humidities. The data plotted in Fig. 2 show thresholds extrapolated from pairs of uptake and loss values obtained from the same animals. The rates between individuals of roughly the same size vary widely, but both gains and losses show thresholds grouped around 88–90% RH, considerably lower than the equilibrium humidity of the haemolymph of 99% RH. Since passive losses are unlikely to be affected by anything but physical variables, we may conclude with greater confidence that the pumping mechanism actually does involve a separate and quantitatively distinct fluid compartment within the animal. Steady-state exchanges show that the absorbing mechanism has two impressive kinetic properties, a five-fold asymmetry between mean rates of gain and loss and linear increase in uptake rate with humidity. This means that the absorption mechanism shows no sign of saturation at least up to maximum measured absorption rates of 380 μg h^{-1} in 100 mg animals.

The most likely explanation of the transient exchanges following humidity change is that the active absorption mechanism is located some distance beneath the rectal surface and is loosely coupled to it by one or more passively exchanging fluid compartments. These exchanges can be used to estimate the size of the intermediate compartment, which in turn will help to locate the site of the uptake mechanism. The mathematical basis for calculating solvent volumes of compartments from pairs of vapour pressure values is given fully in Machin (1972). The solute concentration of a solution is given by its vapour pressure lowering or saturation deficit:

$$vp_0 - vp_x,$$

where vp_0 is the vapour pressure of pure solvent and vp_x the vapour pressure of the solution at the same temperature. Where changes in concentration are brought about by gains or losses of the solvent only, solute concentration and solvent volume, V is related by:

$$V = \frac{1}{vp_0 - vp_x}.$$

Using an expanded version of this equation (Machin, 1972) it was possible to calculate solvent volumes from pairs of solute concentrations knowing

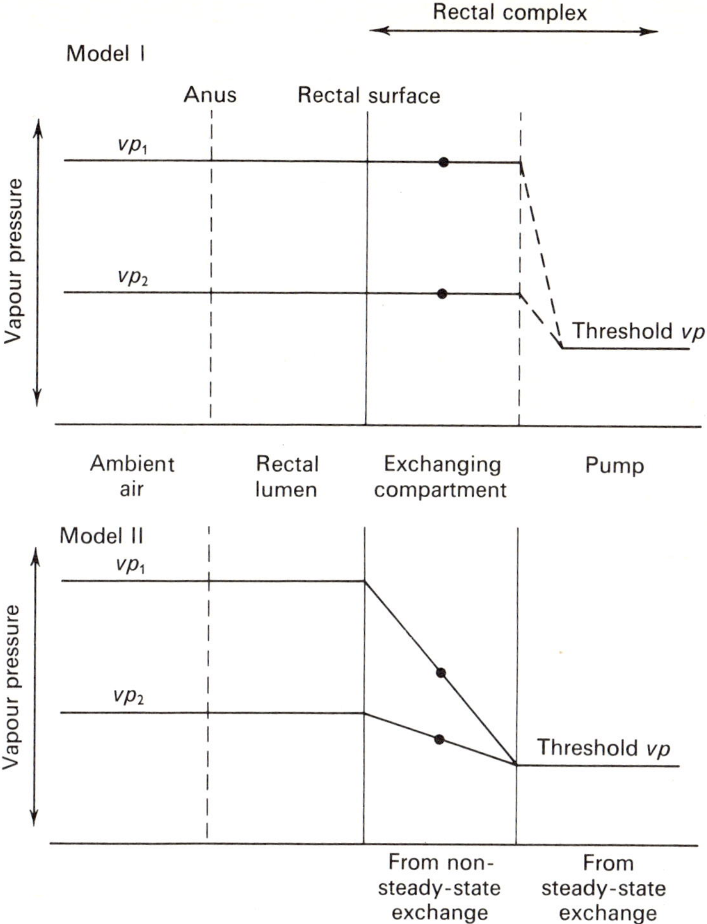

Fig. 3. Diagrammatic representation of models I and II indicating two possible relationships between ambient and exchanging compartment vapour pressures (closed circles).

the amount of water gained or lost from the compartment to bring about the concentration change. Gains and losses in the animal were obtained from the continuous weight records by subtracting surface absorption. Unfortunately the precise relationship between ambient vapour pressure and the vapour of the solvent compartment is unknown. The two most likely relationships are shown in Fig. 3. Model I describes a superficial exchanging compartment located close to the rectal surface and assumes that the difference between ambient and compartment vapour pressures are negligible. Model II is consistent with a deeper exchanging compart-

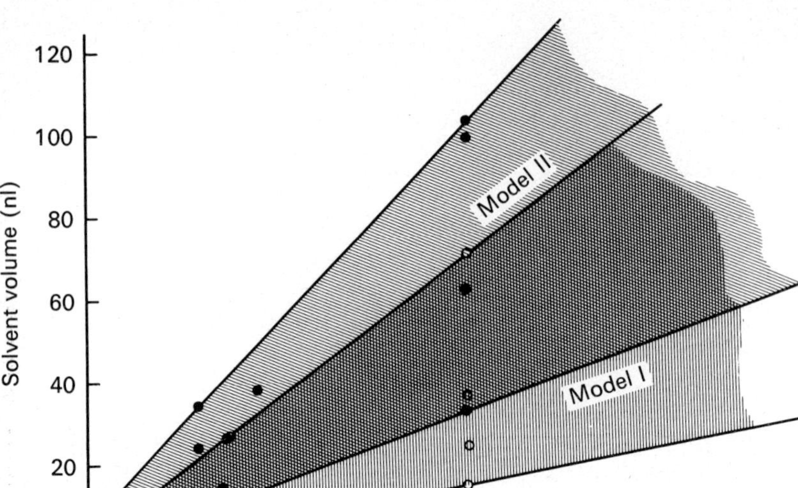

Fig. 4. Graph showing relationship between exchanging solvent volumes and the reciprocal of vapour pressure lowering. Areas relating to the two models are shaded.

ment whose vapour pressure is midway between the lumen and that of the threshold of the uptake mechanism. When the data for absorbing mealworms are analysed (Fig. 4), it can be seen that differences between the quantitative predictions of the two models are not very great in the light of considerable variation in solvent volumes between individual mealworms of the same general size. It should be noted that the justification for drawing straight regression lines between two points comes from other values based on a whole series of different humidity changes using the same animal (Machin, 1976). The passing of the regression line close to the graph origin means that the exchanging compartment behaves like a simple solution.

Turning now to the actual structures involved, it is possible to reconstruct the concentrations existing in the intact, posterior rectal complex by photographing the progressive melting of frozen transverse sections in controlled temperatures. Despite the fact that resolution is low and there is some uncertainty about the precise meaning of melting point in a continuous osmotic gradient, some useful comparisons can be made within the same section. The distribution of melting points seen in Fig. 5 suggests that osmotic pressure increases radially from the rectal lumen,

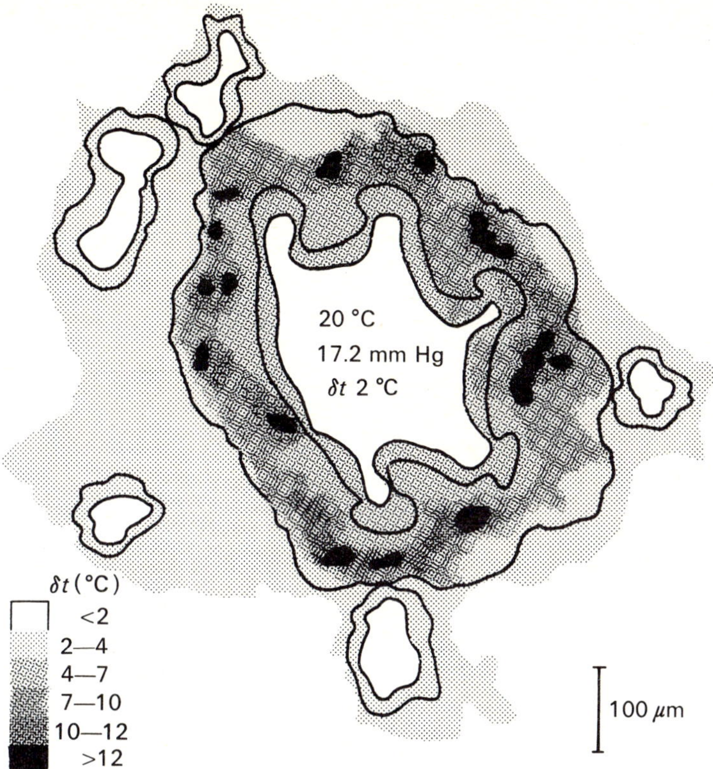

Fig. 5. Melting patterns in a quick frozen, transverse section of the posterior rectal complex in an absorbing mealworm. Free Malphigian tubules, rectal lumen, epidermis, and the outer limit of the complex are outlined. Concentration equivalents of the humid air from which the animal was absorbing are indicated in the lumen.

reaching maximum values in the Malphigian tubules applied to the rectum. This distribution of osmotic pressure is similar to that found by Ramsay (1964) using micropuncture techniques, but giving higher maximum values particularly in the case of the tubules. Since at ambient temperature (20 °C) a solution with an equilibrium humidity of 88% RH would have a melting point of -12.4 °C, the distribution of osmotic pressures is consistent with a tubular absorbing mechanism. It would appear that no other compartment in the complex generates solute concentrations high enough to absorb water vapour with a threshold of 88% RH. By contrast, in control sections from premoult mealworms which are unable to absorb, all concentration gradients disappear and the rectal complex melts uniformly at about -1.1 °C. (0.6 osm kg^{-1}).

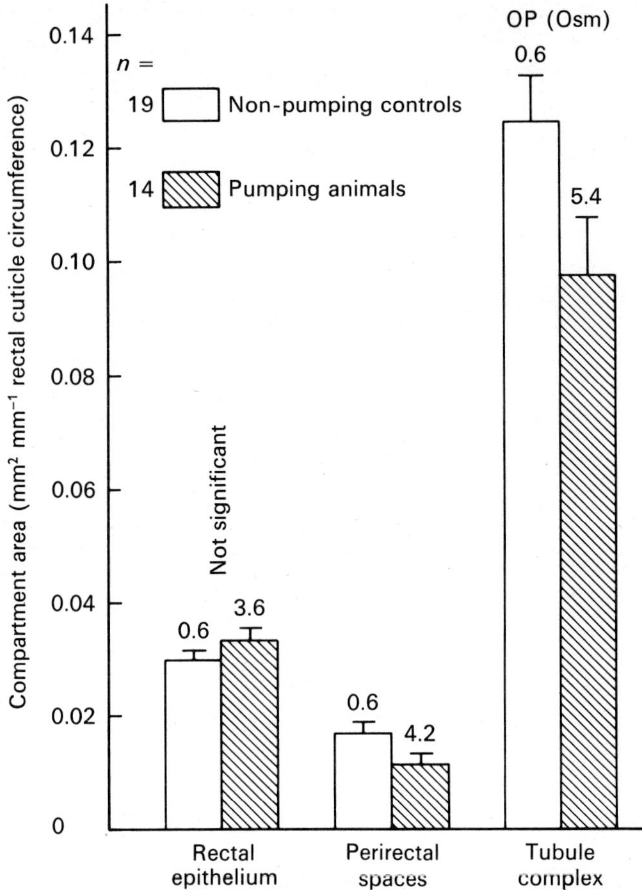

Fig. 6. Histogram showing rectal compartment areas in pumping and non-pumping animals together with their estimated osmotic pressures (OP). Areas are expressed per mm of cuticle circumference to correct for small differences in individual size. Calculations based on cuticular water contents in Fig. 7 indicate a maximum change in circumference of 5 % from complete dehydration to haemolymph equilibration. Maximum errors resulting from swelling during freezing, based on the density of ice and water at 20 °C, are 6 %.

Frozen transverse sections may also be used to establish what compartments in the rectal complex respond to changes in concentration. Although it is extremely laborious and difficult to obtain serial transverse sections in order to determine compartment volumes, compartments are rather uniform along their length and area measurements can provide evidence of relative volume change when two concentrations are compared. With this method single transverse sections per animal can be easily obtained and statistically valid comparisons made. Since area differences are proportionately smaller than actual volumes, comparisons were made only between gross concentration changes. Accordingly the areas of compartments in absorbing and non-absorbing premoult mealworms were compared in order to locate the source of passively exchanging water. The areas measured, together with the observed osmotic pressures in the two groups are summarised in Fig. 6. It can be seen that the tubule complex and perirectal spaces swell significantly when these compartments become iso-osmotic to haemolymph in moulting animals. The swelling, however, is considerably smaller than would be predicted from the seven- to nine-fold change in osmolarity. This is to be expected in the tubules and perirectal space where any radial resistance to swelling can be readily compensated by longitudinal displacement of fluid. In completely cellular structures such as the rectal epithelium, longitudinal fluid displacement is impossible. It is remarkable, therefore, that despite a six-fold change in osmolarity no significant swelling was found in this structure. It would appear that the rectal epithelium may be excluded as the source of passively exchanging water, which must therefore originate from other rectal structures such as the cuticle or perirectal spaces.

More direct estimates of compartment sizes within the rectal complex, possible only with few specimens, may be obtained from three dimensional reconstructions of transverse sections taken at carefully measured intervals along the posterior third of the body. Unfortunately even when analysis is complete there is a major uncertainty about the values. Sections even of starved animals show variable amounts of faecal material in the rectal lumen. This makes it difficult to estimate with accuracy the proportion of the rectum engaged in atmospheric absorption. It seems likely in fact that the previously described variability in uptake rates and exchanging compartment volumes is caused by the obstruction of faecal material in the rectal lumen. In one particular animal, in which uptake rates and hence exchanging compartment volumes are known, it has been possible to estimate the extent of the rectum involved in absorption (the posterior 1.8 mm) from the extent of air spaces in the faecal material. These values may then be used to estimate the total amount of exchangeable water, compartment by compartment. Because of the presence of faeces within the rectum, they must also be included among possible sources of passively exchanging water.

Unfortunately water contents of faecal material cannot be determined directly in the serially reconstructed animals. However, the rectal contents of starved mealworms are almost certainly less than those of feeding animals and an upper limit of 7 nl exchangeable water, based on faecal water content data in Ramsay (1964) and Machin (1975), seems likely. It is possible to obtain better estimates of the exchangeable water in the rectal

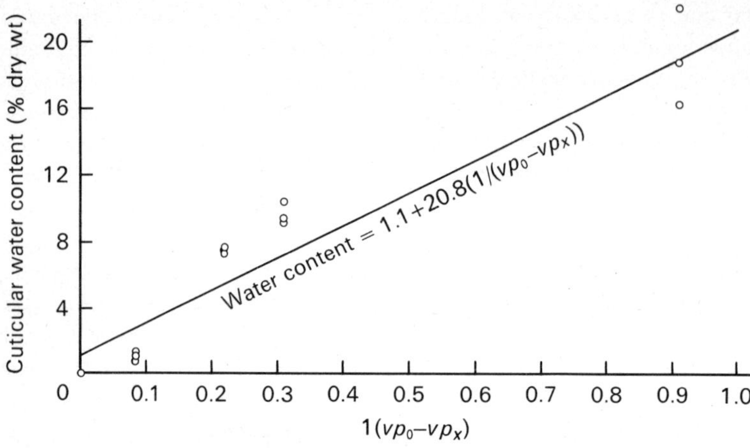

Fig. 7. Graph showing relationship between cuticular water content and the reciprocal of vapour pressure lowering.

cuticle. Samples of cuticle, free of epidermal cells, can be obtained by pulling the dissected complex through the closed tips of fine forceps. Using the cuticles from ten animals, changes in water content in five humidities were followed on a microbalance. Humidity dependent exchanges proceeded rapidly to equilibrium showing no histeresis. Again cuticular water contents were found to be proportional to the reciprocal of vapour pressure lowering (Fig. 7). A value of 1.4 for cuticle density, based on chitin (Richards, 1951), was used to convert water content per dry weight to water content per wet volume. The water content of 14.6 nl of rectal cuticle in the reconstructed animal was calculated to be 2.9 nl.

Melting point data (Fig. 5) have suggested that the absorption mechanism is located in the Malphigian tubules. If this were so, compartments between these and the rectal surface would be subject to humidity-induced concentration changes. Since the perirectal spaces and possibly the tubule walls have been shown to change their volumes in response to concentration changes, passive water exchange seems unlikely to be restricted to the superficial faecal and cuticular compartments. If deeper compartments are involved, model II provides the most accurate experimental estimate of exchanging compartment solvent volume. For the rectum of the reconstructed animal model II predicts that at the mean concentration ($\delta t = 9.7\,°C$) there are 20 nl of water in the passively exchanging compartment. This fits fairly well with the sum of the estimated faecal and cuticular water contents and perirectal volume ($7+2.9+11 = 20.9$ nl), assuming the perirectal space to be filled principally with solvent.

Clearly final conclusions regarding the location of the absorption mechanism must still heavily depend on the results of future experiments. The confirmation of the validity of model II will depend on further studies of the volume-regulating capacity of the rectal epidermis and of the volume–concentration characteristics of the perirectal fluid.

References

Beament, J. W. L. (1965). The active transport of water: evidence, models and mechanism. *Symposia of the Society for Experimental Biology*, **19**, 273–98.

Buxton, P. A. (1930). Evaporation from the mealworm (*Tenebrio*: Coleoptera) and atmospheric humidity. *Proceedings of the Royal Society*, **106**, 560–77.

Dunbar, B. S. & Winston, P. W. (1975). The site of active uptake of atmospheric water in the larvae of *Tenebrio molitor*. *Journal of Insect Physiology*, **21**, 495–500.

Grimstone, A. V., Mullinger, A. M. & Ramsay, J. A. (1968). Further studies on the rectal complex of the meal worm *Tenebrio molitor*, L. (Coleoptera, Tenebrionidae). *Philosphical Transactions of the Royal Society* B, **253**, 343–82.

Kalmus, H. (1936). Die Verwendung der Tracheenblasen der Corethralarve als Mikrohygrometer. *Zeitschrift für Wissenschaftliche Zoologie*, **53**, 215–19.

Machin, J. (1972). Water exchange in the mantle of a terrestrial snail during periods of reduced evaporative loss. *Journal of Experimental Biology*, **57**, 103–11.

Machin, J. (1975). Water balance in *Tenebrio molitor* L. larvae: the effect of atmospheric water absorption. *Journal of Cellular and Comparative Physiology*, **101**, 121–32.

Machin, J. (1976). Passive exchanges during water vapour absorption in mealworms (*Tenebrio molitor*): a new approach to studying the phenomenon. *Journal of Experimental Biology*, **65**, 603–15.

Mellanby, K. (1932). The effect of atmospheric humidity on the metabolism of the fasting mealworm (*Tenebrio molitor* L., Coleoptera). *Proceedings of the Royal Society* B, **111**, 376–90.

Noble-Nesbitt, J. (1970). Water uptake from subsaturated atmospheres: its site in insects. *Nature, London*, **225**, 753–4.

Noble-Nesbitt, J. (1975). Reversible arrest of uptake of water from subsaturated atmospheres by the firebrat, *Thermobia domestica* (Packard). *Journal of Experimental Biology*, **62**, 657–69.

Okasha, A. Y. K. (1971). Water relations of an insect, *Thermobia domestica*. I. Water uptake from sub-saturated atmospheres as a means of volume regulation. *Journal of Experimental Biology*, **55**, 435–48.

Ramsay, J. A. (1964). The rectal complex of the mealworm *Tenebrio molitor* L. (Coleoptera, Tenebrionidae). *Philosophical Transactions of the Royal Society* B, **248**, 279–314.

Richards, A. G. (1951). *The Integument of Arthropods*. Minneapolis: University of Minnesota Press.

Rudolph, D. & Knülle, W. (1974). Site and mechanism of water vapour uptake from the atmosphere in ixodid ticks. *Nature, London*, **249**, 84–5.

7. Uptake of water vapour by mites and mechanisms utilized by the Acaridei

By G. W. WHARTON

Mites comprise seven of the eight orders of Acari (van der Hammen, 1970); representatives of the eighth order are ticks. Most mites gain sufficient water by imbibition or ingestion of food with high moisture contents. They adapt to the ambient humidity of their habitats by frequency of intake or restriction of water loss (McEnroe, 1961; Winston, 1963; Madge, 1964, 1965*a,b*, 1966; Winston & Nelson, 1965; van de Vrie, McMurtry & Huffaker, 1972). Ticks (Rudolph & Knülle, this volume) and some mites can use water vapour of unsaturated air to replenish their body water and thus maintain water balance. The literature on water balance of mites has been reviewed recently (Vannier, 1976). I will restrict this paper to a consideration of the net uptake of water from unsaturated air. The minimum humidity from which a species can extract more water than is lost, expressed as the critical equilibrium water vapour activity (a_v) or CEA, will be discussed first. The kinetics of water exchange when it appears that all of the water in the mite behaves as a single compartment will be presented second followed by a consideration of two-compartment kinetics. Mechanics of the water uptake system and evidence that favours it will be given fourth. Finally, I will attempt to explain why uptake of water vapour from unsaturated air is so restricted in its occurrence among organisms.

Critical equilibrium activity (CEA)

Those mites that can remove water from unsaturated air, such as the spiny rat mite, *Laelaps echidnina* Berlese (= *Echinolaelaps echidninus*), that I have studied extensively (Wharton & Cross, 1957; Cross & Wharton, 1966) can do so only when the humidity of the air is greater than their CEA. The activity of the water in the blood and tissues of mites, 0.99 a_w (Davidson, 1972, in Wharton, 1976), is similar to that of other air-breathing organisms. A plane water surface containing water at a given a_w will be in equilibrium with the water vapour in the air over the surface at the same a_v. The activity of the water vapour is numerically equal to one-hundreth of the relative humidity ($a_v = $ RH/100). Activity is the proportion of the chemical potential of a pure substance that is characteristic of a solution containing that substance. It is one of the units used for the colligative properties of solutions and as such can be converted into any of many units such as osmoles, atmospheres of osmotic pressure, freezing point depression in Δ °C or others. Activity is used here because it is the one unit that is able to describe the potential for either a gas or a liquid. Most biologists use relative humidity for gases and osmoles, freezing point depressions, or NaCl equivalents for liquids, but few recall the humidity in equilibrium with a particular osmolarity:

$$a_w = 55.508/(55.508+x), \tag{1}$$

where 55.508 is the moles of water and x the osmoles of solute in 1 kg water. Use of equation (1) will convert a_w to osmoles to three significant decimal places or milliosmoles.

The CEAs of only five species of fasting mites are known. (1) *Acarus siro* L.: 0.62–0.65 a_v at 20 °C (Solomon, 1962, 1966), 0.71 a_v at 22 °C (Knülle, 1965). (2) *Tyrophagus putrescentiae* Koch: 0.75–0.84 a_v at 25 °C (Cutcher, 1973). (3) *Dermatophagoides farinae* Hughes: 0.70 a_v at 25 °C (Larson, 1969). (4) *Dermatophagoides pteronyssinus* Trouessart: 0.73 a_v at 25 °C (Arlian, 1975*a*). (5) *Laelaps echidnina* Berlese: 0.90 a_v at 15–25 °C (Wharton & Kanungo, 1962). When food is available, mites ingest water with their food and for species that feed on essentially dry food such as stored products mites, the humidity at which they can survive is reduced slightly, *A. siro* about 0.60 a_v at 10 °C (Solomon, 1965) and *T. putrescentiae* 0.65 a_v at 15–25 °C (Cunnington, 1969). In the case of mites that feed on food with high water contents, such as blood at 0.99 a_v, survival at a_vs much below their CEAs is possible (e.g. *L. echidnina*, 0.73 a_v; Knülle, 1967).

Biological membranes are more permeable to water than they are to respiratory gases (Waggoner, 1967); therefore, water will be able to diffuse in and out of mites more rapidly than O_2 and CO_2 per unit of partial pressure. If the air has a humidity of 0.70 a_v and the haemolymph of the mite has an aqueous activity of 0.99 a_v, 41 % more water will transpire or diffuse out of the mite than will be sorbed or diffuse in. It is clear then that a CEA below the a_v of the body fluid of a mite requires that energy be expended to transport water from the vapour phase at a low a_v to the liquid phase at high a_v. Hence, oxygen would be required by the spiny rat mite if it were to move water from 0.90 a_v to 0.99 a_w. Mites in pure N_2 failed to maintain their weight at 0.95 a_v, but did so if the gas mixture contained as little as 2.5 % O_2 (Kanungo, 1963). Studies on carefully standardized female mites that had completed digestion of their blood meals (Kanungo, 1964) determined that the mites consumed some 0.17 μl of O_2 h^{-1} at 27 °C, whether they were losing water at 0.80 a_v or gaining it at 0.98 a_v at the rate of 1 μg h^{-1}. The energy needed to convert

1 μg of water per hour at 0.90 a_v to 0.99 a_w requires a respiratory rate of only 6×10^{-4} μl O_2 h^{-1} (Kanungo, 1965). That the energy required to actively sorb water from unsaturated air is not a large part of a mite's energy budget is confirmed by the fact that mites deprived of food and water and held at 0.93 a_v or alternated daily on a 8–16 h schedule from 0.73 a_v to 0.93 a_v survived for about the same period, 24 days and 21 days, respectively (Knülle, 1967).

One-compartment kinetics

The actual amounts of water moving in and out of female spiny rat mites were found to exchange with the air as if all of the mite's water were in a single compartment. It is possible to label water with ^{3}H and follow its movement across an interface by measuring its radioactivity (T^*) in counts per minute (c.p.m.) in a liquid scintillation spectrometer. As labelled water moves into an unlabelled compartment, the specific radio-activity (S) of the water in the compartment will change from 0 to an equilibrium value such that S on both sides will be identical. The specific radioactivity of such water is

$$S = T^*/m, \tag{2}$$

where $m =$ the mass of water. Spiny rat mites exposed to ^{3}H-labelled water vapour at 0.93 a_v and 25 °C gained ^{3}H by diffusion of ^{3}H vapour from the air until the specific radioactivity (S) of the mite was equal to that of the water vapour. The actual ^{3}H concentration or specific radioactivity at any time after exposure to the ^{3}H-labelled water vapour could be calculated by

$$\ln\left(1 - S_t/S_\infty\right) = -kt, \tag{3}$$

where S_t is the specific radioactivity at any time t, S_∞ is specific activity at equilibrium, k is a rate constant and t is time. The amount of ^{3}H lost by a mite to a large ^{3}H sink, so that the water vapour surrounding the mite had a negligible amount of ^{3}H, fits

$$\ln\left(T_t^*/T_o^*\right) = -kt, \tag{4}$$

where T_o^* is equal to the initial c.p.m. at $t = 0$. As long as mites were held above the CEA, the water mass did not change significantly so that as mites lost tritiated water their specific radioactivity decreased according to:

$$\ln\left(S_t/S_0\right) = -kt. \tag{5}$$

If tritiated water is in fact a good tracer for water in these studies, then when $\dot{m} = 0$,

$$\dot{m}_S = \dot{m}_T = km, \tag{6}$$

where $\dot{m}$ is the rate of change of water mass with time, $\dot{m}_S$ is the rate of sorption and $\dot{m}_T$ is the rate of transpiration. Using these models, k was found to be 0.0373 ± 0.0012 h^{-1} at 0.93 a_v and 25 °C. At this rate of aqueous exchange, mites with water contents $m = 100$ μg, lost 90 μg day^{-1} and gained a like amount so that $\Delta m = 0$. At 0.90 a_v, the CEA for spiny rat mites, at least 9 μg per day must be moved by an active mechanism (Wharton & Devine, 1968).

Aerobic respiration of 1 μg of carohydrate produces 0.56 μg of metabolic water while 1 μg of fat will yield 1.29 μg of water. Respiration at the determined rate of 0.17 μl O$_2$ h^{-1}, if used to metabolize a reasonable mix of carbohydrates and fat, would produce about 2.5 μg of metabolic water a day or 2.8% of total water sorbed in that time. On the other hand, 2.5 μg represents 28% of the water required by active uptake and is probably significant in overall water balance.

Spiny rat mites lose water when in an atmosphere below their CEA and the rate of water loss is

$$\dot{m} = \dot{m}_S - \dot{m}_T. \tag{7}$$

The actual water loss has been observed under a number of environmental conditions and the value of water mass at time t has been found to fit

$$\ln \left[(m_t - m_\infty)/(m_0 - m_\infty) \right] = -kt, \tag{8}$$

where m_∞ is the equilibrium water mass or water mass at infinite time. Below the CEA, except at 0.00 a_v, an equilibrium would be achieved if the mites could survive at water contents at equilibrium. When mites die they lose water rapidly and their transpiration rates increase dramatically. Equilibria below the CEA are therefore virtual rather than real, but they can be estimated by solving equation (8) with approximated values for the virtual equilibrium water mass (m_∞) until a minimum correlation coefficient (r) is arrived at. In estimating m_∞, inspection of a plot of log m_t versus t is helpful. The curve should approach m_∞ asymptotically (Fig. 1). A similar plot of $(m_t - m_\infty)$, when m_∞ has the appropriate value, will be a straight line of slope $-k$. Using ^{3}H-labelled water to follow the net and total transpiration concurrently, it was found that k of equation (8) and k of equation (4) are identical. Thus, using the value of k calculated from observations of ^{3}H loss to a large ^{3}H sink so that ^{3}H in the vapour surrounding a mite is essentially 0, m_∞ can also be calculated from observations of m_t by substitution of k in equation (8).

Sorption rates at any one a_v appear to be constant in time; in other words,

$$\dot{m}_S = C. \tag{9}$$

This circumstance is required because in a constant environment neither the permeability of the exchange surface, nor the potential of vapour driving sorption, change. The total sorption rate of these mites above the

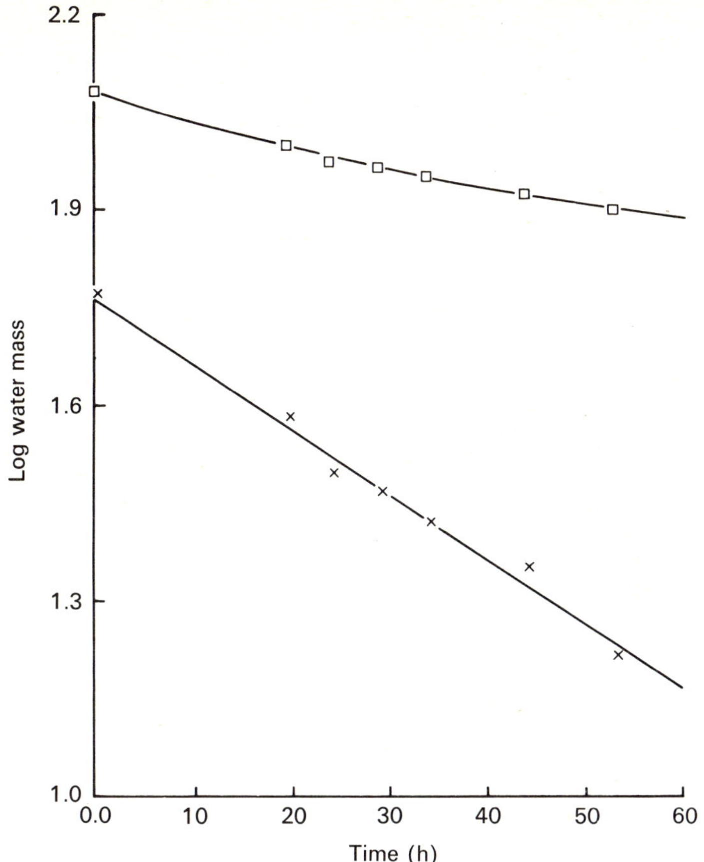

Fig. 1. Plots of water mass, m, and duration of exposure, t, data for *Laelaps echidnina* exposed to 0.75 a_v at 25 °C (Devine, 1969). Top points ($\square$) are for observed ms and top line fits the model t, $\log((m_0 - m_\infty)\exp - kt) + \log m_\infty$. Bottom points ($\times$) are for $(m - m_\infty)$ where $m_\infty = 61.71$ μg and bottom line fits t, $\log[(m_0 - m_\infty)\exp - kt]$. The bottom line is a plot of the data fitted to equation (8); the slope of the line is equal to $-k$.

CEA is equal to total transpiration and when exposed at 25 °C to a humidity of 0.93 a_v, it is 3.75 μg h^{-1} rather than the maximum net value of 0.42 μg h^{-1} observed (Wharton & Kanungo, 1962). As k and the constant sorption rate $\dot{m}_S$ in μg h^{-1} increase with a_v, an a_v will be reached where $\dot{m}_S$ equals km at the equilibrium value. Therefore, m_∞ can be defined by

$$m_\infty = \dot{m}_S\, k^{-1}. \tag{10}$$

Because body water, in so far as sorption and transpiration are concerned, behaves as a single compartment, exchanges of body water from compartment to compartment within the mite are almost infinitely rapid compared with exchanges across the air–mite interface (Devine & Wharton, 1973).

In much previous work on water loss net transpiration rates have been expressed as zero order rather than first order and net transpiration has frequently been considered as being gross transpiration or that sorption of water vapour during transpiration was negligible. We owe a debt of gratitude to Devine (1969) for obtaining enough ^{3}H and water loss data to demonstrate that concurrent flows and first order kinetics are appropriate for water loss by the spiny rate mite.

Two-compartment kinetics

Flows of water between the air and mites of the order Acaridei include studies on protonymphs (Ellingsen, 1976) and females of *Dermatophagoides farinae* Hughes, the American house dust mite (Arlian & Wharton, 1974) and the European house dust mite, *D. pteronyssinus* Trouessart, as well as males of the latter species (Arlian, 1975*a*).

When female American house dust mites were labelled with tritiated water, ^{3}H loss above the CEA was described by equation (4) and gave decreasing values of k with decreasing a_vs. In dry air 0.00 a_v, ^{3}H loss and water loss did not fit equations (4) and (8). Both water and ^{3}H were lost more rapidly at first and then after at least 14 h their loss was characterized by being a constant percentage of the content (Fig. 2). Another anomaly was the increase of the specific activity as water and ^{3}H were lost. Such a phenomenon could only be produced by water transpiring at a greater rate than ^{3}H. Both anomalies can be overcome by assuming a two-compartment in place of a single-compartment system for the water content of the American house dust mite. For the two-compartment system, $m = m' + m''$ and $T^* = T' + T''$. Each compartment will fit equations (4) and (8). The fast compartment m' is characterized by a lower ^{3}H concentration than the slower compartment m''. The regression required to determine the percentage loss from the slow pool k'' also gives the values of m_0'' and T_0'' so that m'' and T'' values can be calculated for any time and by subtraction from m and T^* the values for the fast pool can be obtained. Above the CEA only a single water compartment was observed as would be the case if $k = k' = k''$. If this assumption is accepted, then the 'fast' compartment must be the proportion of water used in an active water-sorbing system or pump, while the slow pool represents the proportion of water remaining in the body of the mite (Arlian & Wharton, 1974).

If the values of k obtained by Arlian & Wharton (1974) for a_vs above

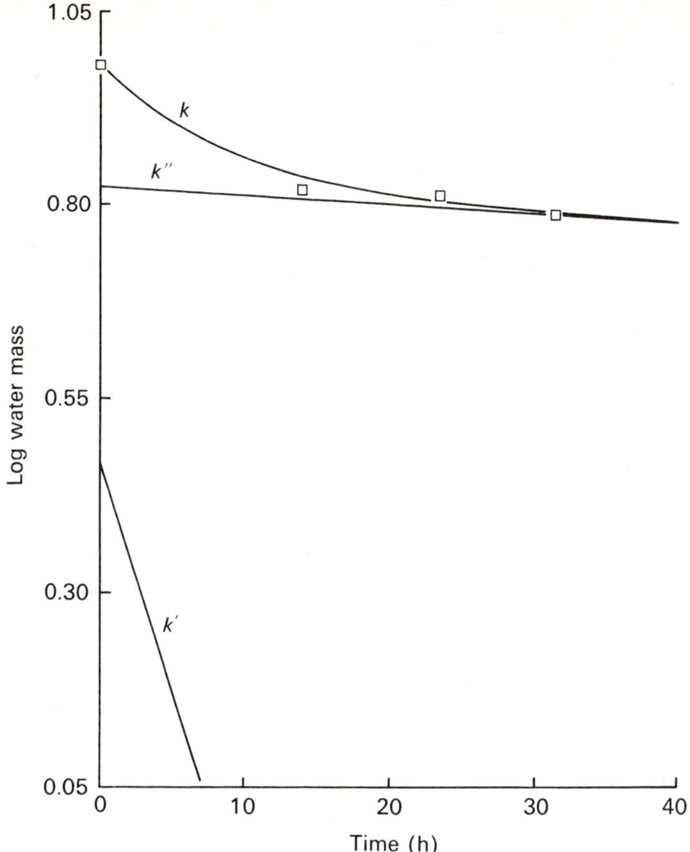

Fig. 2. Plots of water mass, m, and duration of exposure, t, data for *Dermatophagoides farinae* exposed to 0.00 a_v at 25 °C (Arlian & Wharton, 1974). Middle line calculated according to equation (8), using times 14 h and greater. Bottom line calculated according to equation (8), using times 14 h and less. Top line is the sum of middle and bottom lines. Points are observed values. The slope of the bottom line is $-k'$ for the pump below the CEA and the slope of the middle line is $-k''$ for the body water.

the CEA and at 0.00 a_v are regressed on a_v, a formula for calculating k'' for any a_v that explains 99.45 % of the sum of the squares of the differences between observed and calculated values is obtained:

$$k''_{a_v} = 0.02427 a_v + 0.00268 \text{ h}^{-1}. \tag{11}$$

Using equation (10), sorption $\dot{m}_S$ can be calculated if k and m_∞ are known. The minimum viable water mass is 48 % of the equilibrium water mass

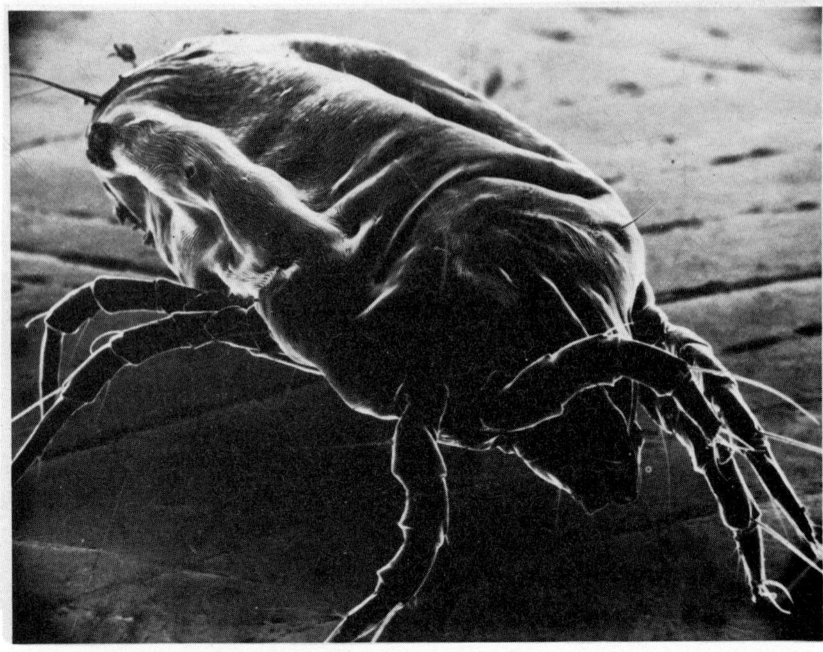

Fig. 3. (*a*) Dehydrated *Dermatophagoides farinae*, pump water has been lost. (*b*) Rehydrated *D. farinae*, pump water has been regained. Length of mite = 0.5 mm. Picutres reproduced by permission of L. G. Arlian (Arlian, 1972).

Table 1. *Values for water masses and rates of exchange for* Dermatophagoides farinae *calculated from observations (Arlian & Wharton, 1974), using a zero order rather than a first order relationship between k and a_v*

Activity a_v	Observed total, $\bar{k}$ (h^{-1})	Calculated body, k'' (h^{-1})	Calculated pump, k' (h^{-1})
0.000	0.00257	0.00268	0.13833
0.225	—	0.00814	0.13833
0.525	—	0.01542	0.13833
0.700	—	0.01820	0.01820
0.750	0.02225	0.01967	0.01967
0.850	0.02347	0.02331	0.02331
0.925	0.02522	0.02513	0.02513
1.000	0.02634	0.02695	0.02695

Activity a_v	Observed total, $\bar{m}_\infty$ (μg)	Calculated body, m''_∞ (μg)	Calculated pump, m'_∞ (μg)
0.000	0.000	0.000	0.000
0.225	—	1.480	0.000
0.525	—	3.452	0.000
0.700	9.589	4.603	4.986
0.750	9.589	4.932	4.657
0.850	9.589	5.589	4.000
0.925	9.589	6.083	3.506
1.000	9.589	6.576	3.013

Activity a_v	Calculated body, $\dot{m}''_S$ (μg h^{-1})	Calculated pump, $\dot{m}'_S$ (μg h^{-1})	Calculated total, $\dot{m}_S$ (μg h^{-1})
0.000	0.000	0.000	0.000
0.225	0.012	0.000	0.012
0.525	0.053	0.000	0.053
0.700	0.084	0.091	0.175
0.750	0.097	0.092	0.189
0.850	0.130	0.093	0.223
0.925	0.153	0.088	0.241
1.000	0.177	0.081	0.258

For symbols, see text.

above the CEA (9.589 μg) or 4.603 μg. The pump stops operating below the CEA and mites lose considerable water in dry air but can regain it when placed above the CEA (Fig. 3). The water that is lost first is the fast pool and this must be that associated with an active uptake mechanism that from now on I shall refer to as a 'pump'. It should also be emphasized that losing water from the pump effectively waterproofs it. It therefore seems reasonable to assume that 4.603 μg is the upper limit for the virtual m_∞'' below the CEA. It is also obvious that m_∞'' at 0.00 a_v will be 0.000 μg. If we calculate m_∞ for any $m_{\infty_{av}}''$ by

$$m_{\infty_{av}}'' = 4.603 \; \mu\text{g} \cdot a_v \cdot \text{CEA}^{-1}, \tag{12}$$

then $\dot{m}_S''$ can be calculated for any a_v by combining equations 10, 11 and 12. It is also possible to calculate $\dot{m}_S$ above the CEA by equation (6). The difference between $\dot{m}_\infty''$ and $\dot{m}_\infty$ is equal to the amount of water in the pump circuit as contrasted with the water in the body (haemolymph and tissues) (Table 1).

Pump mechanics

The size of the water mass associated with the pump of *D. farinae* requires that the volume of the pump be a considerable proportion of the total volume of the mite. The fine structure of female *D. farinae* has been studied (Brody & Wharton, 1970; Wharton & Brody, 1972; Brody, McGrath & Wharton, 1972, 1976). On the basis of these studies the most likely candidate for the production of hygroscopic material that could achieve net uptake of water from 0.70 a_v appeared to be the paired supracoxal glands. Each gland pours its secretion onto the supracoxal plate of coxa I by its own side, the right-hand gland opening on the right and the left hand on the left. The common duct is formed by the confluence of four ducts. Three of the four ducts have a large diameter. Each of the ducts that have significant diameters is associated with a pair of cells. One cell, A-cell, surrounds the greater part of the cell (B-cell) that contains the duct. The A-cell is in intimate contact with the haemolymph by means of a series of mitochondria-rich membranes that surround the B-cell. The B-cell has numerous microtubules as well as layers of cisternae around the duct. It also possesses a few mitochondria-rich elements that project into the haemolymph. The cytoplasm of the B-cell appears to be most closely associated with the lumen of the duct by means of a sieve-like disk that closes one end of the duct. A similar disk is present on the arm of the Y that is in the section of the B-cell in contact with the main body of cytoplasm of the A-cell. That disk is in close contact with the cytoplasm of the A-cell. The other two large ducts are associated with A and B cells as well. The small duct is enclosed in a small elongated D-cell that connects a secretory C-cell with the main duct of the supracoxal gland.

On the supracoxal plate is an open groove, the podocephalic canal, that runs down the plate to its base where the salivary duct opens into it. From there the canal runs over the dorsal surface of the base of the palp into the prebuccal cavity.

If the supracoxal gland is in fact the source of hygroscopic material for the pump, then its orifice must be occluded when the pump is water-proofed. When these mites are dehydrated, as a last resort they appear to expel plugs from the supracoxal glands, where they can be seen as stalactite-like formations. Similar plugs, but not so large, were observed closing the supracoxal gland openings of desiccated *T. putrescentiae* and *A. siro*. An electron-probe elemental analysis of the plugs of *D. farinae* and *T. putrescentiae* demonstrated significant concentrations of potassium and chlorine in the plugs. Chemical analysis of superficial wash water from gram quantities of dried *T. putrescentiae* recovered a chloride-rich solution that, when dried down, deliquesced at 0.85 a_v. The actual plugs of all three species were observed to deliquesce when exposed to moist air under the microscope.

When mites are mounted in mineral oil and observed under the micro-scope, their pharyngeal pump is constantly in operation (Wharton & Brody, 1972). The pharyngeal pump can remove fluid from the prebuccal cavity and pump it into the esophagus, which lies in a haemolymph sinus that traverses the brain. The collapsed esophagus is star-shaped in cross-section so that it is capable of expanding sufficiently to completely occlude the sinus in which it lies. Waves of expansion resulting from pumping material through the esophagus to the midgut will propel haemolymph from the anterior mid-ventral region of the haemocoel to the dorsal mid-section.

On the basis of these observations, it appears that the pump consists of (1) the supracoxal glands that secrete a hygroscopic solution rich in KCl, (2) a supracoxal plate where the hygroscopic supracoxal fluid picks up water vapour from air above the CEA, (3) a podocephalic canal in which the supracoxal fluid flows across the supracoxal plate and dorsal proximal margin of the palps into the prebuccal cavity, and (4) a muscular pharynx that pumps the vapour-enriched fluid through the esophagus into the midgut. The water contained in each of the above structures is the water involved with the pump and could easily amount from about 30–50% of the total water (Table 1, Fig. 4).

From the midgut, the salt and water from the supracoxal fluid must be moved into the haemolymph where the excess salt is removed and used in the manufacture of more hygroscopic supracoxal fluid by the supracoxal

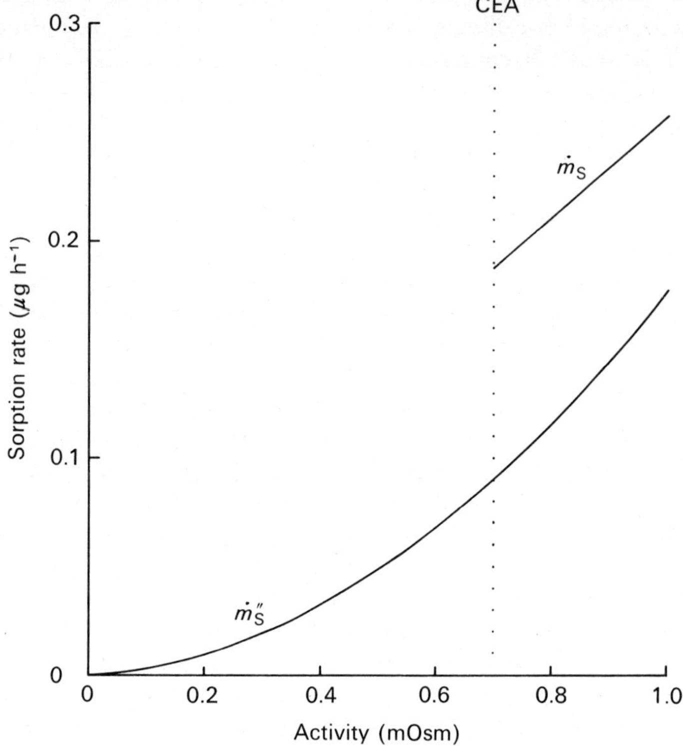

Fig. 4. *Dermatophagoides farinae*, females. The relationship between sorption rate ($\dot{m}_S''$) and water vapour activity (a_v) for mites held in a specific water vapour activity. Diffusion and active sorption are calculated by equation (10). Sorption by diffusion is calculated by equations (12) and (10). (Modified from Arlian & Wharton, 1974.)

glands. The blood flow induced by the movement of the supracoxal fluids through the esophagus will produce a flow from the midgut past the supracoxal glands and back to the anterior mid-ventral region from which it can be recirculated through the cephalic sinus (Wharton & Furumizo, 1977).

When this pump is exposed to air below the CEA, a net loss of water from the fluid to the air causes precipitation of solutes in the fluid. The water in the podocephalic canal and at the orifice of the duct is lost first and, as a consequence, solutes are precipitated out of the fluid. A plug forms over the orifice of the gland and when sufficient water has been lost, sufficient salt builds up in the duct of the gland so that it is waterproofed at least to the same extent as the rest of the body. It is the extensive rapid loss of water involved in waterproofing the gland that is responsible for two-compartment kinetics. Even so, the energy required for the conversion of water from the CEA to the equilibrium a_w of the tissues and haemolymph is only a small fraction of that provided by the mite's metabolism (Arlian, 1975b).

One-compartment kinetics require a pump system that can be waterproofed without the rapid loss of a considerable portion of the body water. The anatomy of *L. echidnina* has been studied at the level of light microscopy (Jakeman, 1961) and two sets of glands could be candidates for the role of producer of hygroscopic fluid for the pump. The salivary glands have already been suggested as unlikely candidates (Wharton & Devine, 1968), but were not completely ruled out. Spiny rat mites have a pair of coxal glands that open between coxae I on the ventral surface of the body anterior to the sternal plate and lateral to the tritosternum. In some predacious relatives of *L. echidnina*, fluid from their prey has been observed to collect in this region and then it is moved anteriorly via the deutosternal groove by the tritosternum into the prebuccal cavity (Wernz & Krantz, 1976). Given a mechanism for moving fluid from the anterior ventral surface of the mite to the prebuccal cavity, the probability that the coxal glands are the producers of hygroscopic fluid for the pump is greatly increased. Structures resembling lips are seen at the gland's orifice, thus it is possible that they can be waterproofed merely by closing their openings. Fine-structural studies of the spiny rat mite are needed.

One feature of the type of gland described here is that the hygroscopic material serves as a reserve water supply that can function almost like a canteen as soon as the mite is again in air above its CEA. This is especially true of the pump characteristic of the Acaridei. The hygroscopic material is dried by air below the CEA and is thus able to sorb more water when it deliquesces above the CEA. Also, in the dry state the salt is effectively isolated from participating in the osmotic balance of the internal fluids.

Generalities

All mites have a large surface to volume ratio and, consequently, are more prone to desiccation than larger organisms. Despite this fact, only two acarine life styles are known to lead to the development of adaptations that permit the utilization of water vapour from unsaturated air as a source of fresh water. These are the gorgers and fasters such as the spiny rat mite and those that feed on dry food such as *D. farinae*. Gorgers and fasters have CEAs mostly between 0.85 and 0.90 a_v. They feed on food, usually blood, that has a water activity of 0.99 a_w so that their food can supply their water needs when it is available. The fasting periods that they undertake are required either by the seasonal availability of their hosts or the requirements of their life cycles. During their periods of fasting, they can sequester themselves in niches that are relatively free of other organic material. In their fasting sites, they need only avoid condensation of water in their vicinity. Mites that consume dry food such as many of the Acaridei have CEAs between 0.62 and 0.84 a_v. They have quite a different problem from that of the gorgers and fasters. They must be able to obtain water from concentrations of organic material that are too dry to support excessive bacterial and especially fungal growth. Their CEAs and rates of population growth are such that fungi do not destroy them in their preferred habitats. Both of the life styles that appear to favor the development of water uptake mechanisms probably have a significant history in geological time, and competitors without the ability to get water from unsaturated air would be unable to cope with those that can. With the exception of the sea, the activity of which is 0.98 a_w, the surface of the earth through climatic changes and plate tectonics changes its humidity too rapidly to promote the development of water uptake mechanisms. Special life styles are required. For example, the feather mites that live on the barbs and never utilize the living tissues of their hosts will someday be found to use water vapour as a source of fresh water. They are known to have well-developed supracoxal glands (Dubinin, 1951).

It appears that the major requirement for obtaining water from unsaturated air is the development of a gland that can produce a substance that is hygroscopic above the CEA demanded by the life style of the organism in question. The salt glands of mites have the same sort of structure as those of other animals (Schmidt-Nielsen, 1960; Copeland, 1967) and even plants (Campbell, Thomson & Platt, 1974). These salt glands are needed to remove salt from water that is sorbed or imbibed. Because so many organisms in so many different groups have developed salt glands, it is predictable that should the earth become much drier than it now is, life as we know it will be able to adapt to the drier conditions by developing techniques for removing water from the air, such as those used by gorgers and fasters or inhabitants of nests, homes, or stored products (Solomon, 1965).

Following this line of reasoning, I think that if life is found on Mars, the organism will have a salt gland that can produce a salt capable of extracting water from the Martian atmosphere.

Contribution from the Department of Entomology, College of Biological Sciences, The Ohio State University.

References

Arlian L. G. (1972). Equilibrium and non-equilibrium water exchange kinetics in an atracheate terrestrial arthropod, *Dermatophagoides farinae* Hughes. Ph.D. Dissertation, The Ohio State University, Columbus, Ohio.

Arlian, L. G. (1975*a*). Dehydration and survival of the European house dust mite, *Dermatophagoides pteronyssinus*. *Journal of Medical Entomology*, **12**, 437–42.

Arlian, L. G. (1975*b*). Water exchange and effect of water vapour activity on metabolic rate in the dust mite, *Dermatophagoides*. *Journal of Insect Physiology*, **21**, 1439–42.

Arlian, L. G. & Wharton, G. W. (1974). Kinetics of active and passive components of water exchange between the air and a mite, *Dermatophagoides farinae*. *Journal of Insect Physiology*, **20**, 1063–77.

Brody, A. R., McGrath, J. C. & Wharton, G. W. (1972). *Dermatophagoides farinae*: the digestive system. *Journal of the New York Entomological Society*, **80**, 152–77.

Brody, A. R., McGrath, J. C. & Wharton, G. W. (1976). *Dermatophagoides farinae*: the supracoxal glands. *Journal of the New York Entomological Society*, **84**, 34–47.

Brody, A. R. & Wharton, G. W. (1970). *Dermatophagoides farinae*: ultrastructure of lateral opisthosomal dermal glands. *Transactions of the American Microscopical Society*, **89**, 499–513.

Campbell, N., Thomson, W. & Platt, K. (1974). The apoplastic pathway of transport to salt glands. *Journal of Experimental Botany*, **25**, 61–9.

Copeland, D. E. (1967). A study of salt secreting cells in the brine shrimp (*Artemia salina*). *Protoplasma*, **63**, 363–84.

Cross, H. F. & Wharton, G. W. (1966). Feeding tests with several species of mites on different kinds of blood and blood components. *Annals of the Entomological Society of America*, **59**, 182–5.

Cunnington, A. M. (1969). Physical limits for complete development of the copra mite, *Tyrophagus putrescentiae* (Schrank) (Acarina, Acaridae). In *Proceedings of the 2nd International Congress on Acarology*, ed. G. O. Evans, pp. 241–8. Budapest: Akadémiai Kiadó.

Cutcher, J. (1973). The critical equilibrium activity of nonfeeding *Tyrophagus putrescentiae* (Acarina: Acaridae). *Annals of the Entomological Society of America*, **66**, 609–11.

Devine, T. L. (1969). A systematic analysis of the exchange of water between a mite *Laelaps echidnina* and the surrounding vapor. Ph.D. Dissertation, The Ohio State University, Columbus, Ohio.

Devine, T. L. & Wharton, G. W. (1973). Kinetics of water exchange between a mite, *Laelaps echidnina*, and the surrounding air. *Journal of Insect Physiology*, **19**, 243–54.

Dubinin, V. B. (1951). Feather mites (Analgesoidea). *Fauna USSR*, **6**, 1–363.

Ellingsen, I. J. (1976). Permeability to water in different adaptive phases of the same instar in the American house-dust mite. *Acarologia*, **17**, 734–44.

Jakeman, L. A. R. (1961). The internal anatomy of the spiny rate mite, *Echinolaelaps echidninus* (Berlese). *Journal of Parasitology*, **47**, 329–49.

Kanungo, K. (1963). Effect of low oxygen tensions on the uptake of water by dehydrated females of the spiny rat-mite *Echinolaelaps echidninus*. *Experimental Parasitology*, **14**, 263–8.

Kanungo, K. (1964). Disappearance of blood from the gut of engorged *Echinolaelaps echidninus* (Acarina: Laelaptidae). *Annals of the Entomological Society of America*, **57**, 427–8.

Kanungo, K. (1965). Oxygen uptake in relation to water balance of a mite (*Echinolaelaps echidninus*) in unsaturated air. *Journal of Insect Physiology*, **11**, 557–68.

Knülle, W. (1965). Die Sorption und Transpiration des Wasserdampfes bei der mehlmilbe (*Acarus siro* L.). *Zeitschrift für vergleichende Physiologie*, **49**, 586–604.

Knülle, W. (1967). Significance of fluctuating humidities and frequency of blood meals on the survival of the spiny rat mite, *Echinolaelaps echidninus* (Berlese). *Journal of Medical Entomology*, **4**, 322–5.

Larson, D. G. (1969). The critical equilibrium activity of adult females of the house dust mite *Dermatophagoides farinae* Hughes. Ph.D. Dissertation, The Ohio State University, Columbus, Ohio.

McEnroe, W. D. (1961). The control of water loss by the two-spotted spider mite (*Tetranychus telarius*). *Annals of the Entomological Society of America*, **54**, 883–7.

Madge, D. S. (1964). The water-relations of *Belba geniculosa* Oudms. and other species of oribatid mites. *Acarologia*, **6**, 199–223.

Madge, D. S. (1965a). The effects of lethal temperatures on oribatid mites. *Acarologia*, **7**, 121–30.

Madge, D. S. (1965b). Further studies on the behaviour of *Belba geniculosa* Oudms. in relation to various environmental stimuli. *Acarologia*, **7**, 744–57.

Madge, D. S. (1966). The significance of the sensory physiology of oribatid mites in their natural environment. *Acarologia*, **8**, 155–60.

Schmidt-Nielsen, K. (1960). The salt-secreting gland of marine birds. *Circulation*, **21**, 955.

Solomon, M. E. (1962). Ecology of the flour mite, *Acarus siro* L. (= *Tyroglyphus farinae* DeG.). *Annals of Applied Biology*, **50**, 178–84.

Solomon, M. E. (1965). The ecology of pests in stores and houses. In *Ecology and the Industrial Society*, pp. 345–66. Oxford: Blackwell Scientific.

Solomon, M. E. (1966). Moisture gains, losses and equilibria of flour mites, *Acarus siro* L., in comparison with larger arthropods. *Entomologia experimentis et applicata*, **9**, 25–41.

van der Hammen, L. (1970). La phylogenèse des Opilioacarides, et leurs affinités avec les autres Acariens. *Acarologia*, **12**, 465–73.

van de Vrie, M., McMurtry, J. A. & Huffaker, C. B. (1972). Ecology of tetranychid mites and their natural enemies: A review. III. Biology, ecology, and pest status, and host-plant relations of tetranychids. *Hilgardia*, **41**, 343–432.

Vannier, G. (1976). Principaux modes d'étude de la balance hydrique chez les Acariens. *Acarologia*, **18**, 3–19.

Waggoner, P. E. (1967). Moisture loss through the boundary layer. *International Journal of Biometeorology*, **11** (Supplement), 41–51.

Wernz, J. G. & Krantz, G. W. (1976). Studies on the function of the tritosternum in selected Gamasida (Acari). *Canadian Journal of Zoology*, **54**, 202–13.

Wharton, G. W. (1976). House dust mites. *Journal of Medical Entomology*, **12**, 577–621.

Wharton, G. W. & Brody, A. R. (1972). The peritrophic membrane of the mite, *Dermatophagoides farinae*: Acariformes. *Journal of Parasitology*, **58**, 801–4.

Wharton, G. W. & Cross, H. F. (1957). Studies on the feeding habits of three species of laelaptid mites. *Journal of Parasitology*, **43**, 45–50.

Wharton, G. W. & Devine, T. L. (1968). Exchange of water between a mite, *Laelaps echidnina*, and the surrounding air under equilibrium conditions. *Journal of Insect Physiology*, **14**, 1303–18.

Wharton, G. W. & Furumizo, R. T. (1977). Supracoxal gland secretions as a source of fresh water for Acaridei. *Acarologia*, **19**. (In press.)

Wharton, G. W. & Kanungo, K. (1962). Some effects of temperature and relative humidity on water-balance in females of the spiny rat mite, *Echinolaelaps echidninus* (Acarina: Laelaptidae). *Annals of the Entomological Society of America*, **55**, 483–92.

Winston, P. W. (1963). Humidity relations in the clover mite, *Bryobia praetiosa* Koch. *Ecology*, **44**, 669–78.

Winston, P. W. & Nelson, V. E. (1965). Regulation of transpiration in the clover mite *Bryobia praetiosa* Koch (Acarina: Tetranychidae). *Journal of Experimental Biology*, **43**, 257–69.

8. Uptake of water vapour from the air: process, site and mechanism in ticks

By DIETER RUDOLPH & WILLI KNÜLLE

In their natural habitat ticks are faced with long periods without ingesting food or imbibing liquid while they are awaiting a passing host. During these periods it is a special problem for them to maintain their water balance against the drying power of an atmosphere in which the activity of water is normally much below that of the body. In numerous ixodid and argasid ticks, active uptake of water vapour has been shown to occur above *c.* 85% RH, as an effective means of maintaining the water balance in their bodies for many months.

To obtain information in ticks on the process, site and mechanism of water vapour uptake from the atmosphere, we performed experiments on the common African tick *Amblyomma variegatum* (Fabricius).

In adults of *A. variegatum* the discharge of faeces and excreta ceased about 10 weeks after the final moult and the weight of the ticks remained practically constant up to the twentieth week at 32 mg for the females and 23 mg for the males. The water content was similarly constant at $60.3 \pm 1.0\%$ of total weight for the males and $60.2 \pm 0.7\%$ for the females. During this period, some 5–6% of the body solids were metabolized in both sexes. Assuming that this occurred entirely by the oxidation of fats – since Less (1946) established that lipids and lipoproteins are virtually the only energy reserves of non-feeding *Ixodes ricinus* – the maximal production of metabolic water would amount to some 800 μg.

On the evidence that the metabolic rate of the ticks proceeded at only 0.03% of their body weight each day and that the average proportion of body water was constant during the above period, it could be concluded that changes in the body weight might simply be identified with changes in the weight of body water that occurred through transpiration or sorption of atmospheric water vapour.

To obtain information on the dynamics of the water balance the weight changes of individual adults of *A. variegatum* exposed to 93% RH were observed at intervals of two days during a period of seven weeks. Weighing at shorter time intervals was inadvisable since frequent handling of the test animals disturbed their natural water balance. Neither CO_2 nor any other anaesthetic was used during weighing. The course of variations of the body

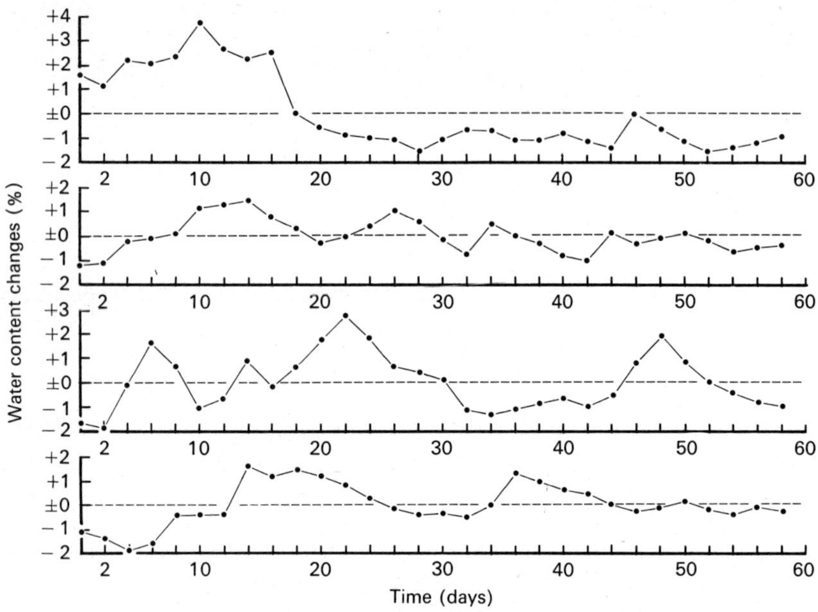

Fig. 1. Changes in the body water content (expressed as percentage of the mean water content over a 9 week observation period) on 4 individual adults of *Amblyomma variegatum*. Broken horizontal lines indicate individual mean water contents.

water contents of individuals from their mean values are illustrated in Fig. 1. The mean value of water content of each individual was calculated from its particular weight at each observation and from its water content at the end of the test.

It is apparent that the water content of adults of *A. variegatum* during prolonged exposure to high relative humidities was not maintained at a constant value, but instead longer periods of 8 to 14 days of slight water loss alternated with shorter periods of 2 to 6 days of rapid uptake. Overall, the water content varied from about −2% to +2% about the mean value. A regular pattern for variations of the water content was not recognized in either the time course or the amplitude of these variations. The losses proceeded at an average of 0.31% of the mean water content each day while the gains were about twice as rapid, at 0.66% per day. In some individuals both the loss rate and the recovery rate decreased with the age of the tick.

We can conclude from these data that there is neither a consistent lower threshold of water content necessary to initiate vapour uptake nor is there an upper limit at which uptake consistently ceases.

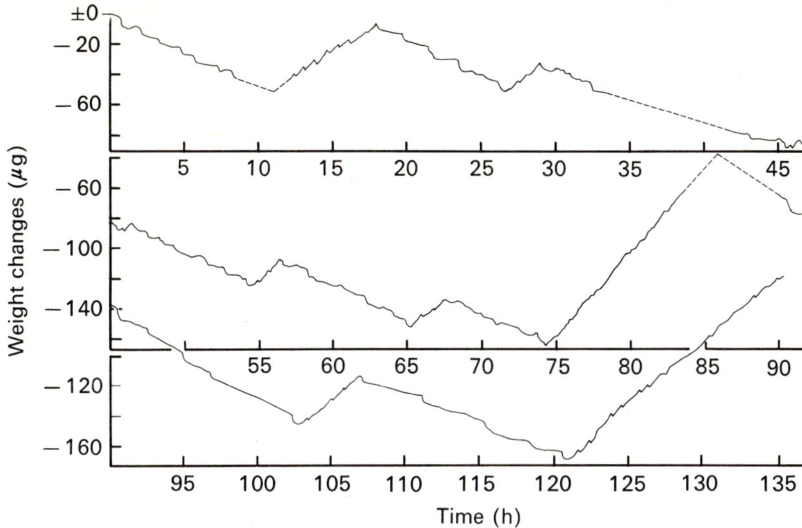

Fig. 2. A continuous recording of weight on an electrobalance of an individual female of *Amblyomma variegatum*. Broken lines indicate periods during which the recorder ran out of scale.

Since it could not be excluded that the observed variations of the body water content could have resulted from handling the test animals, in spite of the relatively long intervals of two days between weighings, we also made continuous observations of undisturbed individuals at constant humidity and temperature conditions on a recording balance with a sensitivity of 0.4 μg. In Fig. 2, weight traces from the recording balance were copied point by point for an individual female of *A. variegatum* that was continuously exposed to 93 % RH and 20 °C.

The traces confirm those variations of the water content, that were observed in intermittently weighed individuals. Furthermore these traces show that the long continued loss phases are not an uninterrupted process, as appeared in the data taken at two day intervals, but instead were composed of a sequence of long loss periods alternating with brief uptake periods in which the water loss of the preceding phase was seldom recovered. This thereby resulted in an overall weight loss over extended periods of time. In the female illustrated in Fig. 2, a continuous 10 h phase of vapour uptake, resulting in an almost full recovery of the water deficit, only occurred after a 3 day period of overall water losses. During uptake, brief interruptions of only a few minutes, accompanied by rapid water loss, could be observed. Although this gives the impression that the operation of the uptake mechanism is discontinuous, this possibility may be excluded,

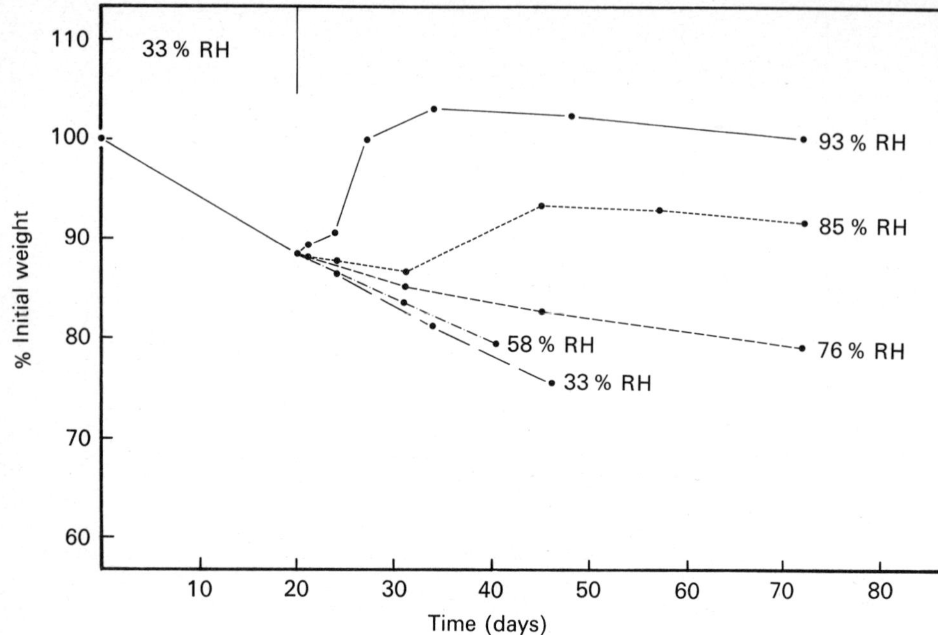

Fig. 3. Mean weights (expressed as percentage of initial weight) of adults of *Amblyomma variegatum*. Following 20 days of partial dehydration at 33% RH and 20 °C, groups of 20 adults were exposed to various humidities.

since comparable short phases of rapid loss were also seen during the period of sustained water loss. It is very probable that during these short phases the spiracles were open. Similar rapid water losses also occurred whenever the test animals exhibited locomotory activity; this was registered on the balance as a burst of sharp 'pulses'.

Fluctuations in the equilibrium water level of the proportion observed in *A. variegatum* do not seem to occur in *Xenopsylla cheopis* (Knülle, 1967), *Liposcelis rufus* (Knülle & Spadafora, 1969) and *Thermobia domestica* (Noble-Nesbitt, 1969). Only by use of a recording balance was Noble-Nesbitt (1969) able to observe variations as small as 0.1% of the total weight in the firebrat during prolonged exposure to a high humidity.

It is remarkable that the equilibrium water level of *A. variegatum* depends upon the environmental humidity (Fig. 3). When partially dehydrated ticks were transferred to high humidities, vapour uptake proceeded at 93% RH at a rate more than seven times as fast as that at 85% (117 μg per day), and equilibrium water content was achieved and maintained in 93% after a total vapour uptake of 3.9 mg and that was more than twice the 1.5 mg taken up from 85% RH.

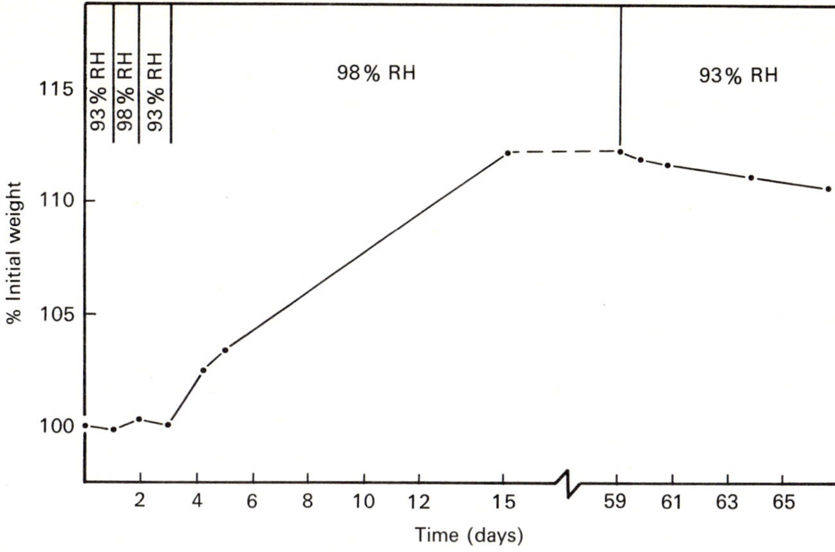

Fig. 4. Mean weights (expressed as percentage of initial weight) of 10 adults of *Amblyomma variegatum* subjected to alternating 93% and 98% RH at 20°C.

The striking dependence of the water content on the environmental humidity is further illustrated when, after a prolonged exposure to 93% RH, ticks were transferred to 98% RH. The water content of these ticks gradually increased and after 12 days attained an equilibrium level 12% higher at 98% RH than that at 93% RH. This level was maintained as long as the animals were held at 98% RH and returned to a lower level only when the animals were subsequently returned to 93% RH (Fig. 4).

Corresponding results were obtained by other investigators in the ixodid ticks *Hyalomma asiaticum* (Balashov, 1960) and *Hyalomma dromedarii* (Hafez, El-Ziady & Hefnawy, 1970), in the argasid tick *Ornithodoros savignyi* (Hafez *et al.*, 1970), in the grain mite, *Acarus siro* (Knülle, 1965) and in the larvae of the oriental rat flea *Xenopsylla cheopis* (Knülle, 1967). On the other hand *Thermobia domestica* (Noble-Nesbitt, 1969) *Ctenolepisma terebrans* (Edney, 1971), as well as Psocoptera of the genus *Liposcelis* (Knülle & Spadafora, 1969) adjusted their equilibrium water content independently of the surrounding relative humidity. We do not know the physiological basis for the adjustments of the water content in either the humidity-dependent or humidity-independent cases.

The site and the mechanism of active water vapour uptake have been under discussion since the phenomenon was discovered. In ticks, as in other arthropods, both the tracheal system and the external integument have been suggested as sites for uptake of atmospheric water vapour. In the 1960s,

Beament (1961, 1964, 1965) and Locke (1964, 1965) proposed detailed models for the transport of water through the cuticle, based on its ultra-structure.

Attention was shifted from the external integument as the site of active uptake to inner surfaces, when in 1970 Noble-Nesbitt demonstrated that occlusion of the anus prevented uptake of water vapour in the firebrat. He inferred that the site of uptake might be in the rectum. The tick *Dermacentor variabilis*, however, was not impaired in this ability when the anus was blocked, as Knülle & Devine and McEnroe demonstrated in 1972.

In order to investigate the region of the tick body at which water vapour is extracted from subsaturated atmospheres, 8 week old unfed ticks (males and females) were first subjected to dehydrating conditions in air of 33% RH and 20 °C. After an exposure of 3–4 weeks to these conditions, they lost, on average, 24% of their initial water mass. Subsequently the following parts of the body were covered with paraffin wax: the anus, the complete dorsal surface of the idiosoma, the ventral surface except the spiracles, the spiracles themselves and the distal part of the capitulum with the mouth proximally up to the insertion point of the palps at the basis capituli. During the ensuing exposure to air of 93% RH and 20 °C, no significant difference in vapour uptake was observed among the control ticks, which remained without paraffin treatment, those with the dorsal surface covered, those with the ventral surface covered and those with the anus covered. However, neither those ticks with the spiracles covered nor those with the mouthparts covered were able to achieve an equilibrium water mass in 93% RH (Table 1).

Of the test-group with covered spiracles, about half of the individuals were able to take up some water vapour, at least within the first 20 h after paraffin treatment, although the net gains were very low, 0.06% of their initial body weight, while the other individuals were able to maintain their body water at a constant level during this time. Over longer periods, however, all individuals of this test group continuously lost small amounts of water with an average rate of 0.2% of the initial weight per day. Following removal of the paraffin cover from the spiracles all ticks began absorbing water vapour from the atmosphere within 48 h.

This experiment demonstrates that basically, ixodid ticks, are able to take up water vapour with the spiracles sealed. The tracheal system therefore cannot be the site of uptake. This is supported by the occurrence of water vapour uptake in the larvae of ixodid ticks (Knülle, 1966) which have no tracheal system. The observed cessation of vapour uptake 1 ½ day following experimental closure of the spiracles apparently resulted from preclusion of gas exchange through the spiracles. Ticks with blocked spiracles soon became inactive and appeared anaesthetized. In this context

Table 1. *Effect of covering different parts of the body with paraffin wax on water vapour uptake in previously dehydrated adults of* Amblyomma variegatum *exposed to 93% RH, 20 °C*

Weight after c. 4 weeks in 33% RH, 20 °C (% initial weight ±s.d.)	Water loss after c. 4 weeks in 33% RH, 20 °C (% initial water mass ±s.d.)	Part of body covered with wax	Rate of uptake or loss per day in 93% RH, 20 °C (% initial water mass ±s.d.)	No. of individuals tested
80.6±3.3	32.4±5.3	Dorsal idiosoma	+3.1±0.8	10
90.6±1.7	15.6±1.1	Ventral idio-soma except spiracles	+2.9±1.4	10
86.4±1.1	22.7±1.9	Anal opening	+3.4±0.6	10
85.3±0.8	24.6±1.4	Spiracles	−0.2±0.02	10
87.5±2.6	20.8±4.3	Distal part of rostrum	+2.9±0.5	10
85.2±4.1	24.7±6.9	Rostrum totally	−0.3±0.07	10
No dehydration	No dehydration	Rostrum totally	−0.2±0.1	10
87.2±4.9	21.3±8.2	Untreated controls	+3.0±0.7	20
82.9±1.5	28.5±2.4	Proximal part of rostrum	+3.2±0.2	10

it is of interest that partially dehydrated adults of *A. variegatum*, when exposed to an anaesthetizing atmosphere of CO_2 at 93% RH ceased to take up water vapour from the atmosphere, as has been shown by other investigators for other species of vapour-absorbing arthropods (Browning, 1954; Knülle, 1967; Noble-Nesbitt, 1969; Knülle & Devine, 1972).

In contrast, those ticks with their mouthparts covered did not show net water gain at any time during an observation period of 4 weeks; instead they lost an average of 0.3% of their body water per day. In order to show that the application of paraffin wax and the contact between the mouthparts and the applied wax had no adverse effects on water uptake, the following experiments were performed.

In one group of ticks, dehydrated as above, only the distal half of the mouthparts was painted with paraffin wax such that both the dorsal space between the chelicerae and the lateral spaces between the hypostome and the chelicerae provided paths between the surrounding air and the buccal cavity (Fig. 5). These ticks were able to increase their body water by an

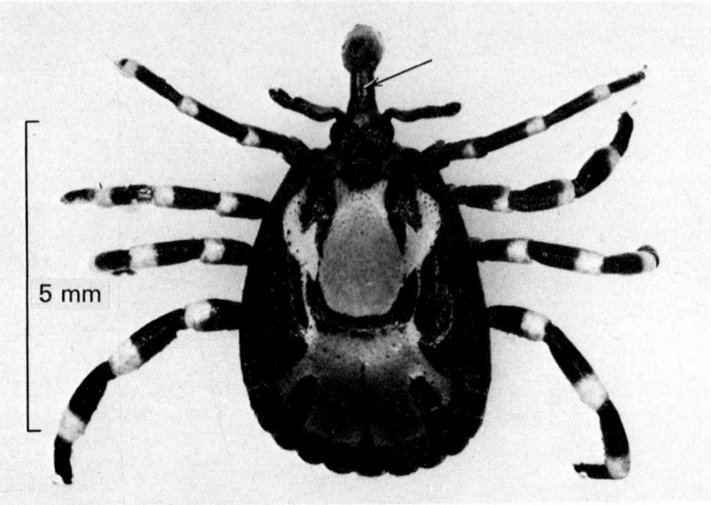

Fig. 5. Male of *Amblyomma variegatum* with the distal part of the rostrum covered with paraffin wax. Arrow pointing to the dorsal space between the chelicerae.

amount comparable to that in the control group. Likewise no effect on uptake was found in ticks covered with wax at the proximal part of the capitulum in the region of the areae porosa.

In another experiment the mouth of partially dehydrated ticks was completely sealed with wax. After the ticks had suffered weight loss for 24 h, the wax was removed from the mouthparts by means of a needle. Although this operation could not be executed in a gentle manner, since the wax adhered tightly to the chelicerae and hypostome, all of these ticks subsequently began an uptake of vapour. Moreover in most ticks with sealed mouthparts the wax cover progressively softened from the inner surface, beginning after approximately 4 weeks – perhaps because of ejected salivary secretion – the cover gradually detached itself from the mouthparts, and at *c.* 5 weeks could be removed in one piece. Accompanying this process, the rate of gain of body water mass increased gradually from *c.* 0.5% per day in the early stages of loosening to *c.* 3% per day after the removal of the wax cover. From these experiments we can conclude that blockage of the mouth prevents net water gain from the atmosphere in adults of *A. variegatum.*

That water vapour actually passed through the mouthparts was shown in an experiment which did not require blockage of the mouthparts with

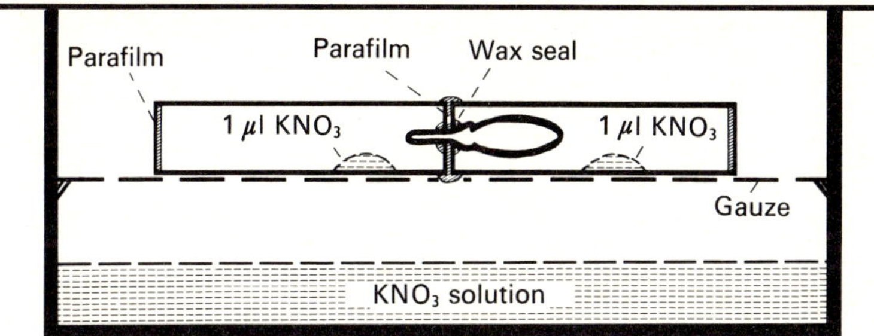

Fig. 6. Apparatus for simultaneous exposure of the anterior and posterior portions of a tick to separated atmospheres. For fuller explanation see text.

paraffin wax. For this purpose the mouthparts of a partially dehydrated tick were inserted through a wax-coated parafilm membrane up to the base of the palps. This membrane was clamped between the open ends of two glass tubes – inner diameter 5 mm, length 20 mm – such that the mouthparts projected into one tube and the rest of the body into the other (Fig. 6). The contact line between the two tubes, and the penetration point of the mouthparts through the membrane, were carefully sealed with melted paraffin wax. The opposite ends of the glass tubes were closed by parafilm membranes coated with paraffin wax. By injecting the needle of a microlitre syringe through these parafilm membranes, one microlitre of saturated KNO$_3$ solution (which provides a relative humidity of 93 % at 20 °C) was placed on the wall of each glass tube, adjacent to but not touching the tick. The membranes were subsequently repaired. After 28 h on average, the drop of KNO$_3$ solution adjacent to the mouthparts had changed to a moist crystal of KNO$_3$, while that in the tube containing the rest of the body appeared to be unchanged. The possibility that vapour inadvertently escaped to the surrounding air was completely precluded in that the assembly was held in a container over KNO$_3$ solution. This experiment gave the same result when repeated with 15 different ticks.

The amount of water which had evaporated from a 1 μl drop of saturated KNO$_3$ solution, at the time when moist KNO$_3$ crystals were observed in the glass tubes, was 575 μg. Only 3.6 μg were required to raise the humidity in one of the glass tubes from 40 % RH (room humidity) to 93 % RH, i.e. 570 μg of water vapour must have been taken up by the tick via the mouth. From this an average uptake rate of 1.7 % of the initial weight was calculated. This is in good agreement with the net uptake rates as previously determined for partially dehydrated ticks in 93 % RH and 20 °C.

The results of these experiments demonstrate unequivocally that sealing

Table 2. *Effect of blocking the mouth of different species of ixodid ticks on water vapour uptake*

Species	Initial weight (93% RH, 20 °C) mg±s.D.	Dehydration time (0% RH, 25 °C) days	Weight after dehydration % initial weight±s.D.	Treatment	Rate of uptake or loss per day (93% RH, 20 °C) % initial weight±s.D.	No.
A. variegatum larvae	0.083±0.004	1	88.3±1.0	Untreated	c. +8.1	5
				Mouth blocked	c. −4.3	
A. variegatum nymphs	1.33±0.19	4	92.7±0.6	Untreated	+3.96±0.3	5
				Mouth blocked	−0.79±0.2	
R. bursa adults	3.81±0.63	12	80.0±3.0	Untreated	+2.96±0.4	10
				Mouth blocked	−0.33±0.04	
H. schulzei adults	22.85±2.06	19	83.1±3.7	Untreated	+2.06±0.5	10
				Mouth blocked	−0.38±0.07	
H. anatolicum excavatum adults	13.28±2.25	14	86.0±3.9	Untreated	+3.95±0.8	10
				Mouth blocked	−0.43±0.06	

the mouth with paraffin wax directly interferes with uptake of water vapour and does not mediate prevention of uptake elsewhere. Since similar results had been obtained when these experiments were performed on both larvae and nymphs of *A. variegatum* and adults of *Rhipicephalus bursa*, *Hyalomma anatolicum excavatum* and *Hyalomma schulzei* (Table 2), and since, moreover, Bull & Smyth (1973) reported that applying wax to the capitulum of *Aponomma hydrosauri*, *Amblyomma albolimbatum* and *Amblyomma limbatum* inhibited water uptake, it may be that active uptake of water vapour proceeds via the mouthparts in all ixodid ticks.

Concerning the mechanism of vapour uptake, evidence for a significant role of salivary secretion is shown by the following observations. Severely dehydrated adults of *A. variegatum* which were desiccated to about 50% of their initial water mass were observed to secrete through their mouthparts a clear fluid which accumulated between the the hypostome and the palps and quickly dried in the desiccation chamber to a white crystalline solid (Rudolph & Knülle, 1974) (Fig. 7); but, when these ticks were transferred to humidities above the critical equilibrium humidity, the crystalline solid dissolved in water taken up from the air and the liquid was subsequently imbibed.

The crystalline solids, described above, were collected from the mouthparts of dehydrated ticks and exposed to a series of different humidities. The substance remained unchanged in humidities less than 75%; at 75% RH it became moist without dissolving. Above 80% RH, however, it took up substantial amounts of water within a few seconds and finally dissolved; recrystallization occurred when the dissolved substance was transferred to relative humidities below 75%. It should be noted that the lower limit of water uptake by the crystalline solids corresponded with the critical equilibrium humidity of the species.

To examine whether the ejected fluid originates in the salivary gland, the chelicerae and the dorsal part of the capitulum were removed in order to observe the buccal cavity, the salivarium and the part of the salivary ducts where they enter the salivarium postero-laterally. In all ticks, regardless of water status, a clear salivary secretion could be seen in the salivarium and the salivary ducts. Furthermore, aperiodic pulsations 2–40 per minute, could be seen in the region of the salivarium. The salivary secretion was ejected forward on the dorsal surface of the hypostome, frequently up to its distal point.

Subsequently the liquid was withdrawn into the pharyngeal orifice located below the salivarium. After each ejection small amounts of fluid crystallized and accumulated on the dorsal surface of the hypostome. The crystalline substance resembled that seen in extremely dehydrated ticks, as described above. Pump movements in the region of the pharynx could

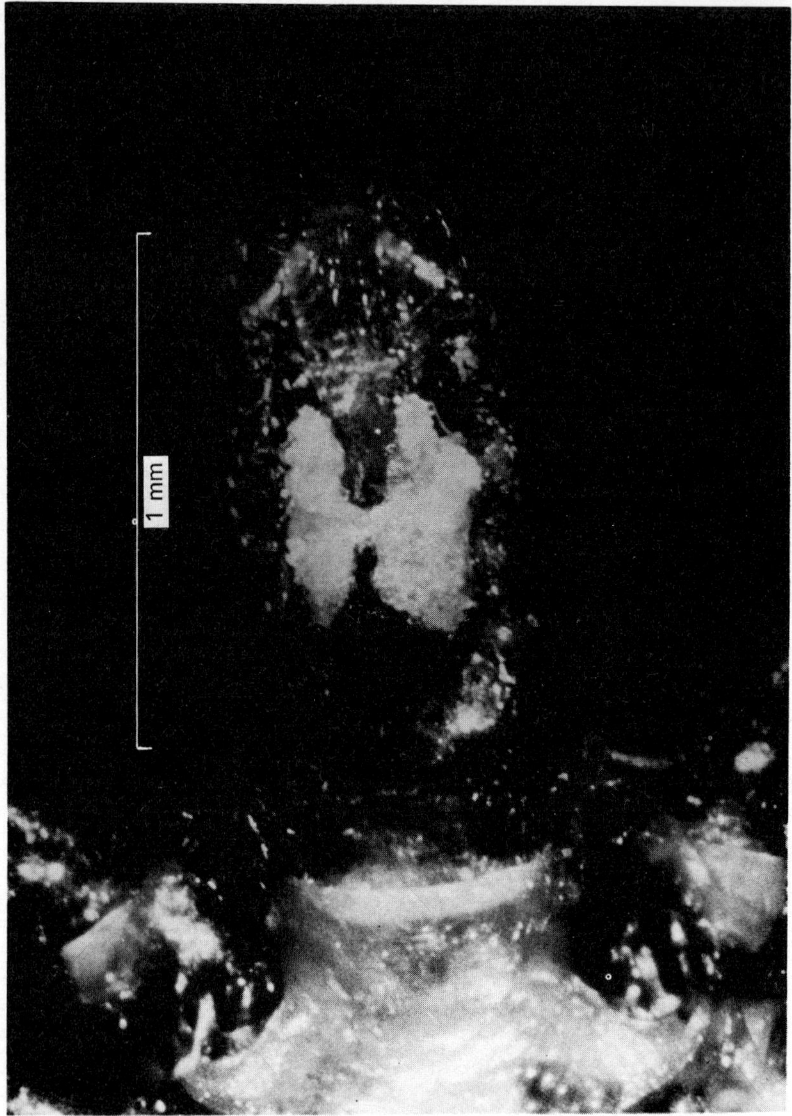

Fig. 7. Ventral view of the mouthparts of an extremely dehydrated male of *Amblyomma variegatum* showing the crystallized salivary secretion. For fuller explanation see text.

also be seen in untreated ticks and fluid could be observed in the buccal channel.

There is no doubt that the oral secretion of extremely dehydrated ticks which accumulated as a white crystalline mass at the rostrum is a secretion of the salivary glands, and that this secretion is present in both the salivarium and the buccal cavity in ticks with a high water content.

The extremely small quantities of salivary secretion made the chemical analysis difficult. A preliminary analysis identified sodium and potassium, with sodium in a higher concentration than potassium. Anions have not been identified as yet. The pH of the aqueous solution was neutral.

To obtain greater amounts of salivary secretion we injected 0.5 μl of a 3% isotonic solution (345 mOsm) of the parasympathetic stimulant pilocarpine into the haemocoel of unengorged adults which were in equilibrium with 93% RH and 20 °C. Pilocarpine has been used successfully to stimulate oral secretion in ticks of different species (Howell, 1966; Tatchell, 1967a; Purnell, Branagan & Radley, 1969; Barker *et al.*, 1973). Only 2 or 3 min after the injection, unengorged adults of *A. variegatum* were observed to perform alternate backward and forward movements of the chelicerae – normally only observed in ticks attaching to a host – accompanied by intermittent ejection of a copious salivary secretion. However, the secretion obtained was different from that described above in that its water evaporated only very slowly at low humidities and it did not exhibit the dramatic water uptake at high humidities. Furthermore, secretion started with the release of a milky white viscous substance which very quickly hardened into a latex-like material. This substance seems to be identical to the so-called 'cement' that has been described by several authors as secreted by a variety of ixodid ticks during their attachment to a host.

Clearly, the salivary secretion obtained by pilocarpine stimulation in unfed ticks is not comparable to that described above for untreated ticks. A modificatory effect of pilocarpine and other drugs on the composition of oral secretion has also been described for engorged ticks (Tatchell, 1967a; Barker *et al.*, 1973).

The functional morphology and the role of the salivary glands in engorging ixodid ticks have been investigated in a series of papers. In contrast to non-feeding individuals, such ticks must eliminate excess water and ions taken up with the host's blood. In argasid ticks this function is performed by coxal glands which are not present in ixodid ticks. In ixodid ticks the salivary glands are the primary organs for the maintenance of water and ion balance during bloodsucking; large quantities of ingested water as well as excess ions are reinjected along with the salivary secretion into the host animal, as has been shown by various authors (Gregson, 1967;

Tatchell, 1967*a,b*, 1969; Kitaoka & Morii, 1970; Kaufman & Phillips, 1973*a,b,c*).

Recent studies on the fine structure of the salivary glands of ixodid ticks by Kirkland (1971), and Coons & Roshdy (1973) have supported the osmoregulatory and ionic regulatory role of the salivary glands. In these studies, attention is drawn to the extensive infoldings of the plasma membrane of the non-granular type I alveoli. These infoldings seem to be a constant characteristic of fluid-transporting epithelia. Extensive infoldings of the basal plasma membrane have also been described by Meredith & Kaufman (1973) in epithelial cells of the granule-secreting alveoli of type III in *Dermacentor andersoni*, the so-called 'water cells'. These authors suggested this alveolus to be a better candidate for producing the bulk of the salivary secretion. Recently, similar cells have been described in the granule-secreting alveoli of the argasid tick *Argas arboreus* (Roshdy & Coons, 1975).

There is as yet no information on the role of the different salivary gland alveoli in non-feeding ticks. The results of our experiments provide good evidence that the salivary glands in non-feeding ixodid ticks are concerned with water and ion balance and that the type I alveolus is evidently of importance is supported by data from the following experiment. In adults of *Hyalomma anatolicum excavatum* heavily infected with *Theileria annulata*, the alveoli of types II and III were completely destroyed, leaving only the two main salivary ducts with the non-granular type I alveoli intact. Such infected ticks, partially dehydrated, started vapour uptake at 93% RH within 2 days as did the uninfected controls, the rate, however, remained with $1.1\pm0.2\%$ of the original weight per day, somewhat behind that of $3.7\pm0.3\%$ obtained for the controls.

In extremely dehydrated non-feeding ticks, the observed ejection of large quantities of a salivary secretion apparently containing considerable amounts of sodium and potassium seems best regarded as an elimination of excess ions during periods when osmotic and ionic stress results from severe water losses.

Secretion of highly concentrated hypertonic salt solutions occurs in other terrestrial animals. Nasal salt glands have been found in several birds and reptiles (Schmidt-Nielsen, 1960; Schmidt-Nielsen *et al.*, 1963; Ohmart, 1972; Peaker & Linzell, 1975). Secretions containing salts at a concentration surpassing that of a 1 M NaCl solution have been demonstrated in several marine birds (Schmidt-Nielsen, 1960). Balashov (1972) called attention to the similarity in ultrastructure of the cells in the non-granular alveoli of ticks and those of the nasal salt glands.

From our experiments on ticks it may be concluded that the first step in vapour uptake from subsaturated atmospheres is accomplished by a salivary secretion exhibiting a water activity below that of the atmosphere. It is likely that the saliva containing the extracted vapour is subsequently imbibed via the pharynx and transferred to the gut. From the gut the ions of the salivary secretion, in particular Na^+ and K^+, along with the water could pass into the haemolymph, possibly without active transport along a concentration gradient maintained by the secretory activity of the salivary glands.

Vapour sorption from subsaturated atmospheres in order to restore a deficit of body water is an adaptive process of decisive significance to survival in non-feeding ticks during the interval between bloodmeals. It will appear too restrictive to view this phenomenon as a simple consequence of an ion-eliminating process; selection pressure for survival might well have perfected the device as an integral part of maintaining water balance.

References

Balashov, Yu. S. (1960). Water balance and behaviour of *Hyalomma asiaticum* in the desert. *Meditsinskaya parazitologiya i parazitarnye bolezni*, **29**, 313–20.

Balashov, Yu. S. (1972). Bloodsucking ticks (Ixodoidea) – Vectors of diseases to man and animals. *Miscellaneous Publications of the Entomological Society of America*, **8**, 161–376.

Barker, R. W., Burris, E., Sauer, J. R. & Hair, J. A. (1973). Composition of tick oral secretions obtained by three different collection methods. *Journal of Medical Entomology*, **10**, 198–201.

Beament, J. W. L. (1961). The water relations of insect cuticle. *Biological Reviews*, **36**, 281–320.

Beament, J. W. L. (1964). The active transport and passive movement of water in insects. *Advances in Insect Physiology*, **2**, 67–129.

Beament, J. W. L. (1965). The active transport of water: evidence, models and mechanisms. *Symposia of the Society for Experimental Biology*, **19**, 273–98.

Browning, T. O. (1954). Water balance in the tick *Ornithodoros moubata* Murray, with particular reference to the influence of carbon dioxide on the uptake and loss of water. *Journal of Experimental Biology*, **31**, 331–40.

Bull, M. & Smyth, M. (1973). The distribution of three species of reptile ticks, *Aponomma hydrosauri* (Denny), *Amblyomma albolimbatum* Neumann, and *Amblyomma limbatum* Neumann. II. Water balance of nymphs and adults in relation to distribution. *Australian Journal of Zoology*, **21**, 103–10.

Coons, L. B. & Roshdy, M. A. (1973). Fine structure of the salivary glands of unfed male *Dermacentor variabilis* (Say) (Ixodoidea: Ixodidae). *Journal of Parasitology*, **59**, 900–12.

Edney, E. B. (1971). Some aspects of water balance in tenebrionid beetles and a thysanuran from the Namib Desert of Southern Africa. *Physiological Zoology*, **44**, 61–76.

Gregson, J. D. (1967). Observations on the movement of fluids in the vicinity of the mouthparts of naturally feeding *Dermacentor andersoni* Stiles. *Parasitology*, **57**, 1–8.

Hafez, M., El-Ziady, S. & Hefnawy, T. (1970). Biochemical and physiological studies of certain ticks (Ixodoidea). Uptake of water vapour by the different developmental stages of *Hyalomma* (*H.*) *dromedarii* Koch (Ixodidae) and *Ornithodoros* (*O.*) *savignyi* (Audouin) (Argasidae). *Journal of Parasitology*, **56**, 354–61.

Howell, C. J. (1966). Collection of salivary gland secretion from the argasid *Ornithodoros savignyi* Audouin (1827) by the use of a pharmacological stimulant. *Journal of the South African Veterinary Medicine Association*, **37**, 236–9.

Kaufman, W. R. & Phillips, J. E. (1973*a*). Ion and water balance in the ixodid tick *Dermacentor andersoni*. I. Routes of ion and water excretion. *Journal of Experimental Biology*, **58**, 523–36.

Kaufman, W. R. & Phillips, J. E. (1973*b*). Ion and water balance in the ixodid tick *Dermacentor andersoni*. II. Mechanism and control of salivary secretion. *Journal of Experimental Biology*, **58**, 537–47.

Kaufman, W. R. & Phillips, J. E. (1973*c*). Ion and water balance in the ixodid tick *Dermacentor andersoni*. III. Influence of monovalent ions and osmotic pressure on salivary secretion. *Journal of Experimental Biology*, **58**, 549–64.

Kirkland, W. L. (1971). Ultrastructural changes in the nymphal salivary glands of the rabbit tick, *Haemaphysalis leporispalustris*, during feeding. *Journal of Insect Physiology*, **17**, 1933–46.

Kitaoka, S. & Morii, T. (1970). Ionic and water balance in the feeding process of ixodid ticks. *National Institute of Animal Health Quarterly*, **10**, 34–41.

Knülle, W. (1965). Die Sorption und Transpiration des Wasserdampfes bei der Mehlmilbe (*Acarus siro* L.). *Zeitschrift für vergleichende Physiologie*, **49**, 586–604.

Knülle, W. (1966). Equilibrium humidities and survival of some tick larvae. *Journal of Medical Entomology*, **2**, 335–8.

Knülle, W. (1967). Physiological properties and biological implications of the water vapour sorption mechanism in larvae of the oriental rat flea, *Xenopsylla cheopis* (Roths.). *Journal of Insect Physiology*, **13**, 333–57.

Knülle, W. & Devine, T. L. (1972). Evidence for active and passive components of sorption of atmospheric water vapour by larvae of the tick *Dermacentor variabilis*. *Journal of Insect Physiology*, **18**, 1653–64.

Knülle, W. & Spadafora, R. R. (1969). Water vapour sorption and humidity relationships in *Liposcelis* (Insecta: Psocoptera). *Journal of Stored Products Research*, **5**, 49–55.

Lees, A. D. (1946). The water balance in *Ixodes ricinus* L. and certain other species of ticks. *Parasitology*, **37**, 1–20.

Locke, M. (1964). The structure and formation of integument in insects. In *The Physiology of Insecta*, ed. M. Rockstein, vol. 3, pp. 379–470. New York: Academic Press.

Locke, M. (1965). Permeability of insect cuticle to water and lipids. *Science, Washington*, **147**, 295–8.

McEnroe, W. D. (1972). The rectum is not the site of water vapour uptake by *Dermacentor variabilis* Say (Acarina: Ixodidae). *Acarologia*, **14**, 542–3.

Meredith, J. & Kaufman, W. R. (1973). A proposed site of fluid secretion in the salivary gland of the ixodid tick *Dermacentor andersoni. Parasitology*, **67**, 205–17.

Noble-Nesbitt, J. (1969). Water balance in the firebrat, *Thermobia domestica* (Packard). Exchanges of water with the atmosphere. *Journal of Experimental Biology*, **50**, 745–69.

Noble-Nesbitt, J. (1970). Water balance in the firebrat *Thermobia domestica* (Packard). The site of uptake of water from the atmosphere. *Journal of Experimental Biology*, **52**, 193–200.

Ohmart, R. D. (1972). Physiological and ecological observations concerning the salt secreting nasal glands of the road-runner. *Comparative Biochemistry and Physiology*, **43A**, 311–16.

Peaker, M. & Linzell, J. L. (1975). *Salt Glands in Birds and Reptiles. Monographs of the Physiological Society*, **32**, 308 pp. London: Cambridge University Press.

Purnell, R. E., Branagan, D. & Radley, D. E. (1969). The use of parasympathomimetic drugs to stimulate salivation in the tick *Rhipicephalus appendiculatus* and the transmission of *Theileria parva* using saliva obtained by this method from infected ticks. *Parasitology*, **59**, 709–18.

Roshdy, M. A. & Coons, L. B. (1975). The subgenus *Persicargas* (Ixodoidea: Argasidae: Argas). 23. Fine structure of the salivary glands of unfed *A.* (*P.*) *arboreus* Kaiser, Hoogstraal and Kohls. *Journal of Parasitology*, **61**, 743–52.

Rudolph, D. & Knülle, W. (1974). Site and mechanism of water vapour uptake from the atmosphere in ixodid ticks. *Nature, London*, **249**, 84–5.

Schmidt-Nielsen, K. (1960). The salt secreting glands in marine birds. *Circulation*, **21**, 955–67.

Schmidt-Nielsen, K., Borut, A., Lee, P. & Crawford, E. (1963). Nasal salt excretion and the possible function of the cloaca in water conservation. *Science, Washington*, **142**, 1300–1.

Tatchell, R. J. (1967a). A modified method for obtaining tick oral secretion. *Journal of Parasitology*, **53**, 1106–7.

Tatchell, R. J. (1967b). Salivary secretion in the cattle tick as a means of water elimination. *Nature, London*, **213**, 940–1.

Tatchell, R. J. (1969). The ionic regulatory role of the salivary secretion of the cattle tick, *Boophilus microplus*. *Journal of Insect Physiology*, **15**, 1421–30.

9. The site of water vapour absorption in *Arenivaga investigata*

By M. J. O'DONNELL

The desert or burrowing cockroach, *Arenivaga investigata*, is found primarily in the desert regions of southern California. Edney (1966) found that starving animals were able to gain weight in subsaturated humidities down to 85% RH, but not at 80% RH. This process, which was observed in nymphs and adult females but not in adult males, could not be explained by either the production of water by metabolism or the reingestion of faeces equilibrated with high relative humidities. The point at which the animal began to increase in weight was apparently unaffected by temperature in the range 10–30 °C.

This account describes experiments in which a new site of absorption for insects has been positively identified and some characteristics of the uptake mechanism measured. Previous studies with the insects *Tenebrio molitor* (Noble-Nesbitt, 1970a; Dunbar & Winston, 1975; Machin, 1975) and *Thermobia domestica* (Noble-Nesbitt, 1970a, b) have demonstrated the importance of the rectum and its associated structures with water vapour absorption.

If the anus or the dorsal surface of *Arenivaga* is blocked with a beeswax–resin mixture (Krogh & Weis-Fogh, 1951) and the animals are subsequently exposed to 96% RII for at least one week, daily weight measurements show a decline in weight gains relative to non-blocked animals, but not a cessation of absorption (Fig. 1). This decline in absorption rate appears to be caused by an inhibitory effect of wax-blocking upon the absorption behaviour of the animal but not upon the absorption mechanism itself. Continuous weight recordings of anus-blocked animals show that absorption is intermittent but the uptake rate is unchanged. Similarly, the mean humidity at which net uptake ceased, calculated by extrapolating uptake rates at various humidities, was found to be unchanged (82.5±3.0% RH prior to blocking the anus with wax and 82.6±1.2% RH after blockage).

A further confirmation of a non-rectal uptake site was obtained by following weight changes of freshly eliminated faecal pellets after the method of Ramsay (1964). Recently fed, absorbing animals lost weight upon defaecation, then gradually returned to the pre-defaecation rate of

115

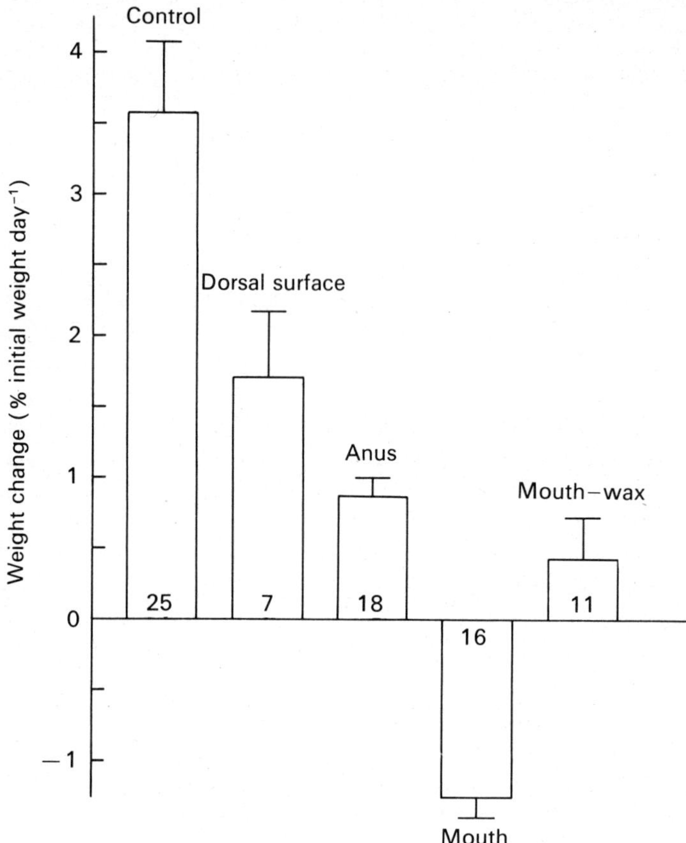

Fig. 1. Histogram of mean weight changes shown by dehydrated animals in 96% RH and 25 °C with different parts of their surfaces blocked with wax. Unwaxed animals serve as the control. The right-hand histogram refers to animals following the removal of wax from the mouth. Vertical bars show 1 s.e.m. Number of animals in each histogram is indicated.

weight gain over a period of 4–6 h. These results indicated that faecal pellets are equilibrated to a higher than ambient humidity prior to voiding, showing that concentrations within the rectum are not sufficiently high to permit absorption.

A dramatic effect upon an animal's weight change in high humidity was seen when the mouth region was blocked with wax. Sixteen such animals at 96% RH showed a mean loss of 1.24±0.15% of their initial weight per day. Further experiments showed it was not necessary to block the mouth, but simply to interfere with its movements by applying wax to the labium or labrum. Similar inhibition of uptake was obtained by applying a strand of surgical thread across the mouth. Most animals resumed gaining weight ($0.44 \pm 0.32\%$ initial weight day^{-1}) when the wax or thread was removed. These results showed that oral structures were involved in the process of water vapour absorption.

Wax-sealing experiments in themselves cannot directly prove the involvement of one anatomical region in the absorption process, since such experimental manipulation could interfere with the sensory input or behaviour of the animal in such a way as to indirectly impede the absorption process (Okasha, 1971). Observation of the mouthparts of normal unimpeded nymphs and adult females in above-threshold humidities implicated two bladder-like structures which seemed to be protracted from the mouth whenever absorption of water vapour was taking place (Fig. 2*a*). These sacs are ventro-lateral diverticula of the hypopharynx which can be quickly inflated or withdrawn by the animal, presumably by alterations of haemolymph pressure. Subsequent studies revealed a significant positive correlation between the proportion of time during which these bladders were out of the mouth and the gain of weight in 96% RH for a given period. Animals that did not expose the bladders lost weight. Bladders are reduced in size in males.

Observations during continuous weight recordings showed that gains occurred only while the bladders were inflated. Once the bladders were withdrawn into the mouth, the animal immediately began to lose weight. Before protruding the bladders, or after withdrawing them, the animal salivated copiously and groomed the bladder surface with the mandibles and the maxillae. An increased weight loss, presumably the result of evaporation of this saliva, occurred for 30–60 s after puffing out the bladders. This was followed by establishment of a rate of gain which was constant for any given relative humidity. If humidity was quickly reduced, the surface of the bladders dried out; no subsequent increase in weight occurred and the animal withdrew the bladders into the mouth within a minute. Generally the bladders were withdrawn before they dried out. These results suggested a direct involvement of the bladders in the water vapour absorption process, and implied that the bladders could not function if their surface was dry.

Thermocouple measurements of surface temperature were made to determine if the bladders functioned as sites for condensation. The thermocouple (40 gauge copper–constantan) could be manipulated onto the surface of the bladders of an animal mounted ventral surface up within the experimental chamber. Measurements from six animals showed that at 96% RH and 25 °C the bladder surface was warmer than the air temperature by 0.14 °C and warmer than the labium, labrum or foreleg femur

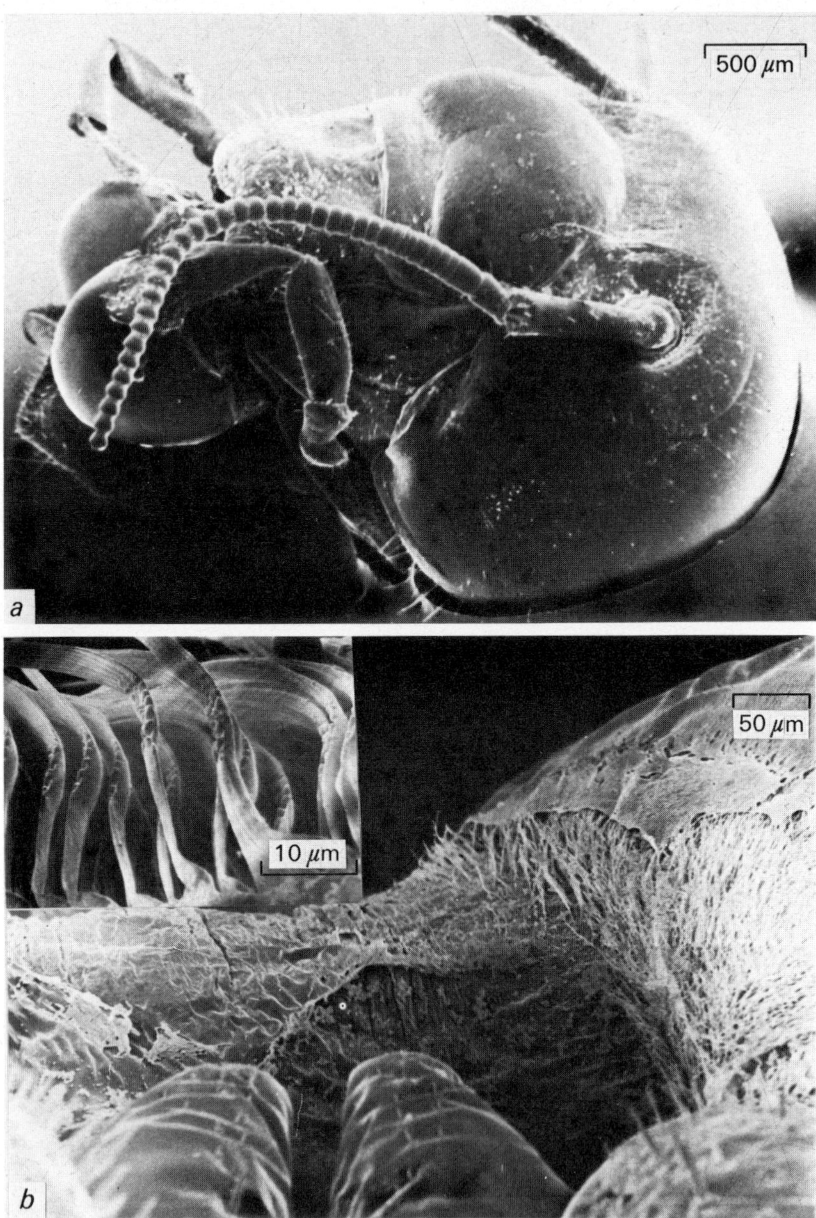

Fig. 2. (*a*) Scanning electron micrograph of the head of *Arenivaga* with inflated bladders (two spherical structures on left). (*b*) Antero-ventral view of bladders showing the fibrils where dried fluid secretion has peeled off during preparation (right centre). The region between the bladders is covered with dried secretion but no fibrils. Inset: basal region of fibrils showing 6–10 fibrils fused together before insertion into the cuticular base.

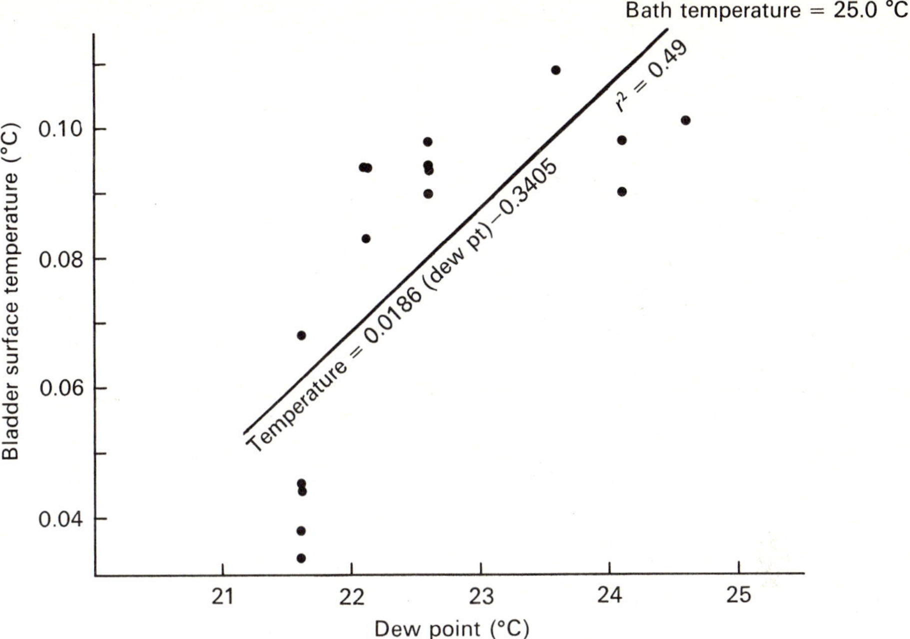

Fig. 3. Graph showing the relationship of bladder surface temperature to humidity (measured as dew point) in one animal.

by 0.09 °C. The region between the bladders (Fig. 2*b*) was warmer than the central part of each bladder (0.21 °C above ambient). If only one of the two bladder surfaces was dried by shining a beam of light upon it or by gradually reducing the humidity, the animal did not usually withdraw the bladders. Dry bladders produced in this way were found not to have warmer surface temperatures than adjacent parts of the head. Temperature measurements made with the thermocouple positioned at one point on the surface of a bladder varied with humidity (Fig. 3).

Studies of the mechanism of absorption require knowledge of the threshold RH above which water vapour uptake occurs. Threshold determinations for the whole animal are inadequate for this purpose, since they represent the humidity at which gains by absorption are balanced by transpiratory losses. Some estimates of the threshold of the absorption site alone have been made in an apparatus which allows exposure of the head to a humidity different from that surrounding the thorax and abdomen. If weight losses through transpiration are minimized by exposing the thorax and abdomen to 96 % RH, the lowest humidity from which absorption can occur is 71.5 % RH. It is important to note that this value is 10 % RH less than the threshold of the whole animal.

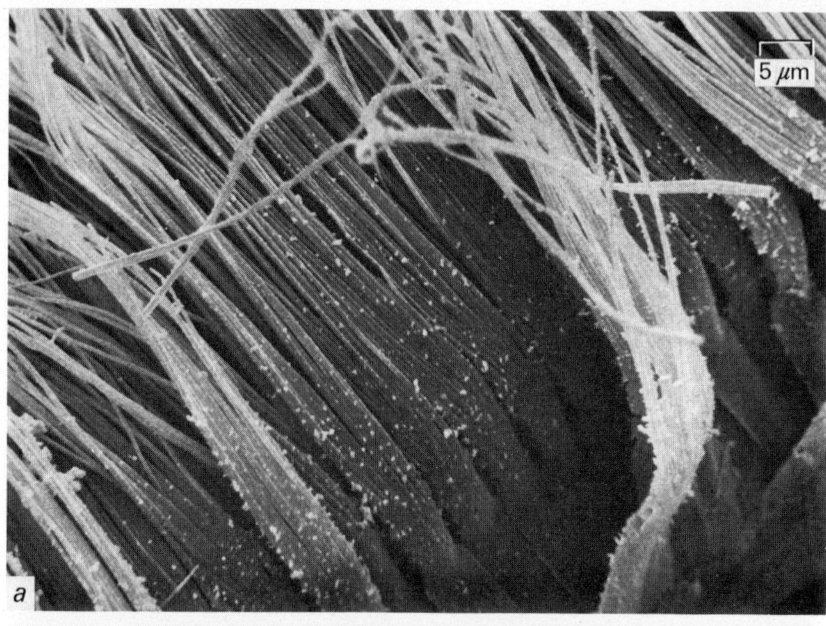

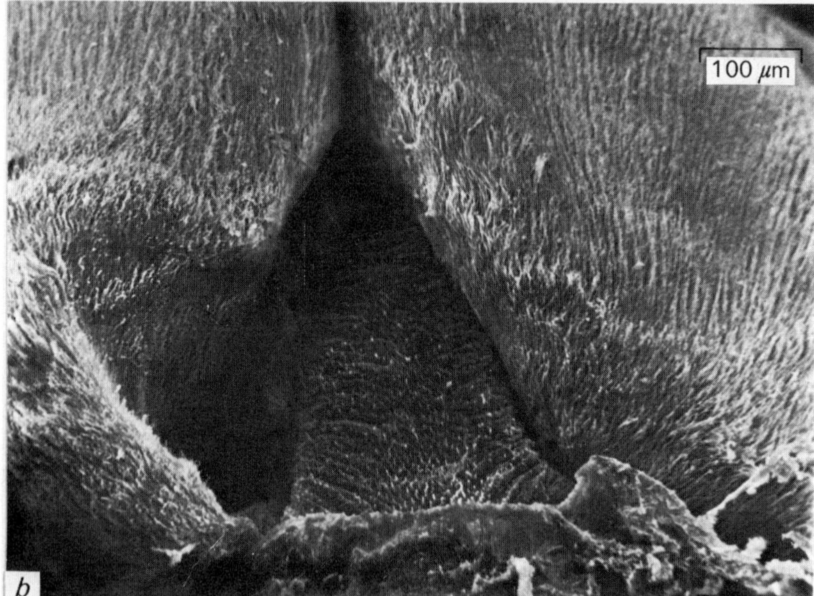

Fig. 4. (*a*) Scanning electron micrograph indicating length of bladder fibrils. (*b*) Ventral view of collapsed bladders, washed free of secretion, indicating density of fibrils. The ventral region between the bladders (centre) is covered with stout spiny projections.

Scanning electron micrographs (Figs. 2 and 4) show that the bladders are covered by densely packed fibrils which vastly increase the surface area of the bladders with an attendant rise in surface tension. The bladder surface may function as a wick, holding to it a layer of fluid which is exposed to the humid air. Presumably this fluid is hygroscopic and causes condensation of water vapour upon the bladder surface, accounting for the above-ambient temperatures of the bladders. The salivary glands, whose duct opens into the mouth at the base of the hypopharynx, are a possible source of the fluid.

References

Dunbar, B. S. & Winston, P. W. (1975). The site of active uptake of atmospheric water in larvae of *Tenebrio molitor*. *Journal of Insect Physiology*, **21**, 495–500.

Edney, E. B. (1966). Absorption of water vapour from unsaturated air by *Arenivaga* sp. (Polyphagidae, Dictyoptera). *Comparative Biochemistry and Physiology*, **19**, 387–408.

Krough, A. & Weis-Fogh, T. (1951). The respiratory exchange of the desert locust (*Schistocerca gregaria*) before, during and after flight. *Journal of Experimental Biology*, **28**, 344–58.

Machin, J. (1975). Water balance in *Tenebrio molitor*, L. larvae; the effect of atmospheric water absorption. *Journal of Cellular and Comparative Physiology*, **101**, 121–32.

Noble-Nesbitt, J. (1970a). Water uptake from subsaturated atmospheres: its site in insects. *Nature, London*, **225**, 753–4.

Noble-Nesbitt, J. (1970b). Water balance in the firebrat, *Thermobia domestica* (*Packard*). *Journal of Experimental Biology*, **52**, 193–200.

Okasha, A. Y. K. (1971). Water relations in an insect, *Thermobia domestica*. I. Water uptake from subsaturated atmospheres as a means of volume regulation. *Journal of Experimental Biology*, **55**, 421–34.

Ramsay, J. A. (1964). The rectal complex of the mealworm *Tenebrio molitor*, L. (Coleoptera, Tenebrionedae). *Philosophical Transactions of the Royal Society* B, **248**, 279–314.

Part 2

Osmotic and ionic regulation in unbalanced environments

10. Some strategies of osmoregulation and ion transport in microorganisms

By D. J. ELLAR

Microorganisms have been found which can grow and reproduce in environments that would rapidly kill the majority of higher organisms. Such hostile surroundings include saturated brine, 65% (w/v) sugar solutions, pH values from 1 to 10 and temperatures ranging from -5 to 99 °C. Brock (1969) has made the general observation that for a range of extreme environments the most successful organisms are those which are morphologically most simple. In citing the above environments it should be emphasised that the microorganisms which colonise them are actively growing and not merely surviving. If we extend our considerations to include the ability of organisms to survive, then the range of ecological and laboratory situations which microorganisms will tolerate is if anything even more dramatic. Here, however, the emphasis will be on some strategies which microorganisms have developed to allow active growth in apparently inimical environments of high solute concentration, although some attention will be devoted to problems of ion transport and dehydration during spore formation in bacteria.

Osmoregulation in non-halophilic bacteria

The necessity for some form of osmoregulation is not confined to a few species of microorganisms growing in media of high osmotic strength as this volume clearly shows. Among the many well-documented examples of osmoregulation in higher organisms are the avian salt gland and adaptations to salinity shown by fish and marine invertebrates (Hochachka & Somero, 1973). Even in those higher animals where the external environment of most cells and tissues is maintained relatively constant, there are nevertheless qualitative differences in the ionic environment on opposite sides of the plasma membrane. Mechanisms of ionic regulation are therefore required to maintain these differences.

Hochachka & Somero (1973) define two strategies by which organisms might realistically regulate their total ionic content. In the first strategy the organism regulates its internal ions qualitatively, but not quantitatively, with respect to the environment. Although the osmolarity of the cell cytoplasm equals that of the surrounding fluid, the cell expends energy to ensure that its internal ionic composition differs qualitatively. However, in such organisms no energy need be devoted to maintain an osmotic difference across the plasma membrane. This is the strategy of osmo-regulation employed by some marine invertebrates and primitive vertebrates.

The second strategy of ionic adaptation calls for the organism to regulate both its internal osmolarity and its qualitative ionic content within precise and narrow limits regardless of changes in the external solute concentration. The cost to these organisms in terms of energy expenditure is obviously higher than was the case for the first strategy, but the result is a cell which possesses a considerable measure of independence of external conditions. This strategy is characteristic of higher animals and advanced inverte-brates.

Neither of these strategies is an accurate description of osmoregulation in microorganisms, as will be seen later. The second strategy is clearly inapplicable, since it has been shown that the internal osmotic pressure of bacteria fluctuates with that of the environment (Epstein & Schultz, 1967). This would appear to favour the first strategy, but this too is not strictly applicable to most microorganisms, since data indicate that bacteria are normally slightly hypertonic with respect to their environment (Epstein & Schultz, 1967). Moreover Christian & Waltho (1961) showed that the extent of this hypertonicity was greater in more osmotolerant micro-organisms growing in media of high osmotic pressure. Osmoregulation in growing microorganisms therefore appears to involve a qualitative control of internal ion content combined with the maintenance of a constant osmotic pressure difference across the plasma membrane in media of differing osmolarity. For microorganisms, this general picture is further modified by the presence of cell walls of varying complexity which lie outside the plasma membrane. These relatively rigid walls provide the organisms with a measure of protection against fluctuations in external osmotic pressure. It should be emphasised again, however, that bacteria are 'osmotic conformers' (Hochachka & Somero, 1973) in the sense that the osmolarity of their cytoplasm fluctuates with that of the growth medium. Probably the most extreme example of this behaviour occurs in the group of bacteria known as obligate halophiles. These halobacteria require 20–25 % (w/v) NaCl in the medium for optimal growth and will not grow at all if the salt concentration falls below about 15 % (w/v). Their natural environment includes the Dead Sea and commercial processes using brine solutions. This requirement for Na^+ and Cl^- is quite specific and no other cation or anion will substitute.

Measurements of chloride levels and freezing points of packed cell pastes have clearly shown that the internal osmolarity of halobacteria equals or exceeds that of the growth medium. Unexpectedly, however, determinations of internal potassium and sodium levels (Christian & Waltho, 1962) gave values of 3.3 M and 1.5 M, respectively, for organisms growing in a medium containing 0.032 M potassium and 4.0 M sodium. Clearly a considerable uptake of K^+ from the medium must be occurring. As will be seen later, it is likely that all bacteria are capable of selectively extracting potassium from their environment, but only in the halobacteria does the internal concentration reach this impressive level.

A comprehensive analysis of the ion balance in the Gram-negative bacterium *Escherichia coli* growing exponentially was carried out by Damadian (1971) and has been summarised by Harold & Altendorf (1974). The current view (Harold & Altendorf, 1974) is that while the cytoplasmic solutes in bacteria are for the most part osmotically active, the situation with cations such as K^+ and Na^+ is somewhat more complex. From measurements of electrical conductivity and osmotic characteristics of two bacterial strains, Marquis & Carstensen (1973) concluded that the mobility of cytoplasmic ions was approximately one-third of the value predicted on the basis of the total cell ion content. In subsequent experiments using cells whose plasma membranes had been ruptured, they observed that in dilute cell suspensions this deviation was not significant, but it increased markedly as the suspensions were made more concentrated. In the intact cell, therefore, it appears that the mobility and osmotic activity of these cations is somewhat restricted as a result of the high viscosity of cell cytoplasm combined with the proximity of macromolecular anions. Despite this restriction, the plasmolysis experiments of Marquis & Carstensen (1973) clearly show these cations to be osmotically active. A similar conclusion is arrived at from the work of Mitchell & Moyle (1956) and Epstein & Schultz (1967), which will be discussed later.

With the halobacteria some uncertainty still prevails regarding the physical state of cytoplasmic cations. In these organisms, as was mentioned above, the internal K^+ concentration may exceed the aqueous solubility of KCl and several workers have concluded that the majority of intracellular K^+ is bound to macromolecules and anions and solvated in semicrystalline water (Cope & Damadian, 1970; Ginzburg, Ginzburg & Tosteson, 1971; Ginzburg, Sachs & Ginzburg, 1971; Ginzburg & Ginzburg, 1975). In the experiments which led to this conclusion, anaerobically incubated bacteria, or bacteria whose oxygen consumption had been reduced to 1 % of control levels by starvation, continued to maintain large ion gradients across the plasma membrane (Ginzburg, Sachs & Ginzburg, 1971). It was concluded that this was not caused by the impermeability of the membrane to these ions, since in experiments where the external Na^+ was replaced by Li^+ or

K^+, the internal Na^+ rapidly equilibrated with the external ion. However, it should be noted that such replacements are likely to disrupt the integrity of the cell envelope of these organisms and could therefore produce an apparent increase in membrane permeability. Further experiments principally with 'starved' halobacteria (Ginzburg, Ginzburg & Tosteson, 1971; Ginzburg & Ginzburg, 1975) appeared to support the view that a fixed proportion of cell K^+ in these organisms is bound in some way. However, the periods of starvation employed in these experiments (up to 8 h) may not have been adequate to deplete the cells' energy reserves. Lanyi & Silverman (1972) have reported that *Halobacterium halobium* can maintain its high intracellular K^+ concentration in a low K^+ medium for up to two days when incubated in the dark without nutrients. Under these conditions the intracellular ATP was maintained at about 30% of the maximal level. From their nuclear magnetic resonance studies with *H. halobium*, Cope & Damadian (1970) also concluded that much of the internal K^+ was complexed by fixed charges and/or was solvated in semi-crystalline water. However, in other experiments with cell pastes of *H. cutirubrum* lysed by freeze-thawing, Lanyi & Silverman (1972) found that intracellular K^+ was readily diffusible, whereas intracellular Mg^{2+} was not. Having in mind reports of the occurrence of a Mg^{2+} salt of a phospholipid from this organism (Rayman, Gordon & MacLeod, 1967) they concluded that their results reflected this bound Mg^{2+} and that the majority of internal K^+ was in a free state, normally retained in the cell by the intrinsic permeability characteristics of the cell membrane. The question of the physical state of internal K^+ in the halobacteria is therefore still a subject for debate, but it should be noted that several studies report that enzymes from these halobacteria are active and remain stable in the absence of large quantities of firmly complexed K^+. As will be seen later, the bacterial spore is a second case in which uncertainty remains concerning the physical state of cytoplasmic cations. The uncertainty results from a combination of factors unique to the spore, including its low water content.

As Table 1 shows, K^+ is the dominant cytoplasmic cation in growing bacteria. Published values for internal concentrations range from 0.2 M in *Escherichia coli* (Harold & Altendorf, 1974) to values as high as 4.6 M in the halobacteria (Epstein & Schultz, 1967). Magnesium is also seen as major cation in vegetative bacteria. A significant difference is seen for dormant bacterial spores where calcium assumes a major role. For various reasons which will be discussed later, estimations of the free calcium content of spores are necessarily imprecise. The general principle which these and other analyses have established is that the cell maintains an ion content qualitatively quite different from its external environment. To accomplish this, bacterial cells are capable of demonstrating dramatic

Table 1. *Inorganic cation content of some vegetative bacteria and bacterial spores[a]*

g per 100 g dried bacteria	K^+	Mg^{2+}	Ca^{2+}	Fe^{3+}	Al^{3+}	Cu^{2+}	Mn^{2+}	Zn^{2+}
Bacillus subtilis	4.9	0.3	0.01	—	—	—	—	—
B. cereus	4.5	1.1	0.03	0.02	0.01	0.01	0.01	0.02
B. megaterium	4.0	0.2	—	—	—	—	—	—
B. megaterium	2.1	1.1	1.1	0.01	0.01	0.01	0.01	—
B. subtilis spores	0.9	0.5	1.6	0.01	0.01	0.01	0.01	—
B. cereus spores	0.2	0.3	1.9	0.02	0.04	0.02	0.01	—
B. megaterium spores	0.6	0.5	1.0	0.01	0.01	0.02	0.01	—
Escherichia coli	1.2	0.6	0.01	0.02	0.01	0.01	0.01	0.01
Aerobacter aerogenes	1.6	0.3	0.02	—	—	—	—	—

[a] From Tempest, 1969.

selectivity in terms of the concentration difference across their plasma membranes. Concentration gradients for potassium as high as $10^6:1$ have been reported for *E. coli* (Weiden, Epstein & Schultz, 1967) and selectivity coefficients of 4000 for potassium over sodium in *Streptococcus faecalis* (Harold & Altendorf, 1974). The selective uptake of these cations is accepted to result in the main from the activity of specific transport proteins residing in the plasma membrane (Harold & Altendorf, 1974). For other cations such as Ca^{2+}, Mg^{2+} and Mn^{2+} the available genetic and biochemical evidence again indicates that accumulation is an energy-dependent process mediated by specific membrane transport proteins. The alternative explanation for selective ion uptake, which would require metabolically dependent binding of the cations to anionic polyelectrolytes in the cell, is generally not held to play a major role, although it continues to be advocated (Damadian, 1973). It is possible that the binding of Mg^{2+} and polyvalent cations to cell wall polymers in bacteria may constitute one exception to this general statement. Baddiley and his colleagues (Baddiley, 1968; Baddiley, Hancock & Sherwood, 1973) have shown that for certain bacteria, binding of Mg^{2+} is related to the presence of negatively charged cell wall polymers known as teichoic acids, but any connection between magnesium bound in this way to polymers outside the plasma membrane and its subsequent transport into the cell remains to be demonstrated. Mention should also be made here of the involvement of specific iron-chelating agents in microbial iron uptake (Lankford, 1973). These are excreted by a variety of microorganisms and comprise the initial phase in the solubilisation and eventual uptake of this element.

A role for K^+ in bacterial osmoregulation was first suggested from the results of Christian & Waltho (1961), who demonstrated a positive correlation between the cellular content of K^+ and the osmolarity of the growth medium. These experiments with non-growing cells were repeated and extended by Epstein & Schultz (1967), using growing cultures of *E. coli*. Similarly with these growing cells, the K^+ content and presumably the level of some balancing anion increased in response to increases in the osmolarity of the growth medium. In other experiments Tempest and co-workers (Dicks & Tempest, 1966; Tempest, Dicks & Hunter, 1966) had shown that, in the absence of any changes in medium osmolarity, the potassium content of bacteria was determined by the cell content of ribosomes, which in turn is determined by growth rate. Recognising that, in the experiments of Epstein & Schultz, variations in growth rate and hence ribosome content may have occurred in media of differing osmolarity, Tempest & Meers (1968) varied the osmolarity of K^+-limited chemostat cultures of the Gram-negative bacterium *Aerobacter aerogenes* to examine this possibility. Although the cell potassium content still varied with growth rate, they were able to confirm that potassium content also varied with medium salinity. From this work it was concluded that cell ribosome content is the major determinant of cellular K^+ levels and that this level of K^+, together with other low molecular weight pool solutes, is sufficient in environments of moderately low ionic strength to provide an adequate internal osmolarity. As the ionic strength of the medium increases above this level, the cell adjusts its internal osmolarity by an additional uptake of K^+ together with a balancing amount of low molecular weight anions. Such a concomitant accumulation of K^+ and accompanying anions therefore provides a possible mechanism for osmoregulation by Gram-negative bacteria in media of increasing osmolarity. However, it is not an adequate explanation of osmoregulation by Gram-positive bacteria, since it has been shown (Measures, 1975) that for *Bacillus subtilis* at least, no measurable change in the intracellular K^+ content occurs in response to variations in the external NaCl concentration. This apparent difference between the two major groups of bacteria will be dealt with below.

The identity of the accompanying anion in this osmoregulation was first suggested by the report of Britten & McClure (1962) that the cellular proline content of *Escherichia coli* was dependent on the medium osmolarity. More recent work by Measures (1975) has confirmed this observation and demonstrated that two other amino acids, closely related biosynthetically, function as balancing anions in this mechanism of osmoregulation. Measures' experiments provide the additional interesting observation that the choice of anion which accumulates in conjunction with K^+ is different for the two major groups of bacteria, viz. Gram-positive and Gram-negative.

In these experiments, pool levels of amino acids were measured in a range of bacteria growing in a sufficiently high concentration of NaCl to produce a 50% reduction in growth rate. The data revealed that for each organism the high medium osmolarity produced an increase in the pool levels of certain amino acids. Glutamic acid was an accompanying anion in all the Gram-negative bacteria examined, with additional contributions from γ-aminobutyric acid and proline in some species. In most Gram-positive organisms, proline was the only amino acid which increased. Confirmation that these changes in amino acid pool levels are indeed osmotically induced was obtained from experiments in which the medium osmolarity was increased by the addition of sucrose instead of NaCl. In this case the same qualitative changes in amino acid levels were recorded, but the extent of the increases per mole of sucrose amounted to only 50% of the increase per mole of NaCl, as would be expected from a non-ionic external solute such as sucrose. At this point it is worth noting that proline accumulation plays an important role in osmoregulation by halophytic plants (Stewart & Lee, 1974) and a limited number of amino acids including glutamate and proline have a similar role in the osmoregulation of marine invertebrates (see Hochachka & Somero, 1973, for review).

To understand osmoregulation in bacteria, it now becomes necessary to explain the link between osmotic stress and these increases in pool levels of particular amino acids. Significantly, Measures (1975) observed that the increases not only occur in a rich growth medium when the bacteria are transporting these amino acids from the medium into the cell, but also when the bacteria are growing in a minimal medium and therefore synthesising their amino acids *de novo*. Under non-stressed conditions, Gram-negative bacteria have comparatively low levels of pool amino acids composed almost entirely of glutamate and also contain low levels of K^+. The response of these organisms to osmotic stress is to increase internal glutamate levels. In contrast, the Gram-positive organisms which Measures showed accumulate proline under stressed conditions have comparatively large free amino acid pools and a high internal K^+ concentration under non-stressed conditions. Glutamate is again the predominant amino acid in the amino acid pools of Gram-positive bacteria under normal circumstances. Against this background, Measures (1975) proposed that the basic response of all bacteria to increasing medium osmolarity is an increase in pool glutamate. This presents no problems to adapting Gram-negative bacteria, with their initially low cellular level of free glutamate. Since glutamate is charged at neutral pH values, however, its accumulation in substantial amounts during osmoregulation would upset the ion balance of the cell. To maintain electrical neutrality, glutamate accumulation in Gram-negative bacteria is accompanied by an equivalent accumulation of cation, and as was shown by the work of Epstein & Schultz and others, potassium fulfils this role. The use of this strategy without modification in Gram-positive bacteria would undoubtedly severely restrict their osmoregulatory capacity, because under non-stressed conditions they already contain high levels of glutamate and K^+, as was seen above. The results of Measures (1975) suggest that Gram-positive bacteria adapt this basic strategy by converting glutamate to the amino acids proline and γ-aminobutyric acid. The use of one or both of these compounds as osmoregulatory solutes instead of glutamate also eliminates any need for these bacteria to increase cellular K^+ above the already high levels which prevail in the absence of any osmotic stress. Since neither acid is highly charged at neutral pH values, the cell will not be required to accumulate a balancing cation when proline or γ-aminobutyric acid levels increase. In fact, Measures (1975) reported that, in those Gram-positive bacteria which accumulated proline, no accompanying change in cellular K^+ was observed in response to osmotic stress. This result does not, however, rule out a role for K^+ in the mechanism of bacterial osmoregulation. Even in Gram-positive bacteria under osmotic stress, there will be an initial rapid increase in the level of most intracellular solutes including K^+. This would be the likely result of the initial loss of water from the cell in response to osmotic stress. The subsequent accumulation of proline or γ-aminobutyric acid would reverse the situation, producing an inward movement of water and restoring the K^+ content to its original value.

In their model for osmoregulation in bacteria, Epstein & Schultz (1967) proposed that the osmotically sensitive site was a K^+/H^+ exchange pump which would effect net potassium uptake in response to an increase in external osmotic pressure. On the basis of Mitchell's chemiosmotic hypothesis, Harold & Altendorf (1974) proposed an alternative explanation in which an intracellular production of acids constituted the osmotic response mechanism. In their words, an increase in external osmotic pressure 'would stimulate production of acids presumably by a shift in metabolic pathways; extrusion of protons is an electrogenic process which is balanced by K^+ uptake'. The wisdom of this suggestion is borne out by experiments carried out by Measures (1975) to investigate the effect of K^+ on the synthesis of glutamate in bacteria. In several bacterial species he observed that glutamate dehydrogenase was activated up to ten-fold in the direction of glutamate by 500 mM K^+, but not at all in the reverse direction. The model of osmoregulation that emerges is one in which an increase in external osmotic pressure produces an increase in cellular K^+ which in turn stimulates glutamate synthesis. A further inflow of K^+ is then required to balance the increased charge produced by the glutamate. This would lead to a further activation of glutamate dehydrogenase and hence to a cyclic increase in glutamate levels. Such a model is supported by the observation that in those bacteria which accumulate glutamate in response to osmotic stress, a concomitant increase in cell K^+ is observed, of the same magnitude as the increase in glutamate concentration (Measures, 1975). As an example of this, Measures quotes the case of *Klebsiella aerogenes* which has a cellular K^+ content of 125 mM under non-stressed conditions. When grown in 1 M NaCl, the K^+ level increases to 625 mM and the observed increase in cell glutamate is 750 mM. For some bacteria, especially those such as bacilli which accumulate principally proline under osmotic stress (see above), additional osmotically sensitive steps in acid production may be involved. These organisms possess a second route of glutamate synthesis involving glutamine synthetase and glutamine: α-ketoglutarate amido-transferase, which Measures found to be unaffected by K^+ ions. Moreover no detectable increase in the total level of these enzymes was observed when organisms were grown in high osmotic strengh media. In these organisms additional mechanisms must be employed and Measures suggests the possibilities that ions may be affecting enzymes of proline synthesis directly or influencing the level of free proline through an effect on normal feedback inhibitory mechanisms. Alternatively the level of internal K^+ may influence the rate of amino acid transport into the cell.

Osmoregulation in other organisms

Reference was made earlier to the fact that certain plants and invertebrates also appear to regulate their internal osmolarity by altering the cell content of amino acids. In the case of marine and euryhaline invertebrates, the work of Schoffeniels & Gilles (1970) indicates that, as with the bacteria, the principle osmoregulatory mechanism involves ion activation of glutamate dehydrogenase. With halophytic plants, the experiments of Stewart & Lee (1974) have suggested that the accumulation of proline is correlated with salt tolerance. In one species, proline represented 10–20% of the dry weight of the shoot and 70% or more of the total amino acid pool. Proline therefore functions as the 'compatible solute' in these plants. This term was coined by Borowitzka & Brown (1974) to describe a substance which at high concentration protects enzyme activity by virtue of its being a very poor inhibitor. As pointed out by Stewart & Lee (1974), proline clearly fits this description, since its solubility is high (162 g per 100 ml at 25 °C) and it is a neutral compound. These workers also showed that proline concentrations as high as 700 mM had no effect on a variety of halophytic plant enzymes. Studies of osmoregulation in algae and osmophilic yeasts have shown that in these organisms the 'compatible solutes' are not amino acids, but instead this role is taken over by polyhydric alcohols. *Dunaliella parva* a photosynthetic halophilic alga accumulates large amounts of glycerol in response to an increase in the external salt concentration (Ben-Amotz & Avron, 1973; Borowitzka & Brown, 1974), while galactosylglycerol has been implicated in *Ochromonas* osmoregulation (Kauss, 1967, 1969). For osmophilic yeasts, arabitol appears to be the 'compatible solute' (Brown & Simpson, 1972). By comparison with the ion activation of glutamate dehydrogenase, it might be expected that similar ion effects would be involved in controlling either the synthesis or degradation of these solutes.

Osmoregulation and ion transport in bacterial sporulation

Spore formation in bacteria is an example of prokaryotic differentiation which has been extensively studied and recently reviewed (Dawes & Hansen, 1972; Freese, 1972). In response to depletion of metabolites in the growth medium, certain bacteria undergo an ordered and well-defined sequence of biochemical and morphological changes which results in the formation of a dormant endospore. The morphological sequence begins when the vegetative cell is asymmetrically divided into two compartments by a transverse division septum (Fig. 1). Since this septum is formed by invagination of the vegetative cell membrane, it is in fact a double-membrane septum. In the next stage the smaller of the two cell compartments resulting from this asymmetric division (forespore) is engulfed within the larger compartment (mother cell) by a continuing invagination of the transverse septum. When the engulfment is complete, the forespore exists within the mother cell cytoplasm as a separate cell bounded by a double membrane (Fig. 1). The development process then continues with the synthesis of various spore-specific components including spore integuments and finally the mature dormant spore is released upon lysis of the mother cell.

In the context of this volume only two aspects of this morphogenesis will be considered, viz. mechanisms of ion transport and osmoregulation during spore formation. As will be seen, bacterial spores present us with a unique problem in considering ion transport and osmoregulation in bacteria, since they are surrounded by two membranes which have opposite polarity relative to each other (Fig. 1; Wilkinson, Deans & Ellar, 1975). It is important at this stage to point out that bacterial spores are highly specialised for their role as dormant resistant cells. Their levels of resistance to heat, radiation and bacteriocidal agents are generally orders of magnitude greater than those of the parent vegetative cell and they exhibit no detectable metabolism. In seeking biochemical/biophysical explanations for these dramatic differences between spores and vegetative cells, detailed comparisons have been made of the ion and solute concentrations and the water content of the two cell forms. From these studies it appears that the

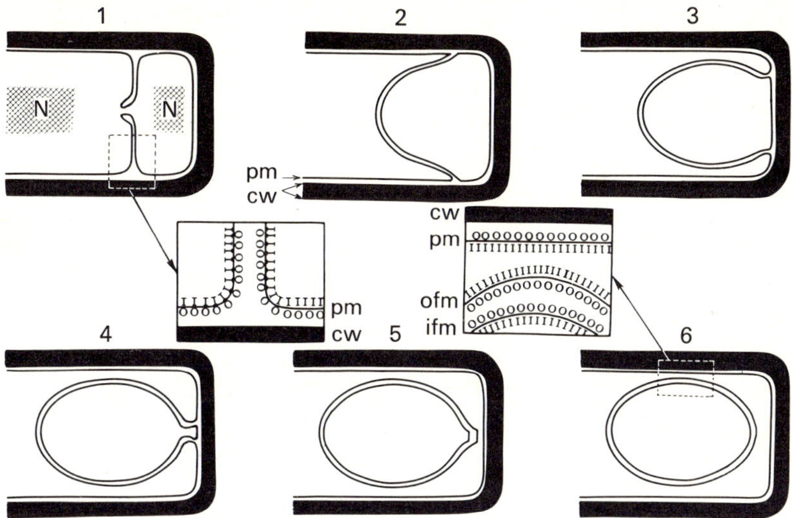

Fig. 1. Formation of the bacterial forespore showing the 'sidedness' of the membranes involved. N, nuclear material; pm, plasma membrane; cw, cell wall; ofm, outer forespore membrane; ifm, inner forespore membrane. In the two insets in the middle of the diagram the symbols O and I represent the 'sidedness' of the membranes; O represents the outer surface of the plasma membrane, I the inner surface.

spore water content may be much lower than that of the vegetative cell (Ross & Billing, 1957; Leman, 1973). While the water content of the latter is typically 70 % or more, values of 20 % or less for spores have been obtained by refractive index measurements. In order to evaluate proposals for the mechanism(s) of this dehydration during spore maturation, it is necessary to know whether this figure of 20 % represents a uniformly low water content throughout the spore, or whether it reflects a typically 'wet' cytoplasm surrounded by dehydrated spore integuments.

The internal contents of ions and solutes in bacterial spores are clearly quite different from those in the vegetative cell. A major difference concerns Ca^{2+}. During their maturation, spores accumulate considerable amounts of this cation from the medium, and at the same time their cell content of K^+ shows a significant decrease (Table 1). At these concentrations, calcium typically accounts for 2 % on average of the spore dry weight and exceeds the mole sum of all the other inorganic spore cations (Murrell, 1969). In contrast to the relatively mobile, osmotically active state of ions in vegetative bacteria, spore calcium is immobilised by chelation with dipicolinic acid (pyridine-2,6-dicarboxylic acid):

$$\text{HOOC} \underset{N}{\bigcirc} \text{COOH}$$

Dipicolinic acid (DPA) does not occur in vegetative cells, but is synthesised during spore formation concomitantly with calcium accumulation. The amount of DPA in spores is generally such as to give a $1:1$ molar ratio with Ca^{2+} (Murrell, 1969). Several studies have suggested that DPA occurs as a chelate in the spore. It is almost invariably extracted from the spores as the Ca–DPA chelate and occasionally chelated with other divalent metals. The order of stability for various DPA chelates is

$$Cu > Ni > Zn > Co > Cd > Ca > Mn > Sr > Ba > Mg,$$

with stability constants ranging from 10 for Cu-DPA, 4.4 for Ca-DPA to 2.4 for Mg-DPA (Murrell, 1969). Of a wide range of DPA analogues tested, only 4H-pyran-2,6-dicarboxylic acid

$$\text{HOOC} \underset{O}{\bigcirc} \text{COOH}$$

was able to substitute for DPA in restoring normal sporulation to a *Bacillus megaterium* mutant unable to synthesise its own DPA (Fukuda, Gilvarg & Lewis, 1969).

Uncertainties about the distribution of Ca^{2+}, DPA, water and other cations within the spore make it extremely difficult to determine the physical state of spore cations. Moreover, the possibility exists that spore solutes and even spore proteins may also be associated in some way with DPA or with a Ca-DPA chelate (Lewis, 1969). A second important difference between spores and vegetative cells which is particularly relevant to considerations of spore osmoregulation is the significant reduction in the level of pool amino acids in spores compared to vegetative cells (Nelson & Kornberg, 1970). The combination of the reduced water content of spores, the high Ca^{2+} and DPA content and the evidence indicating some association between spore amino acids and DPA suggests that the physical state of the cytoplasm in the mature spore is quite different from that in the vegetative cell. Given these conditions it is likely that most of the low molecular weight components in the spore pool will be insoluble, i.e. precipitated, and that the physical state of the protoplasm resembles that of a gel.

The question to be asked now is what are the mechanisms by which the spore produces this atypical internal composition and achieves its controlled dehydration. In approaching this question, two observations must be kept in mind. First, during spore formation no major quantitative change is occurring in the osmolarity of the growth medium, and, secondly, the unusual physical state of spore cytoplasm can revert to the typical vegetative cell condition within minutes, when the spore germinates. Germination is extremely rapid under suitable conditions and is accompanied by loss of calcium and DPA and swelling of the spore protoplast (Gould, 1969).

It was noted earlier that accumulation of DPA and calcium is a unique feature of the sporulating cell. In contrast the evidence indicates that an energy-dependent calcium *efflux* system occurs in vegetative cells (Silver, Toth & Scribner, 1975; Bronner, Nash & Golub, 1975). Therefore one feature of spore ionic regulation might be the development during sporulation of a calcium uptake mechanism. Work in several laboratories has confirmed this prediction (Eisenstadt & Silver, 1972; Bronner & Freund, 1972; Hogarth, Deans & Ellar, 1977). Kinetic and inhibitor studies have shown that at a predictable stage in sporulation, calcium begins to be actively transported into the cell and continues to be accumulated until almost the end of spore maturation. At about the same stage the cell commences DPA synthesis and this continues concomitantly with calcium uptake. Because of the observed association of Ca^{2+} with DPA in an approximately 1:1 molar ratio in the spore it was clearly necessary to discover the sites of both the calcium uptake mechanism and DPA synthesis in order to determine whether the two were linked in some way. It must be remembered that at the time calcium and DPA are accumulating, the spore exists as a discrete cell within the mother cell. Moreover the spore is not only enclosed within the plasma membrane of this mother cell, but is itself bounded by two spore-specific membranes which have opposite polarity (Fig. 1; Wilkinson *et al.*, 1975). This unique feature of spores presents special problems in attempting to understand movement of ions and solutes to and from the spore compartment. If the normal mechanisms of active transport are retained in both spore membranes, they might be expected to oppose each other (Freese, 1972). If this is indeed the situation *in vivo*, it is probable that transport from mother cell cytoplasm into the spore could only occur by passive or facilitated diffusion. In the case of calcium accumulation, this prediction has now been confirmed (Hogarth *et al.*, 1977). In these experiments, the sporulating cell was fractionated in such a way as to permit a study to be made of the ability of isolated developing spores to accumulate calcium (Ellar *et al.*, 1975). The kinetics and inhibitor sensitivity of calcium uptake into these isolated immature spores was then compared to the situation in sporulating protoplasts still containing the forespore (Hogarth *et al.*, 1977). With the sporulating cell, calcium uptake occurred with a K_m of 3.1×10^{-5} M. From the saturability, low K_m value, inhibitor sensitivity and specificity for calcium, it was clear that a carrier-mediated active transport system for calcium uptake exists in the sporulating cell. The task now was to identify which membrane(s) was the site of this active transport. An examination of the isolated immature spores showed that when the external calcium concentration equalled that in the normal sporulation medium (0.1 mM), no measurable calcium uptake occurred. However, in the presence of 7.5 mM external calcium, the isolated spores accumulated calcium at rates comparable to those of the intact sporulating cell. In this case the K_m for isolated spores was found to be 2.1×10^{-3} M. The fact that the rate and extent of calcium uptake by immature spores *in vitro* was influenced by the external calcium concentration, and the knowledge that the concentration of calcium in the mother cell cytoplasm was between 3 and 9 mM, suggested that uptake of calcium by developing spores *in vitro* was accomplished by a carrier-mediated transfer down a concentration gradient. In contrast, the measurements of calcium uptake by intact sporulating cells were made when the external calcium concentration was approximately 0.1 mM. The implications of these observations were confirmed when studies with a range of respiratory inhibitors showed that calcium uptake by intact sporulating cells could be completely blocked by concentrations of inhibitors which had no effect on uptake of calcium by immature spores *in vitro* (Hogarth *et al.*, 1977). These data reveal that the mother cell plasma membrane is the site for the active transport of calcium into mother cell cytoplasm, from which it is accumulated into the spore compartment by a facilitated diffusion mechanism. Values of 3–9 mM for the calcium concentration in mother cell cytoplasm are comparable to the K_m value for in-vitro spore uptake. Up to this point we have not observed any role for DPA in calcium uptake. In this context two observations are relevant: first, in the sporulating cell, the total content of DPA is confined to the spore compartment (Ellar & Posgate, 1974), and, secondly, addition of exogenous DPA has no significant effect on the rate or extent of calcium uptake by intact cells or isolated immature spores (Ellar *et al.*, 1975; Hogarth *et al.*, 1977). The first observation is extremely valuable, since it provides us with a possible mechanism for ensuring that a downhill concentration gradient of calcium is maintained throughout sporulation. The proposed sequence of events which is summarised in Fig. 2 envisages that calcium is first taken up by an active transport system in the mother cell membrane and accumulates to a level of 3–9 mM in the mother cell cytoplasm. From there it moves by facilitated diffusion into the spore compartment where it chelates with DPA, thereby effectively preserving a low free calcium level

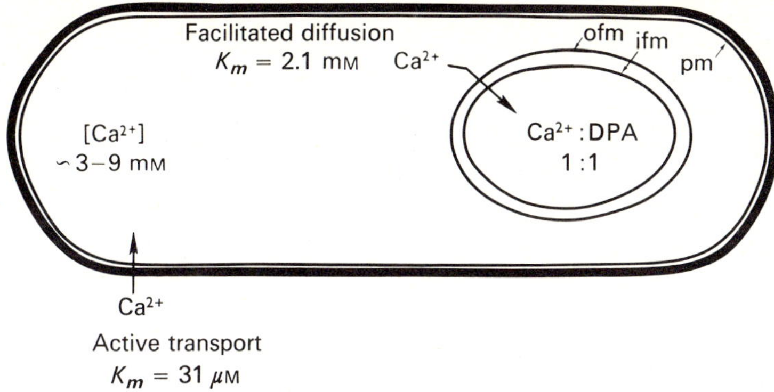

Fig. 2. Model for calcium accumulation during sporulation. pm, plasma membrane of mother cell; ofm, outer forespore membrane; ifm, inner forespore membrane; DPA, dipicolinic acid (pyridine-2,6-dicarboxylic acid).

and allowing continued calcium uptake (La Nauze *et al.*, 1974; Hogarth *et al.*, 1977). This would perhaps explain the unique presence of DPA in bacterial spores in addition to its occurrence in an approximately 1:1 mole ratio with calcium.

In the light of these observations it is now necessary to look again briefly at the question of spore water content. Several mechanisms have been suggested by which dehydration of spore cytoplasm could be accomplished during spore maturation. One of these (Lewis, Snell & Burr, 1960) proposes that dehydration results from the compressive contraction of one of the major spore integuments, viz. the cortex. This is a cross-linked peptidoglycan polymer similar to that which occurs in the walls of vegetative cells (Rogers, 1977). On the basis of the known properties of this cortex, Gould & Dring (1975) reasoned that it was more likely to be expanded than contracted and proposed an alternative novel mechanism in which the cortex functions as an osmoregulatory organelle. These workers observed that spore peptidoglycan is more electronegative and less cross-linked than the similar vegetative cell polymer. Therefore in the absence of high concentrations of cations, spore peptidoglycan might be expanded by electrostatic repulsion of adjacent negatively charged groups. Conversely, contraction of this polymer might then be brought about by exposure to high cation concentrations, especially multivalent cations. In the mechanism proposed by Gould & Dring, the cortex is thought to be electrically neutral and expanded in the dormant spore by virtue of the association of free, positively charged counterions with the free carboxyl groups on the polymer. They suggest that the polymer plus counterions will be osmotically active and capable of exerting osmotic pressures in excess of 30×10^5 N m^{-2}. Provided that the internal osmotic pressure of the spore protoplast is not of this magnitude, it is possible that this osmoregulatory cortex could bring about a dehydration of the spore cytoplasm. In fact, as was discussed earlier, the available evidence indicates that the levels of free ions and low molecular weight solutes in dormant spores are likely to be much less than in the vegetative cell.

Accepting *de facto* that the spore protoplast is dehydrated, the question which might be asked is whether a specialised osmoregulatory organelle is necessary to achieve this. In other words, is it possible to envisage a simpler mechanism of dehydration based on what is known of the biochemistry and compartmentation of the sporulating cell and the membranes which enclose these compartments. It is known for example that extensive protein synthesis occurs in the developing spore compartment throughout spore maturation (Eaton & Ellar, 1974) and it was seen earlier that the levels of free amino acids and low molecular weight solutes in the spore are reduced in comparison to the vegetative cell, either as a result of this synthesis or because of chelation/precipitation with calcium or Ca–DPA. To these observations must be added the fact that the spore is a cell within a cell, i.e. its external environment is the cytoplasm of the mother cell from which it is separated by two membranes. Conceivably then, dehydration could be accomplished when the levels of free ions and solutes in the spore are reduced by the combined operation of the above factors. This progressive reduction in osmotic pressure in the spore compartment could result in a compensatory outflow of water from the spore cytoplasm. According to this model, spore dehydration would be a natural outcome of the macromolecular synthesis and ion movements which characterise spore formation. From this suggestion it would follow that the role of the spore integuments, e.g. cortex, would be to form a rigid casing around the spore protoplast, restricting protoplast expansion and thereby opposing any tendency for water to flow back into the spore when the latter is released from the mother cell cytoplasm into what would inevitably be a more dilute environment. Since spore water content undoubtedly increases rapidly upon spore germination, this line of reasoning leads to the conclusion that initiation of germination must involve reduction or elimination of this rigid component of spore integuments. The collapse of an expanded cortex upon exposure to calcium, proposed by Gould & Dring (1975), provides one possible mechanism for this. Alternatively, the same result could be achieved if the action of germinants is to initiate the activity of latent spore enzymes capable of reducing cortex or coat rigidity by selective hydrolysis.

Proton pumping driven by light in halobacteria

In considering various strategies which microorganisms have evolved to enable them to grow in environments of high osmotic strength, the obligate halophilic bacteria represent a fascinating field for study. Mention has already been made of their natural high salt habitat and the fact that, in one study, their cytoplasmic K^+ and Na^+ levels were shown to be approximately 4.0 M and 2.0 M, respectively (Ginzburg *et al.*, 1970). An osmoregulatory strategy which involves accumulation of ions to these extreme levels in the cytoplasm, suggests that the proteins and perhaps the lipids of these organisms must possess some intrinsic solute resistance. Under these high salt conditions many non-halophile proteins would be denatured or inactive, whereas halophile proteins are active and, in contrast, are frequently found to be denatured at moderate or low ionic strengths. The molecular basis of this ability of halophile proteins to function optimally in high salt is still the subject of investigation and has recently been reviewed by Lanyi (1974). Halobacterial membrane lipids are also quite different from those found in non-halophilic bacteria. Kates and co-workers (Sehgal, Kates & Gibbons, 1962; Kates, Yengoyan & Sastry, 1965) have been largely responsible for showing that halophile phospholipids lack the fatty acid ester linkage found in most other bacteria. Instead their structure is basically that of the diether phospholipid analogue in which dihydro-phytanyl groups are attached to glycerol through an ether linkage. Another interesting feature of these organisms is the presence of substantial amounts of sulphated lipids and sulphated polysaccharide in the cell envelope (Kates *et al.*, 1967; Koncewicz, 1972*a*). The currently popular model for cell membrane structure (Singer & Nicolson, 1972) envisages that a major portion of cell membrane lipid exists in the fluid or liquid crystal phase. That the maintenance of this fluidity in the bilayer is important for most microorganisms is seen in the process termed homeoviscous adaptation by which microorganisms modify their membrane fatty acid content in response to alterations in environmental temperature. In view of these observations it is interesting that recent studies of lipid mobility using spin labels in the cell envelopes of *Halobacterium cutirubrum* (Esser & Lanyi, 1973) led to the conclusion that most of the lipid was in a relatively rigid state. One other unusual feature of halobacteria which should be noted is that the envelopes of these organisms contain none of the rigid peptidoglycan polymer found in other bacteria, although envelope glycoproteins have been isolated (Koncewicz, 1972*b*; Mescher & Strominger, 1976; Newbold, unpublished). The special feature of halobacteria which will be discussed here is their ability to synthesise ATP by a light-driven proton pump. This discovery had its origins in the observation that a fraction containing membrane fragments purple in colour could be isolated from envelopes of *H. halobium* which had been disaggregated by dialysis against distilled water (McClare, 1967; Stoeckenius & Rowen, 1967). These fragments were subsequently found to contain only one protein to which retinal was attached in a Schiff's base linkage with an amino group (Oesterhelt & Stoeckenius, 1971). Because of the obvious similarity between this complex and the visual pigment in higher organisms, this protein was termed bacteriorhodopsin (Oesterhelt & Stoeckenius, 1971). The purple protein accounts for about 75% of the weight of the membrane fragments and has a molecular weight of approximately 26000 daltons. The rest of the purple membrane is composed of lipid and estimates show that this amount of lipid is sufficient to provide ten lipid molecules for every protein molecule (Sumper, Reitmeier & Oesterhelt, 1976). In the isolated purple membrane, bacteriorhodopsin has been shown to exist in a two-dimensional hexagonal crystalline lattice (Blaurock & Stoeckenius, 1971; Blaurock, 1975; Henderson & Unwin, 1975). Bacteriorhodopsin molecules are arranged in threes and asymmetrically shaped. They contain a considerable amount of α-helix and are all oriented in the same direction with lipid in bilayer form occupying the interstices of the protein lattice (fig. 5 in Oesterhelt, 1976*a*). In this structure, bacteriorhodopsin traverses the entire thickness of the purple membrane. Among the evidence from which this picture was constructed is the work of Henderson & Unwin (1975), who used a novel electron microscopical technique to show that bacteriorhodopsin contains seven closely packed α-helical segments which extend roughly perpendicular to the plane of the membrane for most of its width. The individual molecules have dimensions of $250 \times 350 \times 450$ nm and are arranged with the longest axis traversing the membrane width.

A major insight into the in-vivo function of bacteriorhodopsin came with the observation that the purple protein complex is only synthesised when the growth of halobacteria is limited by restricting the oxygen supply (Oesterhelt & Stoeckenius, 1973). Light also stimulates its synthesis, since the amount of purple membrane is six to seven times greater in cells grown in the light than in the dark (Danon & Stoeckenius, 1972). Subsequent experiments have shown (Oesterhelt & Stoeckenius, 1973; Racker & Stoeckenius, 1974; Bogomolni *et al.*, 1976) that the purple membrane complex constitutes a photophosphorylation mechanism mediated by an electrogenic proton pump, which can provide the cell with an alternative source of ATP under conditions of low oxygen tension when the rate of oxidative phosphorylation decreases. Since halobacteria contain no chlorophyll, this represents a novel form of photophosphorylation. The activity of this system is seen clearly when cyanide is added to aerobically

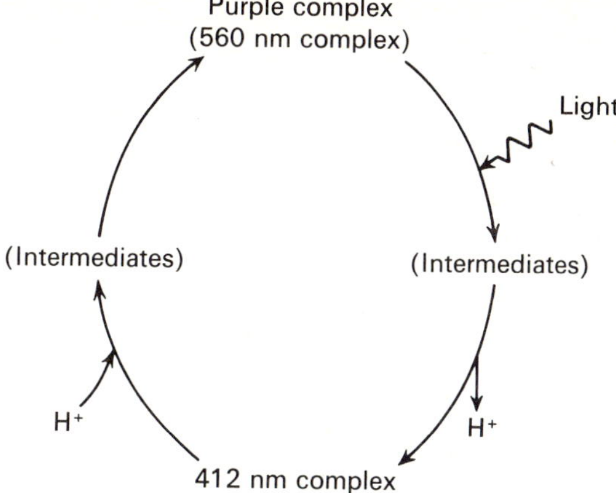

Fig. 3. Bacteriorhodopsin photochemical cycle.

respiring halobacteria in the dark. Under these conditions the internal ATP pool is observed to decrease by about 80%. If the bacteria are now illuminated, there is a rapid rise in ATP to approximately the level observed during uninhibited aerobic resperation (Oesterhelt, 1976a).

How then does bacteriorhodopsin function in the photophosphorylation? Results from a number of laboratories have led to the conclusion that the purple complex functions as a light-driven proton pump whose action is made possible as a result of the photochemical cycle which bacteriorhodopsin undergoes upon exposure to light. This cycle is illustrated in Fig. 3. Oesterhelt & Stoeckenius (1971) found that upon illumination, bacteriorhodopsin responded with a transient shift of its adsorption maximum from 560 nm to 412 nm and the formation of an intermediate product which very rapidly regenerates the purple chromophore. This cyclic change is accompanied by a release and uptake of protons. The absorption of light by the purple complex not only induces changes in the chromophore, but also in the protein itself. It is this latter conformational change which is considered to result in the product which, within milliseconds, regenerates the original chromophore, thereby completing the cycle. When intact halobacteria containing purple membrane were exposed to light, there was a reduction in oxygen consumption together with an increase in cell ATP and an excretion of protons into the medium (Oesterhelt & Stoeckenius, 1973; Danon & Stoeckenius, 1974). In conjunction with the knowledge of the photochemical cycle of bacteriorhodopsin, these observations with intact cells suggested that the purple complex could function as an electrogenic proton pump to supply energy for ATP synthesis in the manner suggested by Mitchell's chemiosmotic hypothesis (Mitchell, 1966). This would require the photochemical cycle to be organised as a vectorial process in the membrane, releasing protons into the medium at the exterior of the membrane and taking up protons from the cytoplasm. This proposal has subsequently been confirmed in several laboratories (for review see Oesterhelt, 1976a) in experiments with intact cells and artificial lipid bilayers. For example, Racker & Stoeckenius (1974) have shown that lipid vesicles which have incorporated purified purple membrane, function as a light-stimulated proton pump. Furthermore, when purified oligomycin-sensitive ATPase was also incorporated into these vesicles, photophosphorylation could be demonstrated. Presumably, in addition to ATP synthesis the halobacteria can use the chemiosmotic gradient produced by the proton pump to drive other energy-requiring processes. Confirmation of this comes from recent reports of light-induced transport of leucine, glutamate and proline (MacDonald & Lanyi, 1975; Lanyi & MacDonald, 1976; Hubbard, Rinehart & Baker, 1976). While a recent proposal which discusses the use of purified purple membrane in the light-driven desalination of sea water (Oesterhelt, 1976b) is certainly a tribute to scientific endeavour, it perhaps also contains a note of irony for halobacteria.

The author is grateful to the *Biochemical Journal* for permission to reproduce Fig. 1 and to Professor S. W. Tempest for permission to use material contained in Table 1.

References

Baddiley, J. (1968). Teichoic acids and the molecular structure of bacterial walls. *Proceedings of the Royal Society, London*, B, **170**, 331–48.

Baddiley, J., Hancock, I. C. & Sherwood, P. M. A. (1973). X-ray photoelectron studies of magnesium ions bound to the cell walls of Gram-positive bacteria. *Nature, London* **243**, 43–5.

Ben-Amotz, A. & Avron, M. (1973). The role of glycerol in the osmotic regulation of the halophilic alga *Dunaliella parva*. *Plant Physiology*, **51**, 875–8.

Blaurock, A. E. (1975). Bacteriorhodopsin: a trans-membrane pump containing α-helix. *Journal of Molecular Biology*, **93**, 139–58.

Blaurock, A. E. & Stoeckenius, W. (1971). Structure of the purple membrane. *Nature, London*, **233**, 152–5.

Bogomolni, R. A., Baker, R. A., Lozier, R. H. & Stoeckenius, W. (1976). Light-driven proton translocations in *Halobacterium halobium*. *Biochimica et Biophysica Acta*, **440**, 68–88.

Borowitzka, L. J. & Brown, A. D. (1974). The salt relations of marine and halophilic species of the unicellular green alga, *Dunaliella*. The role of glycerol as a compatible solute. *Archives of Microbiology*, **96**, 37–52.

Britten, R. J. & McClure, F. T. (1962). The amino acid pool in *Escherichia coli*. *Bacteriological Reviews*, **26**, 292–335.

Brock, T. D. (1969). Microbial growth under extreme conditions. *Symposia of the Society for General Microbiology*, **19**, 15–41.

Bronner, F. & Freund, T. S. (1972). Calcium accumulation during sporulation of *Bacillus megaterium*. In *Spores V, American Society for Microbiology*, pp. 187–90.

Bronner, F., Nash, W. C. & Golub, E. E. (1975). Calcium transport in *Bacillus megaterium*. In *Spores VI, American Society for Microbiology*, pp. 356–61.

Brown, A. D. & Simpson, J. R. (1972). Water relations of sugar-tolerant yeasts: the role of intracellular polyols. *Journal of General Microbiology*, **72**, 589–91.

Christian, J. H. B. & Waltho, J. A. (1961). The sodium and potassium content of non-halophilic bacteria in relation to salt tolerance. *Journal of General Microbiology*, **25**, 97–102.

Christian, J. H. B. & Waltho, J. A. (1962). Solute concentrations within cells of halophilic and non-halophilic cells. *Biochimica et Biophysica Acta*, **65**, 506–8.

Cope, F. W. & Damadian, R. (1970). Cell potassium by ^{39}K spin echo nuclear magnetic resonance. *Nature, London*, **228**, 76–7.

Damadian, R. (1971). Biological ion exchanger resins. 1. Quantitative electrostatic correspondence of fixed charge and mobile counter ion. *Biophysical Journal*, **11**, 739–59.

Damadian, R. (1973). Cation transport in bacteria. *Critical Reviews in Microbiology*, **2**, 377–422.

Danon, A. & Stoeckenius, W. (1972). In *NASA Symposium on Extreme Environments: Mechanisms of Microbial Adaptation*, p. 25. Ames Research Centre, Moffett Field, California, June 1972.

Danon, A. & Stoeckenius, W. (1974). Photophosphorylation in *Halobacterium halobium*. *Proceedings of the National Academy of Sciences*, **71**, 1234–8.

Dawes, I. W. & Hansen, J. N. (1972). Morphogenesis in sporulating bacilli. *Critical Reviews in Microbiology*, **1**, 479–520.

Dicks, J. W. & Tempest, D. W. (1966). The influence of temperature and growth rate on the quantitative relationship between potassium, magnesium, phosphorus and ribonucleic acid of *Aerobacter aerogenes* growing in a chemostat. *Journal of General Microbiology*, **45**, 547–55.

Eaton, M. W. & Ellar, D. J. (1974). Protein synthesis and breakdown in the mother-cell and forespore compartments during spore morphogenesis in *Bacillus megaterium*. *Biochemical Journal*, **144**, 327–37.

Eisenstadt, E. & Silver, S. (1972). Calcium transport during sporulation in *Bacillus subtilis*. In *Spores V, American Society for Microbiology*, pp. 180–6.

Ellar, D. J., Eaton, M. W., Hogarth, C., Wilkinson, B. J., Deans, J. & La Nauze, J. (1975). Comparative biochemistry and function of forespore and mother-cell compartments during sporulation of *Bacillus megaterium* cells. In *Spores VI, American Society for Microbiology*, pp. 425–33.

Ellar, D. J. & Posgate, J. A. (1974). Characterization of forespores isolated from *Bacillus megaterium* at different stages of development into mature spores. In *Spore Research 1973*, ed. A. N. Barker, G. W. Gould & J. Wolf, pp. 21–40. London: Academic Press.

Epstein, W. & Schultz, S. G. (1967). Ion transport and osmoregulation in bacteria.

146   D. J. Ellar

In *Microbial Protoplasts, Spheroplasts and L-forms*, ed. L B. Guze, pp. 186–93. Baltimore, Maryland: Williams & Wilkins.

Esser, A. F. & Lanyi, J. K. (1973). Structure of the lipid phase in cell envelope vesicles from *Halobacterium cutirubrum*. *Biochemistry*, **12**, 1933–9.

Freese, E. (1972). Sporulation of *Bacilli*, a model of cellular differentiation. *Current Topics in Developmental Biology*, **7**, 85–124.

Fukuda, A., Gilvarg, C. & Lewis, J. C. (1969). 4H-Pyran-2,6-dicarboxylate as a substitute for dipicolinate in the sporulation of *Bacillus megaterium*. *Journal of Biological Chemistry*, **244**, 5636–43.

Ginzburg, M. & Ginzburg, B. Z. (1975). Factors influencing the retention of K^+ in a halobacterium. In *Biomembranes*, **7**, 219–51.

Ginzburg, M., Ginzburg, B. Z. & Tosteson, D. C. (1971). The effect of anions on K^+ binding in a *Halobacterium* species. *Journal of Membrane Biology*, **6**, 259–68.

Ginzburg, M., Sachs, L. & Ginzburg, B. Z. (1970). Ion metabolism in a *Halobacterium*. I. Influence of age of culture on intracellular concentrations. *Journal of General Physiology*, **55**, 187–207.

Ginzburg, M., Sachs, L. & Ginzburg, B. Z. (1971). Ion metabolism in a *Halobacterium*. II. Ion concentrations in cells at different levels of metabolism. *Journal of Membrane Biology*, **5**, 78–101.

Gould, G. W. (1969). Germination. In *The Bacterial Spore*, ed. G. W. Gould & A. Hurst, pp. 395–444. New York: Academic Press.

Gould, G. W. & Dring, G. J. (1975). Heat resistance of bacterial endospores and concept of an expanded osmoregulatory cortex. *Nature, London*, **258**, 402–5.

Harold, F. M. & Altendorf, K. (1974). Cation transport in bacteria: K^+, Na^+, and H^+. In *Current Topics in Membranes Transport*, **5**, 1–50.

Henderson, R. & Unwin, P. N. T. (1975). Three-dimensional model of purple membrane obtained by electron microscopy. *Nature, London*, **257**, 28–32.

Hochachka, P. W. & Somero, G. N. (1973). *Strategies of Biochemical Adaptation*. Philadelphia: W. B. Saunders.

Hogarth, C., Deans, J. A. & Ellar, D. J. (1977). Calcium accumulation and membrane morphogenesis in forespore and mother-cell compartments during sporulation of *Bacillus megaterium*. In *Spore Research 1976*, ed. A. N. Barker, J. Wolf, D. J. Ellar, G. J. Dring & G. W. Gould. London: Academic Press, Inc. (In Press.)

Hubbard, J. S., Rhinehart, C. A. & Baker, R. A. (1976). Energy coupling in the active transport of amino acids by bacteriorhodopsin-containing cells of *Halobacterium halobium*. *Journal of Bacteriology*, **125**, 181–90.

Kates, M., Palmeta, B., Perry, M. P. & Adams, G. A. (1967). A new glycolipid sulfate ester in *Halobacterium cutirubrum*. *Biochimica et Biophysica Acta*, **137**, 213–16.

Kates, M., Yengoyan, L. S. & Sastry, P. S. (1965). A diether analog of phosphatidyl glycerophosphate in *Halobacterium cutirubrum*. *Biochimica et Biophysica Acta*, **98**, 252–68.

Kauss, H. (1967). Isofloridosid und Osmoregulation bei *Ochromonas malhamensis*. *Zeitschrift für Pflanzenphysiologie*, **56**, 453–65.

Kauss, H. (1969). Osmoregulation mit α-Galaktosyl-Glyceriden bei *Ochromonas* und Rotalgen. *Bericht der Deutschen botanischen Gesellschaft*, **82**, 115–25.

Koncewicz, M. A. (1972*a*). The presence of a sulphated polysaccharide in the cell envelopes of *Halobacterium halobium*. *Biochemical Journal*, **130**, 40P.

Koncewicz, M. A. (1972*b*). Glycoproteins in the cell envelope of *Halobacterium halobium*. *Biochemical Journal*, **128**, 124P.

La Nauze, J. M., Ellar, D. J., Denton, G. & Posgate, J. A. (1974). Some properties of forespores isolated from *Bacillus megaterium*. In *Spore Research 1973*, ed. A. N. Barker, G. W. Gould & J. Wolf, pp. 41–6. London: Academic Press.

Lankford, C. E. (1973). Bacterial assimilation of iron. *Critical Reviews in Microbiology*, **2**, 273–331.

Lanyi, J. K. (1974). Salt-dependent properties of proteins from extremely halophilic bacteria. *Bacteriological Reviews*, **38**, 272–90.

Lanyi, J. K. & MacDonald, R. E. (1976). Coupling of light-induced chemical and electrical gradients to glutamate transport in *Halobacterium halobium* envelope vesicles. *Biophysical Journal*, **16**, 21A.

Lanyi, J. K. & Silverman, M. P. (1972). The state of binding of intracellular K$^+$ in *Halobacterium cutirubrum*. *Canadian Journal of Microbiology*, **18**, 993–5.

Leman, A. (1973). Interference microscopical determination of bacterial dry weight during germination and sporulation. *Jena Review*, **5**, 263–70.

Lewis, J. C. (1969). Dormancy. In *The Bacterial Spore*, ed. G. W. Gould & A. Hurst, pp. 302–58. New York: Academic Press.

Lewis, J. C., Snell, N. S. & Burr, H. K. (1960). Water permeability of bacterial spores and the concept of the contractile cortex. *Science, Washington*, **132**, 544–5.

Marquis, R. E. & Carstensen, E. L. (1973). Electrical conductivity and internal osmolality of intact bacterial cells. *Journal of Bacteriology*, **113**, 1198–206.

McClare, C. W. F. (1967). Bonding between proteins and lipids in the envelopes of *Halobacterium halobium*. *Nature, London*, **216**, 766–71.

MacDonald, R. E. & Lanyi, J. K. (1975). Light-induced leucine transport in *Halobacterium halobium* envelope vesicles: a chemiosmotic system. *Biochemistry*, **14**, 2882–9.

Measures, J. C. (1975). Role of amino acids in osmoregulation of non-halophilic bacteria. *Nature, London*, **257**, 398–400.

Mescher, M. F. & Strominger, J. L. (1976). Purification and characterization of a prokaryotic glycoprotein from the cell envelope of *Halobacterium salinarium*. *Journal of Biological Chemistry*, **251**, 2005–14.

Mitchell, P. (1966). Chemiosmotic coupling in oxidative and photosynthetic phosphorylation. *Biological Reviews of the Cambridge Philosophical Society*, **41**, 445–502.

Mitchell, P. & Moyle, J. (1956). Osmotic function and structure in bacteria. *Symposia of the Society for General Microbiology*, **6**, 150–80.

Murrell, W. G. (1969). Chemical composition of spores and spore structures. In *The Bacterial Spore*, ed. G. W. Gould & A. Hurst, pp. 215–73. New York: Academic Press.

Nelson, D. L. & Kornberg, A. (1970). Biochemical studies of bacterial sporulation and germination. XVIII. Free amino acids in spores. *Journal of Biological Chemistry*, **245**, 1128–37.

Oesterhelt, D. (1976*a*). Bacteriorhodopsin as an example of a light-driven proton pump. *Angewandte Chemie*, **15**, 17–24.

Oesterhelt, D. (1976*b*). Bacteriorhodopsin as a light-driven ion exchanger. *FEBS Letters*, **64**, 20–2.

Oesterhelt, D. & Stoeckenius, W. (1971). Rhodopsin-like protein from the purple membrane of *Halobacterium halobium*. *Nature, London*, **233**, 149–52.

Oesterhelt, D. & Stoeckenius, W. (1973). Functions of a new photoreceptor membrane. *Proceedings of the National Academy of Sciences, USA*, **70**, 2853–7.

Racker, E. & Stoeckenius, W. (1974). Reconstitution of purple membrane vesicles catalyzing light-driven proton uptake and adenosine triphosphate formation. *Journal of Biological Chemistry*, **249**, 662–3.

Rayman, M. K., Gordon, R. C. & MacLeod, R. A. (1967). Isolation of a Mg^{++} phospholipid from *Halobacterium cutirubrum*. *Journal of Bacteriology*, **93**, 1465–6.

Rogers, H. J. (1977). Peptidoglycans (Mucopeptides), structure, form and function. In *Spore Research 1976*, ed. A. N. Barker, J. Wolf, D. J. Ellar, G. J. Dring & G. W. Gould. London: Academic Press. (In Press.)

Ross, K. F. A. & Billing, E. (1957). The water and solid content of living bacterial spores and vegetative cells as indicated by refractive index measurements. *Journal of General Microbiology*, **16**, 418–25.

Schoffeniels, E. & Gilles, R. (1970). Osmoregulation in aquatic arthropods. In *Chemical Zoology*, vol. 5, ed. M. Florkin & B. T. Scheer, pp. 255–85. New York: Academic Press.

Sehgal, S. N., Kates, M. & Gibbons, N. E. (1962). Lipids of *Halobacterium cutirubrum*. *Canadian Journal of Biochemistry and Physiology*, **40**, 69–81.

Silver, S., Toth, K. & Scribner, H. (1975). Facilitated transport of calcium by cells and subcellular membranes of *Bacillus subtilis* and *Escherichia coli*. *Journal of Bacteriology*, **122**, 880–5.

Singer, S. J. & Nicolson, G. L. (1972). The fluid mosaic model of the structure of cell membranes. *Science, Washington*, **175**, 720–31.

Stewart, G. R. & Lee, J. A. (1974). The role of proline accumulation in Halophytes. *Planta*, **120**, 279–89.

Stoeckenius, W. & Rowen, R. (1967). A morphological study of Halobacterium halobium and its lysis in media of low salt concentration. *Journal of Cell Biology*, **34**, 365–93.

Sumper, M., Reitmeier, H. & Oesterhelt, D. (1976). Biosynthesis of the purple membrane of halobacteria. *Angewandte Chemie (International edition)*, **15**, 187–94.

Tempest, D. W. (1969). Quantitative relationships between inorganic cations and anionic polymers in growing bacteria. *Symposia of the Society for General Microbiology*, **19**, 87–111.

Tempest, D. W., Dicks, J. W. & Hunter, J. R. (1966). The interrelationship between potassium, magnesium and phosphorus in potassium-limited chemostat cultures of *Aerobacter aerogenes*. *Journal of General Microbiology*, **45**, 133–45.

Tempest, D. W. & Meers, J. L. (1968). The influence of NaCl concentration of the medium on the potassium content of *Aerobacter aerogenes* and on the interrelationships between potassium, magnesium and ribonucleic acid in the growing bacteria. *Journal of General Microbiology*, **54**, 319–25.

Weiden, P. L., Epstein, W. & Schultz, S. G. (1967). Cation transport in *Escherichia coli*. VIII. Potassium requirement for phosphate uptake. *Journal of General Physiology*, **50**, 1641–61.

Wilkinson, B. J., Deans, J. A. & Ellar, D. J. (1975). Biochemical evidence for the reversed polarity of the outer membrane of the bacterial forespore. *Biochemical Journal*, **152**, 561–9.

11. Mechanisms of ionic and osmotic regulation in saline-water mosquito larvae

By JOHN E. PHILLIPS,
T. J. BRADLEY & S. H. P. MADDRELL

Habitats and regulatory capacity

Several species of mosquito larvae are known to thrive both in fresh water (FW) and saline waters which may be considerably hyperosmotic (at least two to four times) to the haemolymph (*Aedes detritus*, Beadle, 1939; *A. campestris*, Phillips & Meredith, 1969*a, b*; *Opifex fuscus*, Nicolson, 1972; *A. taeniorhynchus*, Nayar & Sauerman, 1974, and Bradley & Phillips, 1975). The highest salinity tolerance reported to date is the survival of *A. taeniorhynchus* in sea water (SW)* three times more concentrated than normal. Even more remarkable perhaps is the ability of *A. campestris* to thrive in a variety of alkaline salt-lakes which are common in semi-arid regions of North America and in which the dominant salts may be $NaHCO_3$ (at a pH > 10) or equal amounts of $MgSO_4$ and Na_2SO_4 (Scudder, 1969; Phillips & Meredith, 1969*a, b*; Kiceniuk & Phillips, 1974; Phillips & Maddrell, 1974; Maddrell & Phillips, 1975). Recently we have found that larvae from Ctenocladus Pond (700 mOsm (Mg+Na) SO_4) can survive very well in 700 mOsm $NaHCO_3$, 700 mOsm NaCl or in FW; indeed, they survive rather better in these media than in their natural habitat (Bradley & Phillips, 1977*a*). We must conclude that the occurrence of this species in water of various chemical types reflects regulatory adaptability rather than the existence of distinct physiological races.

The regulatory ability of *A. detritus* larvae, described by Beadle (1939), had been questioned (Stobbart & Shaw, 1964, 1974) because it was not shown that a steady-state had been reached. However, Phillips & Meredith (1969*a, b*) found that when *A. campestris* larvae were transferred to tap water, 800 mM NaCl or solutions of intermediate salinity, most of the

* The abbreviation SW indicates fluids similar in ionic composition to sea water; 300% SW is thus a medium containing all the constituents of sea water in their normal proportions but with each present at three times the normal concentration. We use the more general term saline water to indicate media substantially more concentrated than fresh water (FW); it includes sea water (SW) but also other waters in which inorganic ions are found in proportions different (sometimes very much so) from those characteristic of sea water.

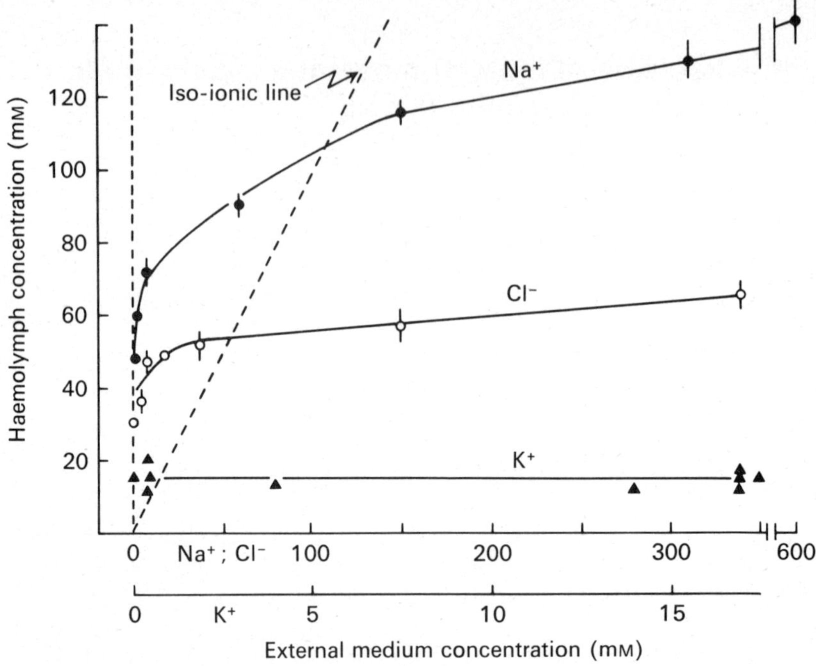

Fig. 1. Ionic concentrations in haemolymph of *Aedes campestris* larvae (4th instar) as a function of external levels. Larvae from a $NaHCO_3$ lake (GR-2; see Phillips & Meredith, 1969a, for composition) were adapted at 10 °C for at least 4 days in various experimental media, prepared by dilution of natural lake-water (1.2 Osm) or by addition of NaCl to natural lake-water (Phillips, unpublished data).

changes in blood Na^+, Cl^-, K^+ and osmotic concentrations occurred during the first day, and that new steady-state values were reached within 2 days. This is also the case for Mg^{2+} and SO_4^{2-} (Kiceniuk & Phillips, 1974; Maddrell & Phillips, 1975). These larvae can adapt to waters varying in ionic concentration over a 500-fold range, without the steady-state haemolymph concentrations of Na^+, Mg^{2+}, K^+, Cl^- and total solutes (osmolarity) changing by more than two-fold, except in very fresh water ($NaCl < 1$–2 mM; Fig. 1). Regulation to changes in sulphate concentration is less successful. The haemolymph concentration of this anion is maintained at low levels (< 10 mM) until the external concentration rises above 100 mM; thereafter, the blood level rises in parallel with external values until it reaches 100 mM. There are, however, no obvious effects detrimental to the larvae over the next several days. Similar results have been reported for *A. taeniorhnychus* larvae reared in various media from FW to 300 % SW (Nayar & Sauerman, 1974; Bradley & Phillips, 1977b). These observations highlight the remarkable regulatory capacity of saline-water mosquito larvae.

To appreciate the load which is imposed upon those organs of euryhaline larvae that are responsible for this regulation, i.e. the Malpighian tubules and rectum of the excretory system, and the extra-renal organs (anal papillae), the net exchange rate of water and major ions between larvae and their environment must be considered (Fig. 2). The total body content of water and salt is turned over relatively slowly in FW. With increasing concentration of the external media, however, a dramatic reduction in turnover times is observed. This is largely caused by the increased ion concentration in the ingested fluid, ingestion being virtually the exclusive means of entry of water and most ions from hypertonic media, as first suggested by Beadle (1939) and reviewed by Phillips & Bradley (1977).

A. campestris larvae respond to rising external salinities by increasing their drinking rate by up to six-fold (Phillips & Bradley, 1977). However, most of this increase occurs only near the upper tolerance limits, when osmotic reduction in body volume, and hence internal hydrostatic pressure, threatens the mobility and therefore the survival of larvae (Kiceniuk & Phillips, 1974). Bradley & Phillips (1977*b*) did not observe any difference in the very high drinking rate of *A. taeniorhynchus* larvae reared in 10%, 100% or 200% SW. Under these conditions drinking rate was found to be proportional to body surface area as expected if it was dictated by the metabolic demands for nutrients. Indeed, larvae can be reared in sterile culture medium in the absence of particulate matter or microorganisms (reviewed by Clements, 1963; Nayar, 1966). Since many natural saline waters in which larvae thrive are undoubtedly very rich in dissolved nutrients and microorganisms, and often contain large numbers of competing filter feeders, larvae may normally indiscriminately ingest the external medium as a method of obtaining nutrients. This would also explain why over 80% of the net uptake of water by *A. campestris* larvae adapted to FW occurs by drinking. This high rate of ingestion would otherwise be surprising since the usual strategy amongst hyper-regulators is to minimize ingestion of fluid.

Kiceniuk & Phillips (1974) and Maddrell & Phillips (1975) have demonstrated that virtually all of the water, SO_4^{2-} and Mg^{2+} ingested by *A. campestris* larvae living in hyperosmotic $(Na+Mg)SO_4$ waters is absorbed into the haemolymph in the midgut. It seems probable that this is also the case for other major ions in various saline waters. It seems very likely that this serves to concentrate a dilute food source in preparation for digestion. The almost complete absorption of Mg^{2+} and SO_4^{2-} in the midgut contrasts sharply with the relatively small amount of intestinal absorption of these divalent ions in marine fish (Conte, 1969).

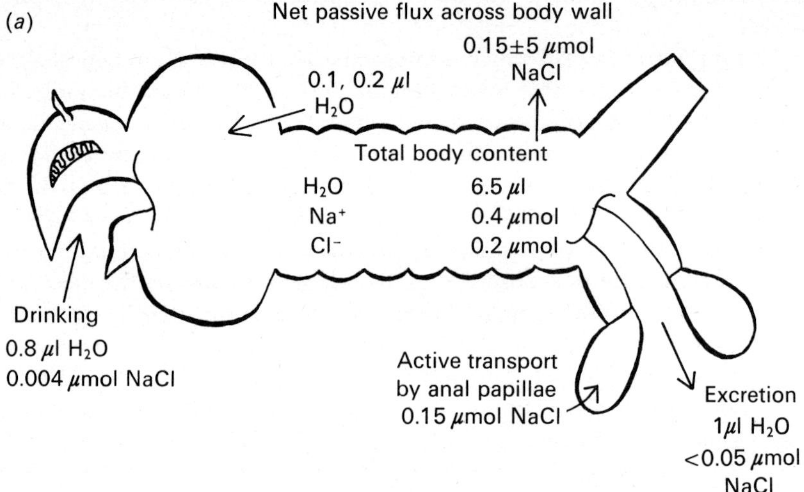

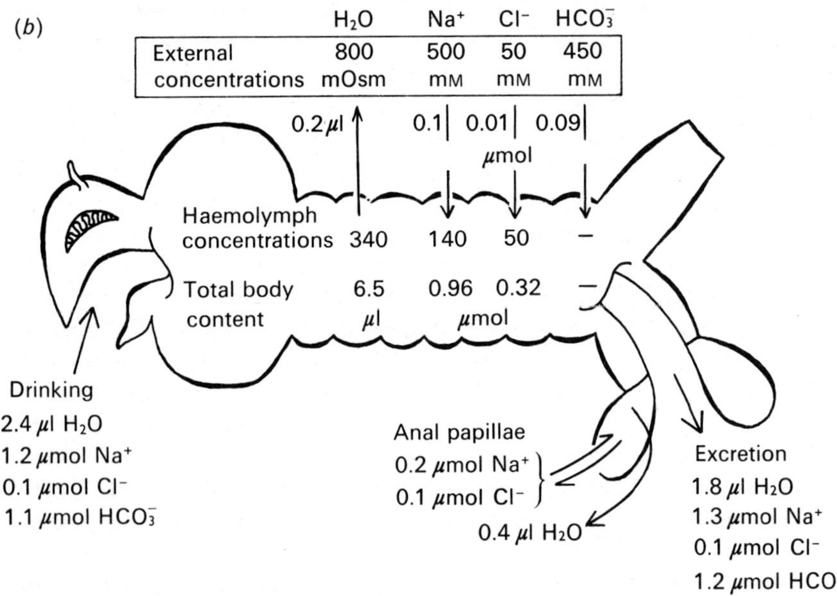

Fig. 2. Net exchanges per day across different body sites for 8 mg larvae of *Aedes campestris* in hypo-osmotic (*a*) or hyperosmotic (*b*) waters (Phillips, unpublished observations). (*a*) Adapted to 5 mM NaCl. Net passive fluxes across the body wall include the anal papillae. The two values for net water movement indicate simple diffsional and osmotic permeabilities, respectively. Excretion was calculated to balance measured rates of net exchange by all other sites. (*b*) Adapted to hyper-

The high rates of drinking in saline-water mosquito larvae may have an additional consequence. If the rate of fluid ingestion, rather than being just large enough to replace the water lost by osmosis across the body wall, is several times larger, then the concentration of the fluid secreted by the rectum no longer needs to be so concentrated to achieve osmotic balance. For example, if twice as much fluid is imbibed as is lost by osmosis, then the rectum must eliminate fluid twice as concentrated as the external medium. If, however, the ratio of imbibition to osmotic loss is increased to 4:1, then the eliminated fluid need only be 4:3 as concentrated as the ingested fluid. Higher rates of drinking do, of course, mean that the rectum must be able to eliminate the extra ions and fluid that are ingested. Presumably the rectal epithelia have sufficient transport capacity for this, but they may be more restricted in the concentration of fluid they can secrete.

Malpighian tubule secretion

Ramsay (1950) showed that the fluid produced by the Malpighian tubules of *A. detritus* reared in FW or SW was approximately iso-osmotic to the haemolymph and that the adjustment of urine osmolarity to achieve osmotic balance in these larvae occurred primarily in the rectum. Further information on Malpighian tubule secretion of saline-water species has only recently been provided by Phillips & Maddrell (1974) and Maddrell & Phillips (1975), using *A. campestris* larvae from $(Mg+Na)SO_4$ waters of moderate salinity (50% of haemolymph osmolarity). Each larvae has five tubules of similar appearance and without obvious divisions into distinct segments at the light-microscope level. These tubules are considerably smaller (< 3 mm long) than those of other insect species which have been successfully studied using the in vitro preparation of Ramsay (1954). However, by suitable miniaturization of the techniques, the same methods could be applied. In the tubules of *A. campestris*, fluid secretion appears to depend on a potassium transport process. There is some evidence that chloride ions may be transported actively, which is unusual in Malpighian tubules (Maddrell, 1971). The rate of fluid secretion appears to be under the control of a diuretic hormone similar to those of other insects (Maddrell

osmotic $NaHCO_3$ waters. Net passive fluxes of water across the anal papillae and the remainder of the body wall are distinguished in this situation; values are for osmotic flow. Values for net movement of HCO_3^- by all routes except drinking, and all of those for excretion, were calculated to balance other measured exchanges. The exchanges of NaCl across the anal papillae, which are indicated by half-arrows ($\rightleftharpoons$), are unidirectional fluxes. Since influx of these ions through terminal segments equalled efflux, no net exchange or transport of salt could be shown experimentally across anal papillae in this situation.

& Phillips, in preparation). Interestingly, this hormone may also act to stimulate sulphate transport; it was found that the rate of secretion of this ion increased substantially after stimulation.

While Malpighian tubules of saline-water mosquito larvae may not be unusual in their mechanism of fluid secretion, they have an unusual ability to secrete divalent ions. *A. campestris* larvae survive in waters in which Mg^{2+} concentration is 100 mM and in which SO_4^{2-}, at 350 mM, accounts for 90% of the total anions. Since these ions are also found in relatively high concentrations in SW (Hoar, 1975), the problem of maintaining low levels of Mg^{2+} and SO_4^{2-} in the haemolymph is probably common to most insect larvae that live in this environment. Active transport of Mg^{2+} amd SO_4^{2-} has been shown to occur against large electrochemical gradients under appropriate experimental conditions in *A. campestris* and also, in the case of SO_4^{2-}, in *A. taeniorhynchus* (Phillips & Maddrell, 1974; Maddrell & Phillips, 1975). These papers appear to be the first reports of the transport of these two ions in insects. Tubules of *Rhodnius* and *Carausius*, for example, are unable to secrete these ions against a concentration gradient. In saline-water mosquito larvae, however, concentration ratios (lumen: haemolymph) across the tubule wall of 12:1 for Mg^{2+} and of 5:1 for SO_4^{2-} were observed and maximum concentrations in the secretion of 45 mM and 9 mM, respectively, were achieved. Both transport mechanisms exhibit saturation kinetics with K_t values of 2.5 mM and 10 mM and V_{max} values of 15 pmol min^{-1} tubule^{-1} for Mg^{2+} and *c.* 50 pmol min^{-1} tubule^{-1} for SO_4^{2-}, respectively (Fig. 3).

Over a wide range of external concentrations, haemolymph levels of Mg^{2+} and SO_4^{2-} are normally maintained at relatively constant values of 1.5–4 and 2–7 mM, respectively. This is near or below the K_t values for the transport processes, i.e. in the region of the curve where the rate of ion transport is more or less directly proportional to Mg^{2+} and SO_4^{2-} levels in the blood (Fig. 3). Thus, rather small increases in haemolymph levels of these ions, resulting from ingestion, automatically give rise to their increased elimination, as a direct consequence of the kinetic properties of the transport mechanisms. At low and moderate salinities, the calculated rates of Mg^{2+} and SO_4^{2-} secretion by Malpighian tubules of *A. campestris* (Phillips & Maddrell, 1974; Maddrell & Phillips, 1975) are adequate to account for elimination of the total ingested amounts of these ions. This may be the first example among insects of ionic regulation as a direct consequence of selective secretion by the Malpighian tubules rather than by selective reabsorption in the rectum.

At higher external concentrations of Mg^{2+} and SO_4^{2-}, the transport capacities of Malpighian tubules reported by the above authors appear to be inadequate to eliminate all the ingested ions. In the case of Mg^{2+}, a more

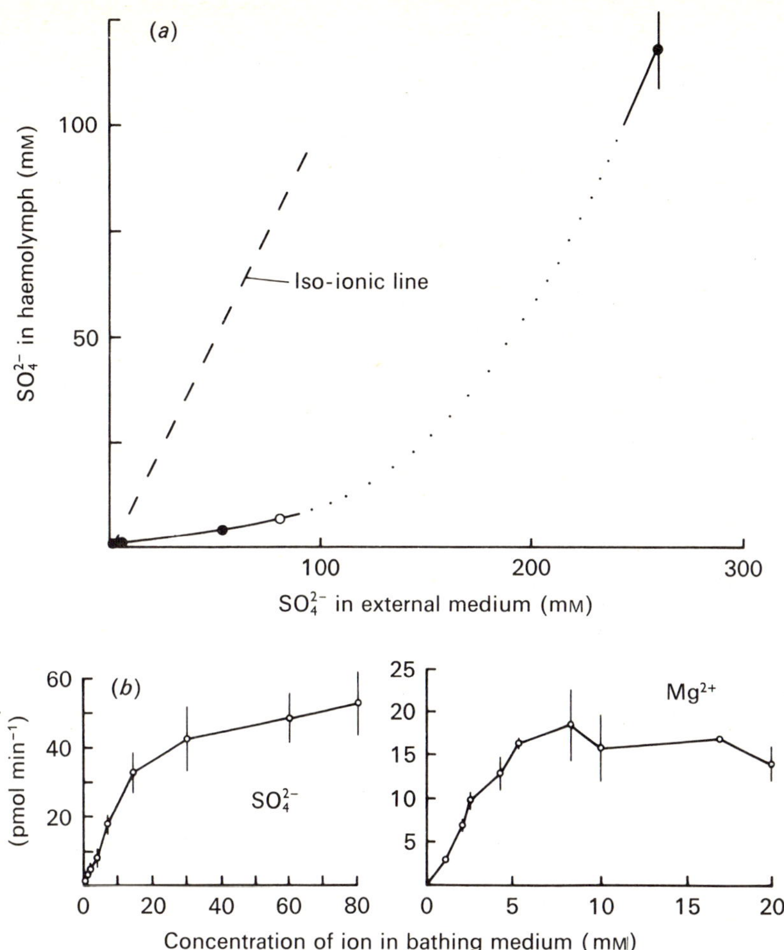

Fig. 3. (a) The influence of external SO_4^{2-} levels on steady-state concentrations of this anion in the haemolymph of the 4th instar *Aedes campestris* larvae. Haemolymph SO_4^{2-} levels are regulated at low external concentrations, but rise sharply when the latter exceeds 100 mM. This may be because of a saturation of the sulphate-secreting mechanism of the Malpighian tubules. (b) The dependence of the rate of secretion of SO_4^{2-} and Mg^{2+} by single isolated Malpighian tubules of *A. campestris* on the concentration of these ions in the bathing medium.

concentrated secretion by the rectum is possible (discussed below) and elimination by the anal papillae cannot be excluded (Kiceniuk & Phillips, 1974; Phillips & Maddrell, 1974). Another form of response has recently been observed by Maddrell & Phillips (in preparation), similar to that

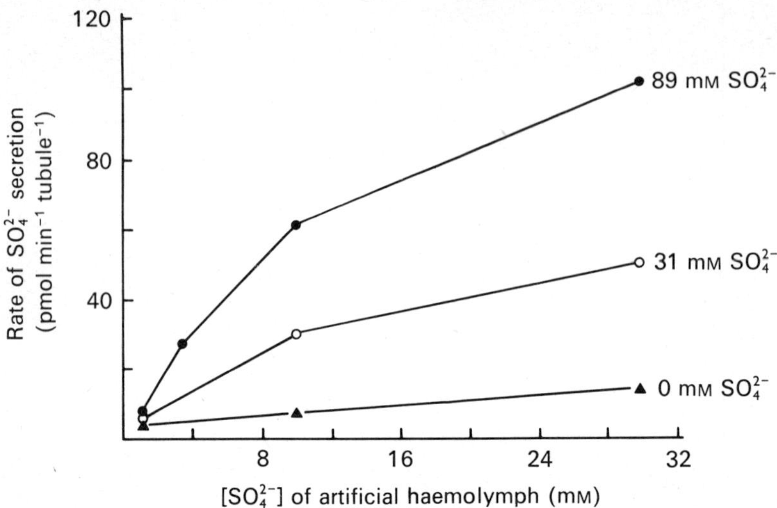

Fig. 4. The influence of external SO$_4^{2-}$ level during early development of *Aedes taeniorhynchus* larvae on the capacity of Malpighian tubules from 4th instar individuals to secrete this anion actively.

Larvae were reared in artificial SW containing 0, 31 or 89 mM SO$_4^{2-}$. Malpighian tubules from all three groups of larvae were exposed to a standard Ringer solution with different SO$_4^{2-}$ concentrations and the rates of secretion of this anion were measured. The results demonstrate that prior exposure to external SO$_4^{2-}$ induces or turns on greater capacity to transport this anion (Maddrell & Phillips, in preparation).

reported for anal papillae by Phillips & Meredith (1969a). If *A. taeniorhynchus* larvae are reared in sulphate-free SW, the tubular secretion contains only low levels of sulphate, well below those of the bathing Ringer solution. Other larvae, reared in normal SW or in SW with its SO$_4^{2-}$ concentration doubled, develop an ability to secrete this ion against large concentration differences, at several times the rate observed for larvae reared in the absence of SO$_4^{2-}$. Indeed, under standard test conditions, the rate of SO$_4^{2-}$ secretion at any haemolymph concentration reflects the external level during adaptation (Fig. 4). Since about 8–16 h exposure to sulphate-rich SW is needed to turn on SO$_4^{2-}$ transport in larvae reared in sulphate-free SW, a synthesis of new protein may well be involved (i.e. induction). This time is much shorter than that required to turn on Na$^+$ transport in anal papillae (see p. 165). Similar regulatory responses also seem to be the rule for ion transport processes in the rectum of saline-water larvae.

Rectal secretion in larvae inhabiting hyperosmotic waters

Until recently, it has been assumed that saline-water insects form hyper-osmotic urine in the same manner as terrestrial insects, i.e. by absorbing water without proportional amounts of solute in the rectum (Stobbart & Shaw, 1964, 1974). However, Meredith & Phillips (1973*a*) questioned this assumption on the basis of their ultrastructural studies of the rectum of saline-water mosquito larva *A. campestris*. They found that the formation of hyperosmotic urine in saline-water larvae is associated with the presence of a posterior rectal segment absent in strictly FW species. This segment consists of a single layer of cells of one type which lack the elaborately developed lateral intercellular spaces of the papillate rectum. There are, however, extensive infoldings of the apical membrane, which are regular and tightly packed and extend 60% of the distance across the cell. Most of the abundant mitochondria are associated with these infoldings. Since saline-water larvae ingest large quantities of the medium in which they live, Meredith & Phillips suggested that these animals do not have to conserve water but only rid themselves of the excess ions so ingested. They proposed that this might be achieved by active secretion of hyperosmotic fluid across the apical plasma membrane into the lumen of the posterior rectum (Meredith & Phillips, 1973*a*).

Indeed, if the animal is ligated at positions anterior and posterior to the rectum, this organ swells with secretion and the osmolarity of the lumen content increases from a value iso-osmotic to the haemolymph to that of the SW in which the larvae are reared (Bradley & Phillips, 1975). As ex-pected if the rectum is the principal site of regulation, the ionic composi-tion of the secretion resembles that of SW, except that the potassium levels are an order of magnitude higher (Fig. 5). Secretion of Na^+, K^-, Mg^{2+} and Cl^- occurs against large concentration differences of 2- to 18-fold. The average transepithelial potential difference across the posterior rectal segment decreases from 10 to 6 mV (lumen positive) during the course of secretion by ligated recta (Bradley & Phillips, 1977*c*). It is therefore necessary to postulate some form of active transport of all four ions to the lumen side. Such secretion only occurs in the posterior rectal segment when the larvae are in hyperosmotic waters. The osmotic concentration of the rectal secretion was found to be proportional to haemolymph Na^+ and Cl^- concentrations, but to be unaffected within the physiological range by haemolymph osmotic concentration (Bradley & Phillips, 1977*b*). The kinetics of the Na^+ and Cl^- transport mechanisms are unusual in that they resemble those of regulatory rather than classical enzymes (Fig. 6). Consequently, if haemolymph concentrations rise because of the abrupt

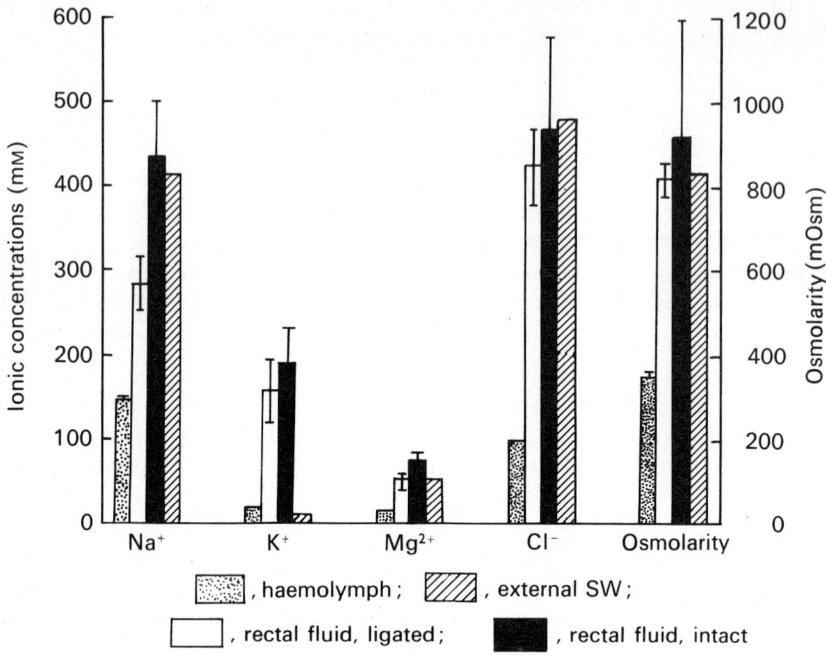

Fig. 5. A comparison of rectal fluid secreted by *Aedes taeniorhynchus* larvae with the haemolymph and the external SW in which these animals were reared. The composition of rectal fluid was determined using both ligated recta exposed to an artificial haemolymph and from intact larvae with the anus blocked for 1 h to permit accumulation of rectal fluid. Vertical bars denote the S.E.M. (Bradley & Phillips, 1975).

changes in external salinity, rates of Na^+ and Cl^- secretion in the rectum automatically increase sharply. This provides an element of automatic or intrinsic regulation. However, the response to long-term changes in external salinity and ionic composition (e.g. in $NaHCO_3$ and $(Na+Mg)SO_4$ lakes) is to vary the number of carrier sites, either by induction of carrier synthesis or by hormonal activation of pre-formed carriers. This is evidence from analyses of secretion from ligated recta of larvae which are reared in three different waters, all 700 mOsm, but differing in ionic composition (SW, $NaHCO_3$ lake water and $(Na+Mg)SO_4$ lake water). The recta from these larvae were all exposed to the same test Ringer solution. When conditions on the haemocoel side were constant, the ionic concentrations of the rectal secretions tended to reflect those prevailing in the external medium to which the larvae had been adapted (Fig. 7).

A model (Fig. 8) for the arrangement of transport processes in this

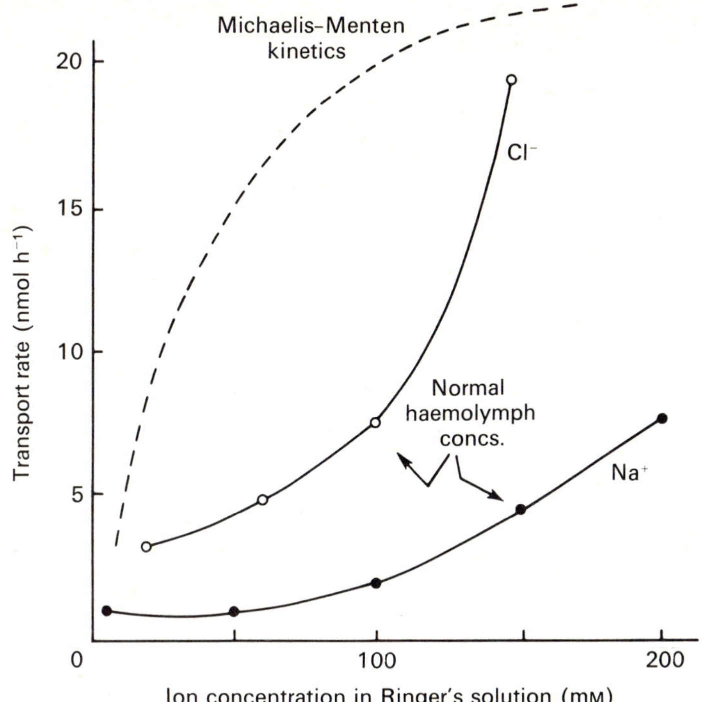

Fig. 6. The rate of Na^+ and Cl^- transport into the lumen of the ligated rectum of *Aedes taeniorhynchus* larvae. The body wall was torn open to expose the rectum to various Ringer solutions. Na^+ and Cl^- concentrations were varied individually by substituting choline and nitrate ions, respectively. Rates were calculated from the volume and ion concentrations of the secretion 2 h after recta had been ligated. Ion transport processes exhibit characteristics of regulatory rather than classical enzymes (Bradley & Phillips, 1977*b*).

epithelium has been proposed, based on the influence of different haemolymph ion ratios on electrical potential differences across the apical border and whole rectal wall (Bradley & Phillips, 1977*c*).

The posterior rectum of the saline-water mosquito larva plays a role analogous to the gills of marine teleosts, the rectal glands of elasmobranchs, and the salt glands of birds and reptiles. Indeed the profile of electrical potential across the posterior recta of mosquito larvae and secreting salt glands of birds are rather similar. The cell interior is negative to both the blood and the lumen so that electrogenic transport of cations is postulated to occur across the apical plasma membrane in both epithelia.

The argument has been repeatedly made for vertebrate transporting epithelia that the relative folding of the plasma membranes is consistent

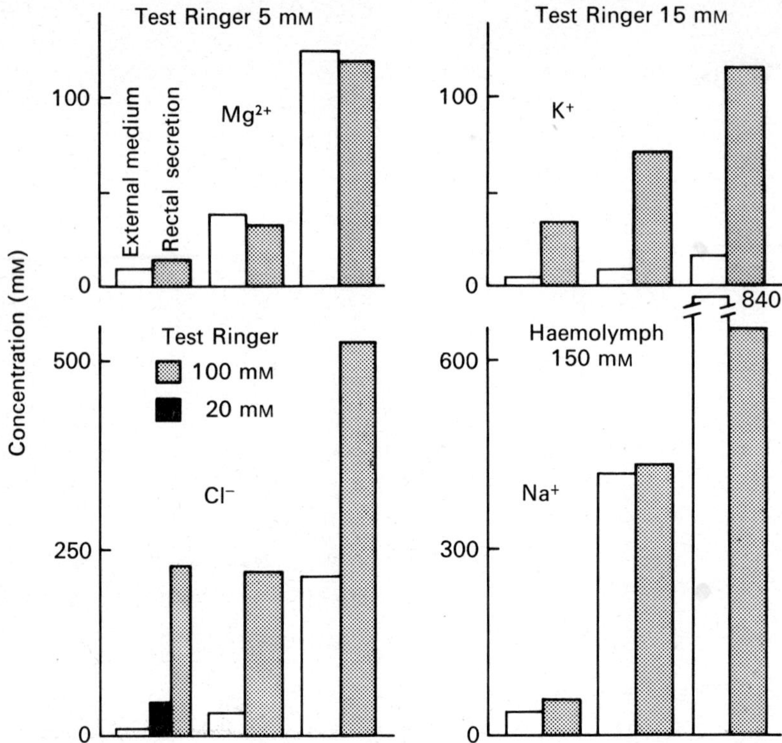

Fig. 7. A comparison of Mg^{2+}, K^+ and Cl^- ion concentrations in rectal secretions of *Aedes campestris* larvae with those in the natural water in which these larvae were reared. All waters were 700 mOsm but the principal salt differed ($NaCl$ or $NaHCO_3$ or $(Mg+Na)SO_4$). The body wall was torn open to expose ligated recta to a single test Ringer's solution, with the ion concentration indicated on each graph. Under these constant test conditions, ion content of the rectal secretions tended to reflect those of the external environment in which the larvae were living. The comparisons of Na^+ levels in external media and rectal secretions are for *A. taeniorhynchus* larvae reared in 10, 100 and 200% SW (Phillips, 1977).

with the model outlined above (e.g. Peaker, 1971). The basolateral membrane is, in general, highly folded in order to allow iso-osmotic exchange between the cell and blood, while the apical membrane is relatively flat so that less water will accompany ions actively secreted into the lumen. In essence it is claimed that it is the lack of folding on the apical side that allows the transfer of very hyperosmotic fluid across this cell border.

Such an interpretation clearly does not hold for the mosquito posterior rectum, which, as we have seen, has a highly folded apical membrane. These apical infoldings are much narrower and longer than those in other

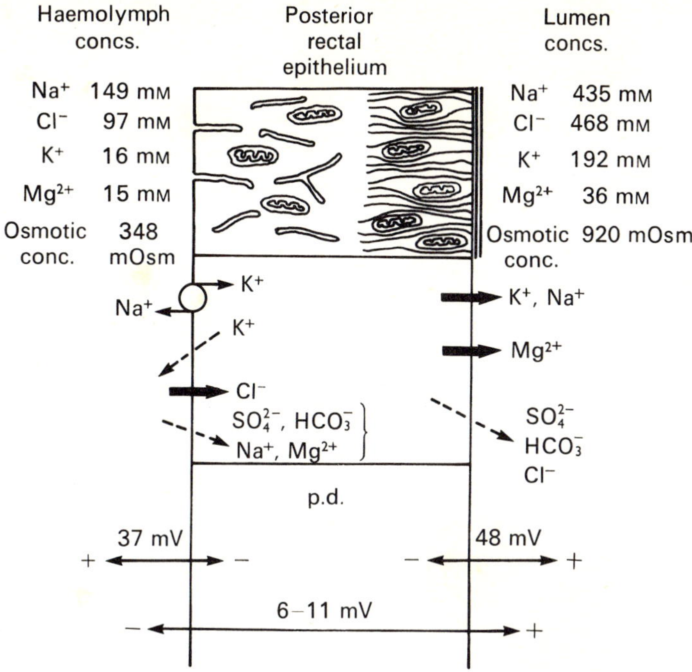

Fig. 8. A model by Bradley & Phillips (1975) for the cellular location of ion transport processes in the posterior rectum of saline-water mosquito larvae. Solid arrows indicate active transport and broken arrows depict passive movements. This model is based in part on measured ion concentration differences developed across this simple epithelium as a result of hyperosmotic secretion to the lumen side (figures shown in upper part of diagram), and on the observed electrical potential differences across the two plasma membranes (shown in the lower part of the diagram).

epithelia where iso-osmotic transport is known to occur (Table 1). Ironically, it is these channels in the mosquito rectum which most closely approach the dimensions that Hill (1975) calculates are necessary for iso-osmotic fluid transport, assuming the standing-gradient model of Diamond & Bossert (1967).

Not only are the extracellular channels very narrow but the cytoplasmic processes are of similar dimension so that the assumption of good mixing on the intracellular side of the membrane is very improbable. Possibly this absence of mixing on the cytoplasmic side, and the 'sweeping-in' effect which would accompany fluid entry into the cytoplasmic processes during secretion, might lead to local accumulation within the processes of major intracellular solutes which are not transported. Moreover, any Na⁺ which

Table 1. *The dimensions of membrane-bound channels associated with solute-coupled water transfer for selected epithelia*

Assuming typical permeability values for animal membranes, Hill (1975) has calculated that the geometric function length^2/radius must exceed 50 cm for the absorbate to achieve iso-osmoticity with the bathing medium. The mosquito posterior rectum most closely approaches this value but secretes a very hyperosmotic fluid.

System	10^4 Length (cm)	10^4 Radius (cm)	L^2/R	Secretion
Rabbit gall-bladder	30	1.5	0.06	Isotonic
Rat small intestine	20	0.7	0.06	Isotonic
Rat proximal tubule	8	0.05	0.12	Isotonic
Avian salt gland	7	0.13	0.04	**Hyperosmotic**
Insect Malpighian tubule	5	0.01	0.28	**Isotonic**
Locust rectal pad	—	—	1.0	Hypo-osmotic
Mosquito posterior rectum	37	0.015	9.0	Hyperosmotic
Required for isosmotic transport (Hill, 1975)	—	—	50	

Values from Hill (1975), except for mosquito (Meredith & Phillips, 1973*a*).

may leak back from the lumen must follow a long path back along the narrow cytoplasmic processes where they would continually be presented to the ion pumps and so re-secreted into the extracellular channels. Rather than enhancing the water transfer, we believe these channels may do the reverse. By an exchange multiplier-type mechanism, an increasing osmotic gradient might be established along the length of the infoldings from haemolymph to lumen side, rather than a decreasing one, an implied by the Diamond & Bossert model. The effect of this arrangement is to reduce the back diffusion (leak) of Na^+ from the lumen and also to limit the amount of water which accompanies ion secretion into the lumen. Clearly, there is a need to investigate this idea, using both mathematical modelling to see if it is feasible and direct analysis to find out whether such large osmotic concentration gradients do exist within these cells.

Anal papillae

Beadle (1939) concluded from some preliminary observations that the external anal papillae of *A. detritus* were rudimentary and impermeable structures which did not function as organs of extra-renal regulation, unlike those of strictly FW species. However, a regulatory function for anal papillae has been demonstrated in the larvae of at least one saline-water species of mosquito, *A. campestris* (Phillips & Meredith, 1969*a*). Indeed, this is the first report of extra-renal regulation among saline-water insects in general. When these larvae are reared in FW and exposed to distilled water for 1 to 2 days prior to experiments, net uptake of NaCl through the anal papillae against large electrochemical gradients occurs from 0.5 though not 0.05 mM NaCl. The larvae of the strictly FW mosquito *A. aegypti* can do the same from external solutions at least one order of magnitude more dilute (Stobbart, 1965). The active uptake by anal papillae of *A. campestris* exhibits Michaelis–Menten-type kinetics, with a K_t of 2 mM and V_{max} of 1.5 nmol h^{-1} mg^{-1} body weight. This K_t value is four times higher than that reported for *A. aegypti* (Stobbart, 1967). In terms of transport per mg wet weight of anal papillae, V_{max} was estimated to be similar for *A. campestris* and *A. aegypti*, but of course the total weight of anal papillae in the latter species, relative to total body weight, is much greater. In all, these differences may account for the regulatory advantages which strictly FW species have over saline-water species in natural waters of low osmotic concentration. When saline-water larvae are transferred from hyperosmotic medium to FW no significant influx of ^{22}Na via the anal papillae is observed initially. However, over 5–15 days, a large influx of the isotope through these organs develops (Fig. 9). Clearly, several days are required to induce Na$^+$ transport in this species. It appears that Na$^+$ and Cl$^-$ transport can be turned on independently, since some net uptake of Cl$^-$ occurs from 5 mM KCl, NH$_4$Cl, LiCl, CaCl$_2$ and choline chloride, while uptake of Na$^+$ will occur from the same concentrations of NaNO$_3$, Na$_2$HPO$_4$ and sodium citrate. In this regard, anal papillae of FW-adapted *A. campestris* appear similar to those of the FW species *A. aegypti* (Stobbart, 1965, 1967).

Ultrastructural studies support the physiological demonstration of active transport by these organs. The papillae of *A. campestris* and *A. togoi* (saline-water species: Meredith & Phillips, 1973*b, c*) and *A. aegypti* (FW species: Copeland, 1964; Sohal & Copeland, 1966) all have similar structures, not unlike the anterior rectum of the two saline-water species. The papillae contain only one cell type, arranged as a single layer covered with cuticle externally. Internally only a basement membrane separates the cell from the haemocoel. Both apical and basal borders are highly folded. Mitochondria are abundant, particularly in the saline-water species, where their profiles may constitute 30% of the cell area seen in electron micrographs. Surprisingly, no major differences in structure were evident between papillae of *A. campestris* larvae adapted for two weeks to FW or to strongly hyperosmotic NaHCO$_3$ water, nor in *A. togoi* which were living in rock pools containing 130% SW. The ultrastructure suggests active

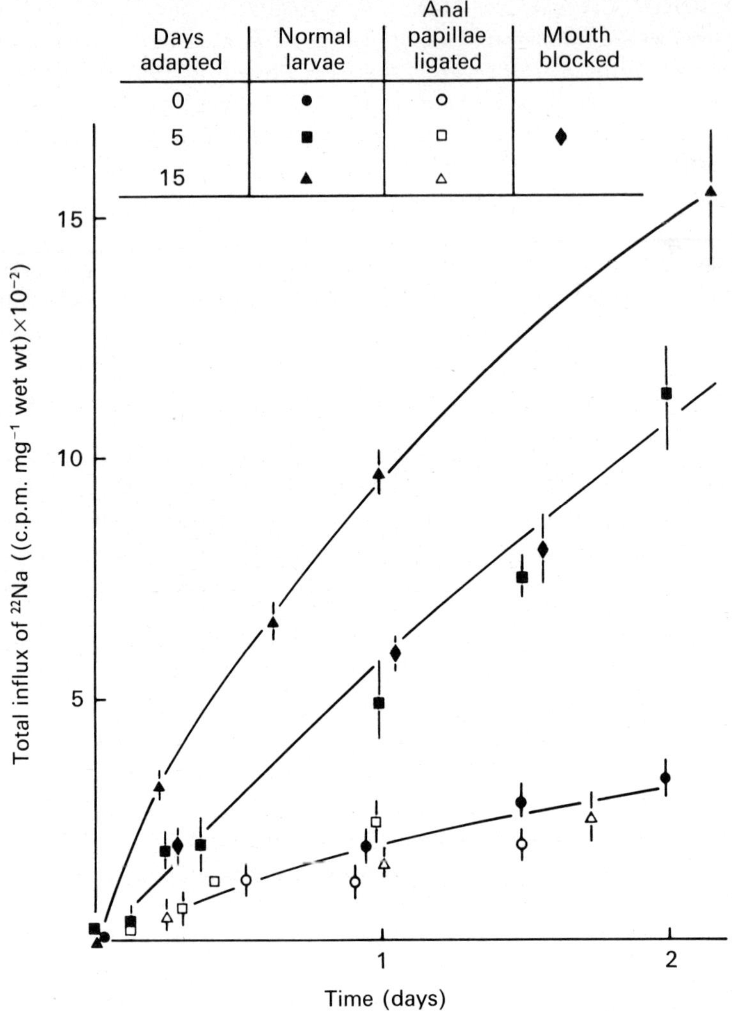

Fig. 9. The unidirectional influx of ^{22}Na into larvae of *Aedes campestris* at various times after transferring animals from hyperosmotic waters (1.2 Osm NaHCO$_3$) to hypo-osmotic media (4 mm ^{22}NaCl; Constant specific activity).

Uptake of ^{22}Na by normal larvae (controls) is compared with that by animals which had their mouths blocked with beeswax-resin to prevent drinking or with others in which the anal segment was ligated to prevent uptake by the anal papillae. Immediately after transfer (0 days) only a small influx of ^{22}Na is observed and this does not occur through the anal papillae. A large increase in influx appears within 5 to 15 days of transfer and all of this increment is the result of uptake by the anal papillae, because blocking the mouth has no effect, while ligating the papillae totally abolishes the increased ^{22}Na influx. (Phillips, unpublished; net uptake of Na$^+$ by the anal papillae, Phillips & Meredith, 1969a.)

transport activity under all conditions. This raises the question of the function of these organs in hyperosmotic waters.

An obvious role for anal papillae in unbalanced hyperosmotic natural water (e.g. $NaHCO_3$ and $(Mg+Na)SO_4$) is to absorb actively those essential ions which are only present in the external environment at relatively low concentrations: e.g. Cl^- and K^+ in the case of the waters just mentioned. The influx of ^{36}Cl via the anal papillae in concentrated $NaHCO_3$ water retains the saturation kinetics and the same K_t of 2 mM observed for FW-adapted larvae. However, the V_{max} (i.e. the amount of carrier) is reduced by an order of magnitude to a value of 0.1–0.2 nmol h^{-1} mg^{-1} body weight (Phillips & Bradley, 1977), another example of long-term adjustment in transport capacity.

There is some other circumstantial evidence to suggest that anal papillae can, like the gills of marine teleosts, reverse the direction of active transport and secrete NaCl to the external medium in SW. The anal papillae of larvae reared in iso-osmotic saline can be destroyed by $AgNO_3$ treatment (Phillips & Meredith, 1969a). When such larvae subsequently were transferred to SW, the level of Cl^- in their haemolymph rose 20% above that of the control larvae. However, some other effect of $AgNO_3$ treatment, in addition to that on the anal papillae, cannot be excluded. As mentioned above, ultrastructural observations have indicated very active anal papillae in *A. togoi* that were living in 130% SW, where both Na^+ and Cl^- were in excess. Moreover, rectal secretion may not be able to account totally for ionic regulation in some hyperosmotic waters (see p. 160). In summary, a considerable body of circumstantial evidence suggests that ions might actively be extruded by the anal papillae of larvae living in hyperosmotic waters but an unequivocal demonstration that this does occur has yet to be provided. Such a demonstration awaits a solution to the technical problem of occluding the anus without damaging the anal papillae.

Conclusions

Having discussed in turn each of the regulatory organs of saline-water mosquito larvae, the nature of the overall process of osmotic regulation, as we presently envisage it, will now be considered. Fig. 10 summarizes the processes which have been demonstrated or suggested for the Malpighian tubules, rectum and anal papillae.

Considerable evidence is now available which points to separate functions for the anterior and posterior rectal segments (Bradley & Phillips, 1977c). The two segments show morphological dissimilarities, the cells of the posterior segment having deeper apical infolds more closely associated with mitochondria. Electrical potential differences relative to the haemolymph are of opposite sign in the two segments, with the anterior being

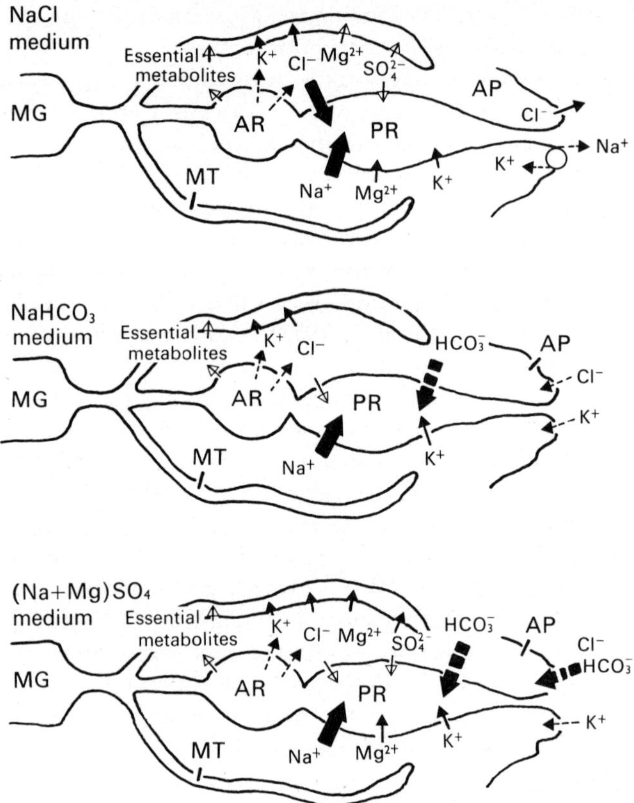

Fig. 10. A diagrammatic representation of the proposed locations of ion transport processes in the Malpighian tubules, rectum and anal papillae of *Aedes campestris* larvae, reared in three media (all 700 mOsm) of different chemical composition. The thickness of the arrows represents the relative rate of net transport as demonstrated by an increased concentration of an ion in the fluid secreted by the Malpighian tubules and rectum. The indicated rates and direction of ion transport in the anal papillae are suggested by the difference in drinking and urine excretion rates. AP, anal papillae; AR, anterior rectum segment; MG, midgut; MT, Malpighian tubule; PR, posterior rectal segment. Dashed arrows represent postulated ion transport pathways which are suggested by observations to date, but which have not been directly demonstrated (Bradley & Phillips, 1977a).

negative and posterior positive. An in-vitro ligated preparation of the posterior rectal segment secretes a strongly hyperosmotic fluid while the anterior does not. These findings, when combined with the observations that rectal cells such as those in the posterior rectal segment are only found in saline-water mosquito larvae, strongly support the hypothesis of

Meredith & Phillips (1973a) that the posterior rectal segment is an important site of hyperosmotic fluid secretion.

We propose that the following events occur in saline-water larvae adapted to FW. The Malpighian tubules secrete an iso-osmotic fluid rich in K^+ and Cl^-. These two ions and possible others (e.g. Na^+), as well as essential metabolites, are reabsorbed in the anterior rectum with a minimum of water. This creates a hypo-osmotic fluid containing largely nitrogenous and other waste products; in other words, the system functions much like that of FW insects (reviewed by Stobbart & Shaw, 1964, 1974). Under these conditions the posterior rectum is inactive and fluid passes through this segment to the exterior without major change. Net uptake of K^+, Na^+ and Cl^- occurs through the anal papillae in any natural FW containing more than 0.1 mM of these ions.

The pattern in larvae adapted to hyperosmotic waters is as follows. The Malpighian tubules continue to produce iso-osmotic fluid rich in KCl. However, if the natural water contains high levels of Mg^{2+} or SO_4^{2-}, active secretion of these ions is induced and accounts for a significant fraction of the total solute in the tubule fluid. An increase in transport capacity (i.e. amount of carrier) for SO_4^{2-}, and possibly for Mg^{2+}, rapidly occurs (within a period of a day) in response to exposure of the larvae to water containing high levels of these ions. In the anterior rectum, most of the KCl and water is reabsorbed in proportional amounts, leaving a smaller volume of approximately iso-osmotic fluid enriched in $MgSO_4$ and other waste products. The posterior rectum is activated and secretes a hyperosmotic fluid. The relative rates at which it eliminates the various ions (Na^+, K^+, Mg^{2+}, Cl^- and HCO_3^-) depend on the concentration of these ions in the external medium to which larvae are adapted, as well as on the haemolymph levels. In more concentrated waters, the total rate of ion transport by the posterior rectum and the volume of its secretion increase. However, proportionally less water follows the ion movement, so that osmolarity of the secretion also increases. This fluid is then eliminated via the anus. Net uptake of individual ions from the external medium may continue in unbalanced waters: e.g. Cl^- uptake from $NaHCO_3$ water poor in Cl^-. In SW, in which both Na^+ and Cl^- are more concentrated than in the haemolymph, uptake of these ions by the anal papillae ceases. Indeed the papillae may now transport Na^+ and Cl^- in the opposite direction, and thus eliminate them from the larva. We suggest that some of the Na^+ secreted in this manner might be linked to a net influx of K^+, as occurs in the gills of marine teleosts. This would help to replace the large quantities of K^+ lost by secretion in the posterior rectum. Two important areas which remain to be considered are HCO_3^- and pH regulation.

We wish to acknowledge the financial support of the North Atlantic Treaty Organization and the National Research Council of Canada during recent studies which were conducted by the authors and which are described in this paper.

References

Beadle, L. C. (1939). Regulation of the hemolymph in the saline water mosquito larvae *Aedes detritus* Edw. *Journal of Experimental Biology*, **16**, 346–62.

Bradley, T. J. & Phillips, J. E. (1975). The secretion of hyperosmotic fluid by the rectum of a saline-water mosquito larva *Aedes taeniorhynchus*. *Journal of Experimental Biology*, **63**, 331–42.

Bradley, T. J. & Phillips, J. E. (1977*a*). Regulation of rectal secretion in saline-water mosquito larvae living in waters of diverse ionic composition. *Journal of Experimental Biology*, **66**, 83–96.

Bradley, T. J. & Phillips, J. E. (1977*b*). The effect of external salinity on drinking rate and rectal secretion in the larvae of the saline-water mosquito *Aedes taeniorhynchus*. *Journal of Experimental Biology*, **66**, 97–110.

Bradley, T. J. & Phillips, J. E. (1977*c*). The location and mechanism of hyper-osmotic fluid secretion in the rectum of the saline-water mosquito larvae *Aedes taeniorhynchus*. *Journal of Experimental Biology*, **66**, 111–26.

Clements, A. N. (1963). *The Physiology of Mosquitoes*, pp. 33. New York: Macmillan.

Conte, F. P. (1969). Salt secretion. In *Fish Physiology*, ed. W. S. Hoar & D. J. Randall, vol. 1, pp. 241–92. New York, London: Academic Press.

Copeland, E. (1964). Mitochondrial pump in the cells of the anal papillae of mosquito larvae. *Journal of Cell Biology*, **23**, 253–64.

Diamond, J. M. & Bossert, W. H. (1967). Standing-gradient osmotic flow. A mechanism for coupling of water and solute transport in epithelia. *Journal of General Physiology*, **50**, 2061–83.

Hill, A. E. (1975). Solute–solvent coupling in epithelia: a critical examination of the standing-gradient osmotic flow theory. *Proceedings of the Royal Society, London*, B, **190**, 99–114.

Hoar, W. S. (1975). *General and Comparative Physiology*. Toronto: Prentice Hall.

Kiceniuk, J. W. & Phillips, J. E. (1974). Magnesium regulation in mosquito larvae (*Aedes campestris*) living in waters of high $MgSO_4$ content. *Journal of Experimental Biology*, **61**, 749–60.

Maddrell, S. H. P. (1971). The mechanisms of insect excretory systems. *Advances in Insect Physiology*, **8**, 199–331.

Maddrell, S. H. P. & Phillips, J. E. (1975). Active transport of sulphate ions by the Malpighian tubules of the larvae of the mosquito, *Aedes campestris*. *Journal of Experimental Biology*, **62**, 367–78.

Meredith, J. & Phillips, J. E. (1973*a*). Rectal ultrastructure in salt- and freshwater mosquito larvae in relation to physiological state. *Zeitschrift für Zellforschung und mikroskopische Anatomie*, **138**, 1–22.

Meredith, J. & Phillips, J. E. (1973*b*). Ultrastructure of the anal papillae of a salt-water mosquito larva, *Aedes campestris*. *Journal of Insect Physiology*, **19**, 1157–72.

Meredith, J. & Phillips, J. E. (1973c). Ultrastructure of anal papillae from a seawater mosquito larva (*Aedes togoi* Theobald). *Canadian Journal of Zoology*, **51**, 349–53.

Nayar, J. K. (1966). A method of rearing salt-marsh mosquito larvae in a defined sterile medium. *Annals of the Entomological Society of America*, **50**, 1283–5.

Nayar, J. K. & Sauerman Jr, D. M. (1974). Osmoregulation in larvae of the salt-marsh mosquito *Aedes taeniorhynchus*. *Entomologia experimentalis et applicata*, **17**, 367–80.

Nicolson, S. W. (1972). Osmoregulation in larvae of the New Zealand salt-water mosquito *Opifex fuscus* Hutten. *Journal of Entomology*, **47**, 101–8.

Peaker, M. (1971). Avian salt glands. *Philosophical Transactions of the Royal Society* B, **262**, 289–300.

Phillips, J. E. (1977). Excretion in insects: function of gut and rectum in concentrating and diluting the urine. *Federation Proceedings*. (In Press.)

Phillips, J. E. & Bradley, T. J. (1977). Osmotic and ionic regulation in saline-water mosquito larvae. In *Transport of Ions and Water in Animal Tissues*, ed. B. L. Gupta, R. B. Moreton, J. L. Oschman & B. J. Wall). London: Academic Press. (In Press.)

Phillips, J. E. & Maddrell, S. H. P. (1974). Active transport of magnesium by the Malpighian tubules of the larvae of the mosquito *Aedes campestris*. *Journal of Experimental Biology*, **61**, 761–71.

Phillips, J. E. & Meredity, J. (1969a). Active sodium and chloride transport by anal papillae of a salt water mosquito larva (*Aedes campestris*). *Nature, London*, **222**, 168–9.

Phillips, J. E. & Meredith, J. (1969b). Osmotic and ionic regulation in a salt water mosquito larva *Aedes campestris*. *American Zoologist*, **9**, 588.

Ramsay, J. A. (1950). Osmotic regulation in mosquito larvae. *Journal of Experimental Biology*, **27**, 145–57.

Scudder, G. G. E. (1969). The fauna of saline lakes of the Fraser Plateau in British Columbia. *Verh. Internat. Verein. Limnol.* **17**, 430–9.

Sohal, R. S. & Copeland, E. (1966). Ultrastructural variation in the anal papillae of *Aedes aegypti* (L.) at different environmental salinities. *Journal of Insect Physiology*, **12**, 429–39.

Stobbart, R. H. (1965). The effect of some anions and cations upon the fluxes and net uptake of sodium in the larva of *Aedes aegypti* (L.). *Journal of Experimental Biology*, **42**, 29–43.

Stobbart, R. H. (1967). The effect of some anions and cations upon the fluxes and net uptake of chloride in the larva of *Aedes aegypti* (L.) and the nature of the uptake mechanisms for sodium and chloride. *Journal of Experimental Biology*, **47**, 35–57.

Stobbart, R. H. & Shaw, J. (1964). Salt and Water Balance: Excretion. In *The Physiology of Insecta*, ed. M. Rockstein, vol. 3, pp. 190–258. New York, London: Academic Press.

Stobbart, R. H. & Shaw, J. (1974). Salt and Water Balance Excretion. In *The Physiology of Insecta*, ed. M. Rockstein, 2nd edn, vol. 5, pp. 362–446. New York, London: Academic Press.

12. Inhibition of magnesium secretion in the prawn *Palaemon serratus* by ethacrynic acid and by ligature of the eyestalks

By S. E. FRANKLIN, BANCHONG TEINSONGRUSMEE
& A. P. M. LOCKWOOD

Magnesium is handled in such a variety of ways by different species of crustaceans that it appears as something of a maverick by comparison with the other major inorganic ions in the body fluids. The more sluggish species of aquatic marine crab have blood magnesium concentrations tending toward the level in sea water and urine concentrations which are little higher; more active crabs usually have lower haemolymph and higher urine values for this ion (Robertson, 1949, 1953). There is a similar correlation amongst semi-terrestrial crabs, the more terrestrial species having the greater capacity to concentrate magnesium in the urine (Gross, 1964); however, by contrast one of the most terrestrial of all crabs, *Gecarcinus lateralis* shows scarcely any ability to concentrate magnesium in the urine above the level in the blood (the U/B for magnesium being only 1.7) (Gross, 1964). The prawn *Palaemon serratus* is comparable to the brachyurans in having low haemolymph, but high urine, magnesium concentrations (Parry, 1954); indeed it outdoes all crabs so far studied in the magnitude of the gradient maintained between haemolymph and urine. However, a low blood magnesium concentration can be associated with a low urine magnesium level, as in the lobster *Homarus americanus* (Burger, 1957) and the freshwater crab *Potamon niloticus* (Riegel & Lockwood, unpublished), while the fiddler crab, *Uca crenulata*, has a high urine concentration but also a relatively high blood concentration (Gross, 1964) (Table 1).

Most of the euryhaline species of crab can vary the U/B for magnesium as the salinity of the medium is changed, though *Gecarcinus lateralis* lacks this ability. In *Pachygrapsus* the rate at which magnesium is lost in the urine is largely determined by the rate at which water passes through the excretory system, so that in magnesium-free sea water the animals rapidly become depleted of this ion (Gross & Marshall, 1960). By contrast, in magnesium-free sea water, the magnesium levels in *Carcinus* show a proportionately greater drop in urine than in blood, resulting in some measure of conservation of magnesium in the body. Both *Cancer antennarius* (Gross, 1964) and *Carcinus* (Webb, 1940) increase the magnesium

Table 1. *Magnesium concentration in the urine and blood of various species of Crustacea acclimatised to sea water[a]*

Species	Mean blood Mg^{2+} (mM)	Mean urine Mg^{2+} (mM)	Mg^{2+} U/B	Authority
Cardiosoma carnifex	13	60	4.4[b]	Gross *et al.* (1966)
Pachygrapsus crassipes	8.5	168	13.6[b]	Gross & Capen (1966)
Cancer antennarius	30.5	51	1.74[b]	Gross (1964)
Hemigrapsus oregonensis	23.5	130	4.96[b]	Gross (1964)
Uca crenulata	30.5	173	5.32[b]	Gross (1964)
Gecarcinus lateralis	14	21	1.74[b]	Gross (1964)
Maia squinado	46	49	1.06[c]	Robertson (1953)
Homarus americanus	6.8	11.5	1.7[c]	Burger (1957)
Carcinus maenas	20.5	90	4.3[c]	Riegel & Lockwood (1961)
Palaemon serratus	6	154	25.6[c]	Present paper
Palaemon serratus	12.5	84	6.7[c]	Parry (1954)

[a] Mg^{2+} in sea water = *c.* 50–53 mM.
[b] Expressed as mean of individual U/B values.
[c] Expressed as the ratio of mean U to mean B.

concentration of the urine when the concentration of magnesium in the external medium is increased and *Carcinus* also increases the urine magnesium concentration and urine U/B for magnesium after injection of a small volume of concentrated $MgCl_2$ into the blood (Lockwood & Riegel, 1969). The excretion of magnesium therefore appears to be under facultative control.

When the concentration of magnesium in the urine of crabs is raised, the sodium concentration is depressed so that the U/B is less than one (Webb, 1940; Prosser, Green & Chow, 1955; Gross, 1957, 1959, 1964; Gross & Marshall, 1960; Riegel & Lockwood, 1961; Gross & Capen, 1966). It does not, however, seem probable that the direct exchange of sodium for magnesium ions across the bladder wall could alone account for the concentration differences observed. Other factors, including water shifts, must also be involved (Lockwood & Riegel, 1969) in concentrating urine magnesium. Flemister (1958) demonstrated that semi-terrestrial crabs are capable of reabsorbing water from the primary urine. The concentrations of magnesium in *Carcinus* and *Pachygrapsus* can be partially accounted for by water reabsorption in the excretory system after formation of the primary urine (Riegel & Lockwood, 1961; Gross & Capen, 1966; Lockwood & Riegel, 1969). Doubts about the importance of this water withdrawal (Binns, 1969*a, b*) have subsequently been, refuted (Riegel *et al.*, 1974).

The ability of species such as *Pachygrapsus* and *Carcinus* to vary the magnesium level in the urine suggests, therefore, that this ion must be actively secreted into the urine, and Gross & Capen (1966) have produced the following evidence supporting the conclusion that this secretion occurs in the bladder.

(1) Crabs with empty bladders when immersed in sea water show increased magnesium concentration in the urine while the bladder fills.

(2) When perfusion fluid is substituted for urine in the bladder, the magnesium concentration of the bladder fluid increases in crabs kept out of water.

(3) When crabs with empty bladders are kept out of water, the fluid initially filling the bladder is low in magnesium but later the concentration rises.

(4) When urine samples from previously blocked and unblocked bladders of the same crab are compared, that from the blocked side has a higher magnesium level.

Remarkably little is known about the actual processes involved in magnesium secretion and control, even in vertebrates. Wen, Evansen & Dirks (1970) obtained results suggesting that renal tubular reabsorption of magnesium in dogs occurs against the electrochemical gradient and that the reabsorption rate declines after infusion of $MgCl_2$ into the blood. Treatment with the diuretic furosemide greatly increases magnesium excretion and Wen *et al.* consider that this could be because of reduced fluid reabsorption. Aldor & Moore (1970) and Behar (1975), working with rat intestines, have obtained evidence compatible with the assumption that magnesium transport is associated with bulk flow and solvent drag effects, the rate of magnesium absorption correlating with the rates of water absorption.

Investigation of magnesium transport in the gut of the phytophagous insect *Cecropia* (silkworm) by Wood, Jungreis & Harvey (1975) has shown that, although in intact preparations of the midgut the potential difference would be adequate to account for the observed gradient of magnesium across the gut wall, magnesium transport continues from the gut to the blood even when the potential difference is short circuited. The transport, which shows Michaelis–Menten kinetics, is independent of sodium and calcium but bears a complex relationship with potassium transport. The presence of a specific transport system for magnesium is therefore implied.

The prawn *Palaemon serratus* has a U/B for inulin which averages between 1.5 and 1.98 (Bryan & Ward, 1962) and also a high magnesium U/B (Parry, 1954). In view of these results in vertebrates, the present experiments were undertaken to investigate whether factors affecting the urine production rates would also influence magnesium transport in the prawn or whether, as in the silkworm, there is evidence of a specific transport process. Two factors were tested: (1) the diuretic ethacrynic acid which is found to decrease water reabsorption in the excretory systems of vertebrates (Proverbio, Robinson & Whittembury, 1969) and crabs (Riegel *et al.*, 1974), and (2) ligature of the base of the eyestalk which increases urine production rate in crayfish (Kamemoto & Ono, 1969). Both injection of ethacrynic acid and ligaturing the eyestalks are found to result in a major decrease in the rate of excretion of magnesium.

Material and methods

Animal maintenance

Palaemon serratus were collected by hand-net from the estuary of the River Hamble and from Weymouth. Prior to use they were maintained in sea water (35‰) in the aquarium in the Department of Oceanography, Southampton, and fed weekly on pieces of fish.

Only males and non-ovigerous females were used for experimentation, as the ionic concentrations of ovigerous females can vary (Panikkar, 1941).

Experimental animals were held in aerated tanks and not fed during experiments. The water was changed daily and the temperature maintained at 15 °C.

All concentrations of the medium quoted are based on dilution or concentration of local sea water with a salinity of 34.5‰ and a magnesium concentration of 53 mM.

Prior to the sampling of blood or urine, animals were allowed to acclimatise to the appropriate medium for 2 days; those being placed into 50 or 25 % sea water were acclimatised to an intermediate salinity before being placed in the experimental media.

Body fluid sampling

Each prawn was blotted dry and adhering medium was removed from around the antennary pore. The tip of a finely drawn Pasteur pipette was then inserted just within the lips of the pore and a urine sample gently sucked up. Urine was always sampled before blood as the animals tended to release urine under stress.

Samples of blood were obtained by inserting the top of a drawn-out pipette through the arthrodial membrane under the hind margin of the cephalothoracic shield, after drying the groove to remove any trapped medium.

Both blood and urine samples were immediately blown out under liquid paraffin in siliconised vessels, and subsamples were taken for analysis. Three replicates of each sample were taken.

Magnesium concentration was measured by flame absorption using a Unicam SP90 spectrophotometer, and sodium by flame emission by a Unicam SP900 spectrophotometer. ^{51}Cr-labelled EDTA, obtained from the Radiochemical Centre, Amersham, was counted using a Panax Reigate Series Scanner and a well-type scintillation probe.

Prawns injected with [^{51}Cr]EDTA solution (3 μl $\equiv$ 3 μCi) were left for 40 h to permit equilibration of the EDTA between blood and urine before sampling commenced. Preliminary experiments suggested that a steady U/B for EDTA is reached after about 12–15 h at a temperature of 15 °C.

Ethacrynic acid for injection was made up to a concentration of 500 p.p.m. by dissolving 5 mg of the material in two drops of *n*-butyl alcohol and then dispersing this in 10 ml of 100 % sea water. Fresh ethacrynic acid solution was made up for each set of injections. The volume injected was 5 μl.

Values for blood and urine magnesium concentrations are given in mM whole blood and whole urine, respectively. The water content of *Palaemon serratus* is some 877 mg ml^{-1} and the protein content is some 12 % by weight according to Parry (1954).

Results

The effect of a range of salinities on the sodium and magnesium concentrations of the haemolymph and urine

For specimens fully acclimatised to a particular saline concentration in the range 25–125 % sea water, the haemolymph magnesium concentration is maintained at an almost constant level of about 5–6 mM, though the urine levels rise from 6.5 to 200 mM over the same salinity range. Consequently the concentration ratio for magnesium between haemolymph and urine increases steeply with salinity (Table 2).

The sodium concentration of haemolymph rises with increased external salinity, though less steeply than the medium. By contrast the urine sodium level declines relative to that of the blood, as the salinity rises (Table 3).

These values for both magnesium and sodium are comparable with those of Parry (1954), although she obtained a significantly higher level of magnesium in the blood and somewhat lower values in the urine in the high salinities.

Table 2. *Magnesium concentration[a] in urine and blood at various external salinities ($N = 8$)*

	Salinity (% sea water)				
	125	100	70	50	25
Urine Mg^{2+} (mM)	200±37	154±26	46±9.3	19±5	6.5±1.1
Blood Mg^{2+} (mM)	6±0	6±0.95	5.7±1.53	6±1.41	5±0
U/B for Mg^{2+}	33.3	25.7	8.1	3.2	1.3

[a] Means ± S.D. are given.

Table 3. *Sodium concentration[a] of blood and urine at various external salinities*

	Salinity (% sea water)				
	125	100	70	50	25
Urine Na^+ (mM)	301±35	282±22	270±35	240±40	190±15
Blood Na^+ (mM)	482±45	395±26	360±20	255±34	200±30
Number	8	8	8	8	8
U/B for Na^+	0.62	0.71	0.75	0.94	0.95

[a] Means ± S.D. are given.

The influence of water withdrawal on magnesium U/B

[51]Cr-labelled EDTA, like inulin, is thought to enter the excretory tubule without restriction during primary urine formation and then to behave as an inert marker, not being reabsorbed, during passage of the urine through the excretory system (Lockwood & Riegel, 1969; Riegel *et al.*, 1974). It is assumed therefore that any difference between the haemolymph and definitive urine concentrations of [51Cr]EDTA, after correction for the differences in water content of blood and urine, will reflect essentially water reabsorption from the primary urine in the excretory system. (The U/B ratio for [51Cr]EDTA does not necessarily indicate solely water reabsorption, since it will also be influenced by time lag effects (Binns, 1969*a*, *b*), although errors from this cause are unlikely to be substantial (Riegel *et al.*, 1974).) The observed U/B for [51Cr]EDTA thus probably reflects a maximum measure of water reabsorption, the true rate being somewhat less.

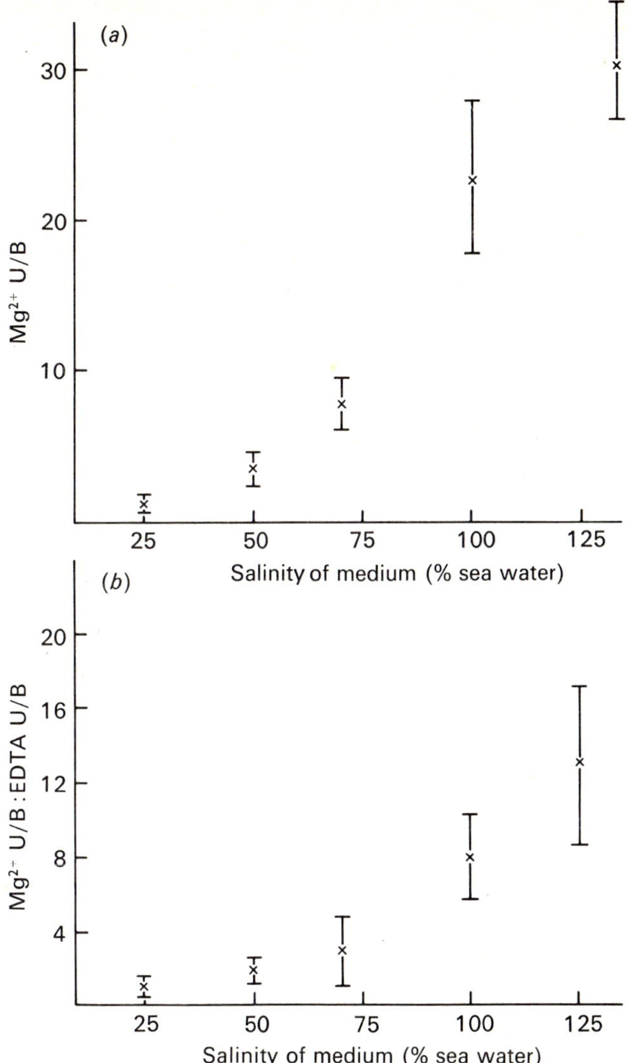

Fig. 1. (*a*) The effect of external salinity on magnesium U/B ratio. Vertical bars indicate the standard deviations. (*b*) The effect of external salinity on the ratio of magnesium U/B to EDTA U/B.

Measurement of the U/B for [^{51}Cr]EDTA suggests that, in the prawn, water reabsorption in the excretory system declines with decreasing concentration of the medium, as it does in crabs (Lockwood & Riegel, 1969) and in *Gammarus duebeni* (Lockwood & Inman, 1973). Reabsorption of water would naturally tend to concentrate magnesium in the urine if this ion was not reabsorbed, but water withdrawal alone cannot account for the concentration of magnesium in the definitive urine (Fig. 1*a*, *b*).

For animals in 100 % sea water, the potential difference recorded between an electrode in the haemolymph and one implanted just within the urinary opening is small (3–9 mV, N = 4, urine positive to blood). This seems to eliminate the possibility that the concentration gradient for magnesium is maintained by the electrical gradient, since a potential difference of some 40 mV would be required to balance the 154 to 6 mM difference between blood and urine in this medium. Some form of active process must therefore be responsible for the establishment and maintenance of the concentration gradient.

Rate of elimination of magnesium in different salinities

Previous experiments (Teinsongrusmee, 1976) have established that almost all the $[^{51}Cr]$EDTA lost from the haemolymph of prawns over a period of 36 h following injection appears in the medium, implying that any uptake by body tissues is small (av. 3 % of haemolymph loss). It is assumed that all the loss is in the urine. Measurements of clearance of EDTA based on this assumption indicate that the rates of primary and definitive urine production change substantially as external salinity varies (Table 4).

Table 4. *Loss of the urinary marker* $[^{51}Cr]$EDTA *in animals exposed to different salinities*

	% Sea water	Wt ranges[a] (g)	Clearance[a] (% body wt day^{-1})	EDTA[a] (U/B)	U/B corrected[b]	Calculated urine flow (% body wt day^{-1})
Medium	100	0.67–1.65	24.7±3.8 (18)	2.7±1.1 (10)	2.3	10.6
	70	0.56–1.21	23.3±5.0 (11)	2.5±0.4 (6)	2.2	10.8
	50	0.44–1.29	39.5±7.9 (15)	2.0±0.3 (6)	1.8	22.3
	10	—	—	1.4±0.6 (5)	1.2	—

[a] Means ± S.D. are given.
[b] The correction for U/B is based on the assumption that blood proteins occupy 12 % of the haemolymph volume (Parry, 1954) and that this space is not available to EDTA.

The rate of excretion of magnesium in the urine, based on the urine Mg^{2+} concentration and calculated urine flow rate, is, for 100% sea water, 16.26 mmol kg^{-1} day^{-1}; 70% sea water, 4.96 mmol kg^{-1} day^{-1} and 50% sea water, 4.08 mmol kg^{-1} day^{-1}. The rate of elimination of magnesium thus rises markedly with increased salinity and this rise is not dependent solely upon haemolymph magnesium concentration and urine flow rate.

Table 5. *Blood and urine* Mg^{2+} *concentrations[a]* (mM) *in prawns with the eyestalks ligatured* $(N = 8)$

| | Medium salinity as % sea water | | | |
	125	100	70	50
Urine	35±6.2	41±4.8	22±2.4	10±1.3
Blood	15±1.6	14±1.3	13±1.9	9±0.4

[a] Means ± S.D.

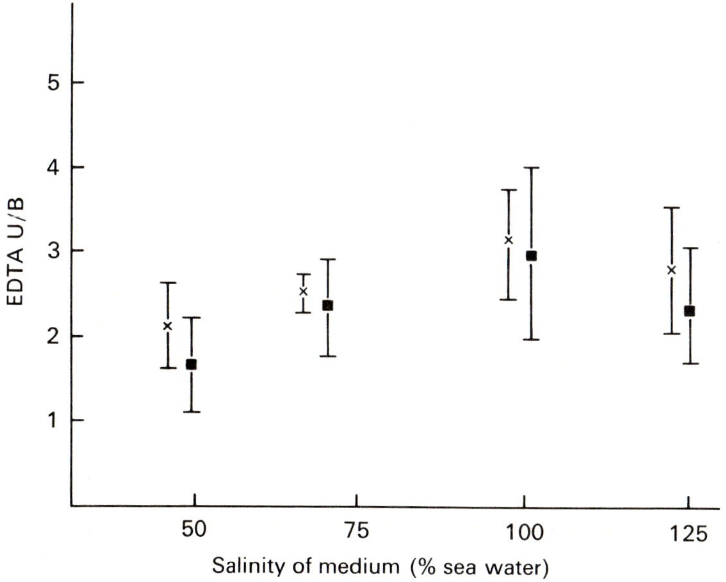

Fig. 2. The effect of ligaturing the eyestalks on EDTA U/B ratio. ×, ligatured; ■, controls. Vertical bars represent standard deviations.

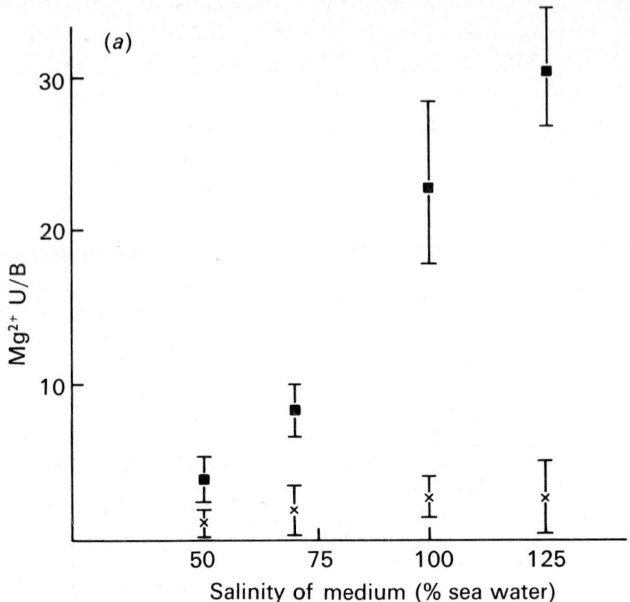

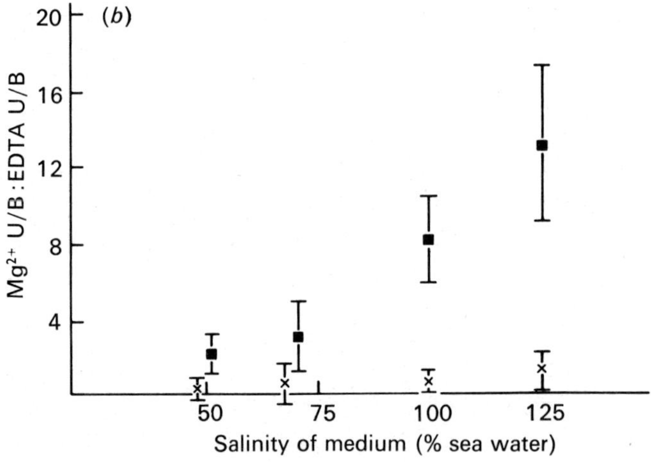

Fig. 3. (*a*) The effect of ligaturing the eyestalks on magnesium U/B ratio. Vertical bars represent standard deviations. (*b*) The effect of ligaturing the eyestalks on the ratio of magnesium U/B to EDTA U/B.

Table 6. *The time course of change in blood and urine magnesium levels*[a] *following eyestalk ligature* ($N = 8$)

	Prior to experiment	1 h	3 h	2 days
Urine	154±26.4	72±19.2	56±14.6	47±8.4
Blood	6±0.95	7±0.45	8±1.06	14±2.34

[a] Means ± S.D. are given.
[b] All values mM Mg^{2+}.

Table 7. *The influence of ligaturing the eyestalks on the clearance of* [^{51}Cr]EDTA *in 100% sea water*

	Weight (g)	Clearance[a] (% body wt day^{-1})	U/B	Urine production (% body wt day^{-1})
Control	0.67±1.65	24.7±3.8 (15)	2.34	10.6±1.6
Ligatured	1.08±1.76	29.1±7.7 (18)	2.94	9.9±2.6

[a] Number of animals given in parentheses.

The effect of ligaturing the eyestalk on magnesium excretion

Prawns from 100% sea water were ligatured with cotton at the base of the eyestalk and then left for a day to recover before being transferred to 125, 100, 70 or 50% sea water. After 3 days in the experimental medium samples of blood and urine were taken for determination of magnesium and 3 μl of [^{51}Cr]EDTA injected. EDTA levels in blood and urine were assayed after a further 40 h.

The haemolymph and urine magnesium concentrations of eyestalk-ligatured prawns differ markedly (Table 5) from those of normal prawns (Table 2), the blood levels rising in all four media and the urine concentrations falling in the more saline media.

The U/B for [^{51}Cr]EDTA is not significantly different between eyestalk-ligatured prawns and normal controls in any of the four media (Fig. 2). The average U/B in each medium is higher than that of the controls, suggesting that there is no evidence for any decrease in water reabsorption of the ligatured individuals. Consequently, it appears that any such ability

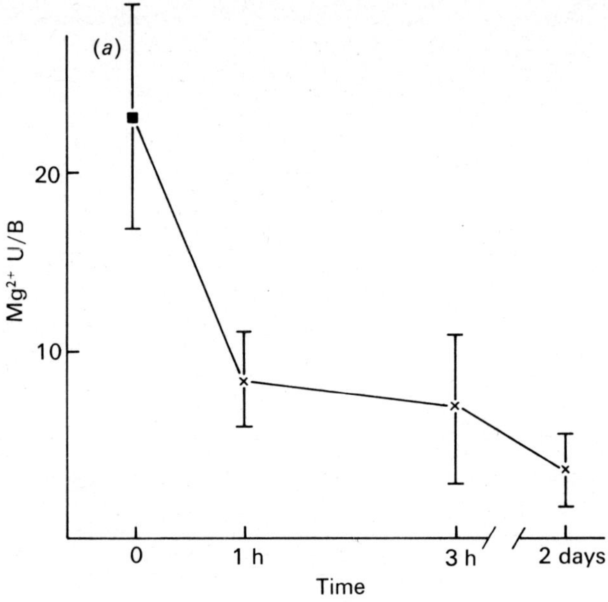

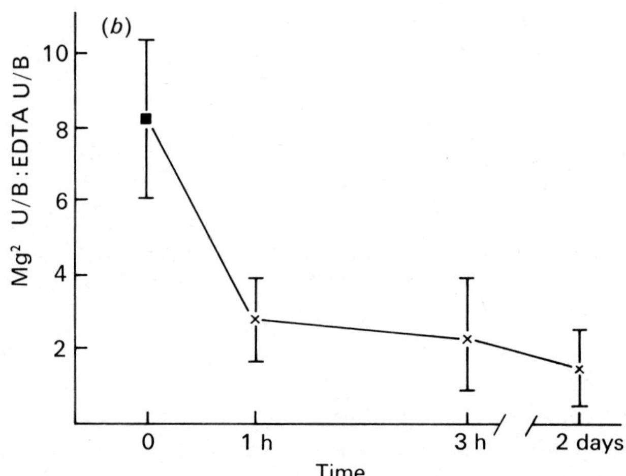

Fig. 4. (*a*) The effect of eystalk ligature with time on the magnesium U/B ratio of prawns in 100% sea water. Vertical bars represent standard deviations. ×, ligatured; ■, control. (*b*) The effect of eyestalk ligature with time on the ratio of magnesium U/B to EDTA U/B. ×, ligatured; ■, control.

that the ligatured animals retain to concentrate magnesium in the urine results essentially from water withdrawal from the primary urine. Correction for this concentrating effect suggests that ligaturing virtually completely blocks net transport of the ion into the urine (Fig. 3*a*, *b*).

Decrease in the excretion rate of magnesium follows fairly rapidly after the eyestalks are ligatured, the urine concentration being halved after 1 h and only *c.* 10 mM above the steady-state value after 3 h (Table 6). The corresponding changes in the magnesium U/B are illustrated in Fig. 4*a*, *b*.

The clearance of [^{51}Cr]EDTA from the haemolymph of ligatured prawns is faster on average than that of controls though the difference is not significant (Table 7).

The urine magnesium concentration and the urine production rate suggests that the rate of elimination of magnesium from the body in ligatured prawns (measured 2 days after the operation) has dropped to about 4.6 mmol kg^{-1} body wt day^{-1} by comparison with the normal rate of 16.3 mmol kg^{-1} body wt day^{-1} despite the higher blood magnesium concentration of the experimental animals and also despite the similarity of the residence time for the urine in the excretory system in the control and experimental animals.

It seems, therefore, that control of magnesium secretion into the urine is exercised directly or indirectly from the eyestalk.

The effect of ethacrynic acid on magnesium excretion

Ethacrynic acid (2,3-dichloro-4-(2-methylene butyryl) phenoxyacetic acid),

$$C_2H_5C-\overset{\overset{O}{\|}}{\underset{\underset{CH_2}{\|}}{C}}\text{—}\langle\text{Cl, Cl ring}\rangle\text{—OCH}_2\text{COOH}$$

Ethacrynic acid

a potent diuretic in vertebrates, has been used in an attempt to elaborate further the processes involved in magnesium secretion by the prawn. Ethacrynic acid tends to bind to thiol groups and acts as a general metabolic poison (Zieve & Solomon, 1968). Among the specific actions attributed to it are the blocking of monovalent cation transport by (*a*) interfering with glycolysis and the production of ATP (Gordon & Hartog, 1971), and (*b*) directly interfering, apparently by competitive interaction with K$^+$, for a transport site (Sachs & Welt, 1968). Duggan & Noll (1965) and Bentley (1969) conclude that it interacts with Na$^+$, K$^+$-ATPase. Poat,

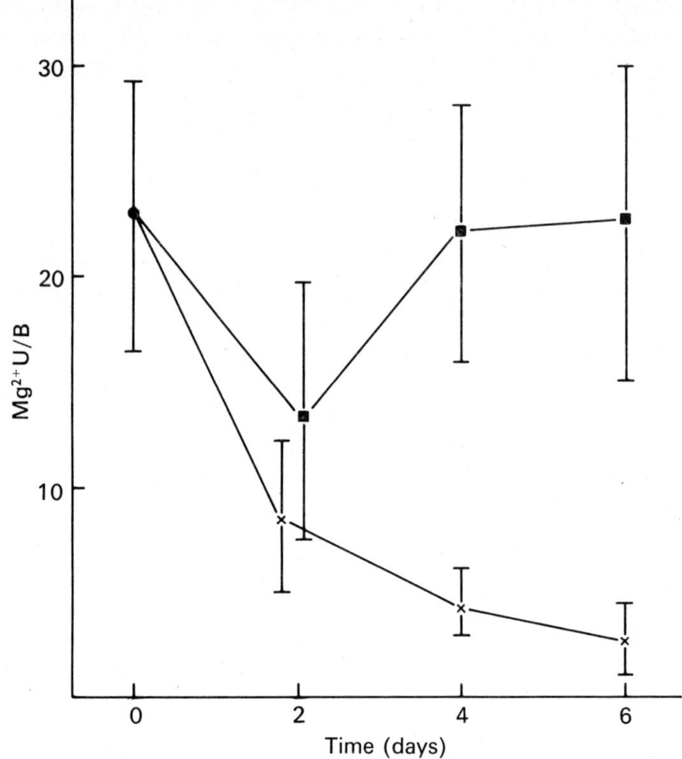

Fig. 5. The effect of repeated injections of ethacrynic acid on magnesium U/B ratio. ●, prior to injection; ×, ethacrynic acid treated; ■, controls. Vertical bars represent standard deviations.

Poat & Munday (1970) find that it inhibits mitochondrial respiration and therefore energy supply but were not able to confirm earlier suggestions that it acts on the alternative (ouabain-insensitive) sodium pump. However, ethacrynic acid does interfere with cell water balance (Zieve & Solomon, 1968). Since Behar (1975) suggests that in the rat intestine magnesium transport correlates well with bulk water movement, experiments were undertaken with the prawn to investigate whether interference with water movement would upset magnesium transport.

Experimental prawns were injected in the haemolymph with 5 μl of 500 p.p.m. ethacrynic acid in 100% sea water; control animals were given 5 μl of 100% sea water. Repeat injections into both control and experimental animals were made following sampling of blood and urine, to ensure that the drug concentration in the body remained at an effective level.

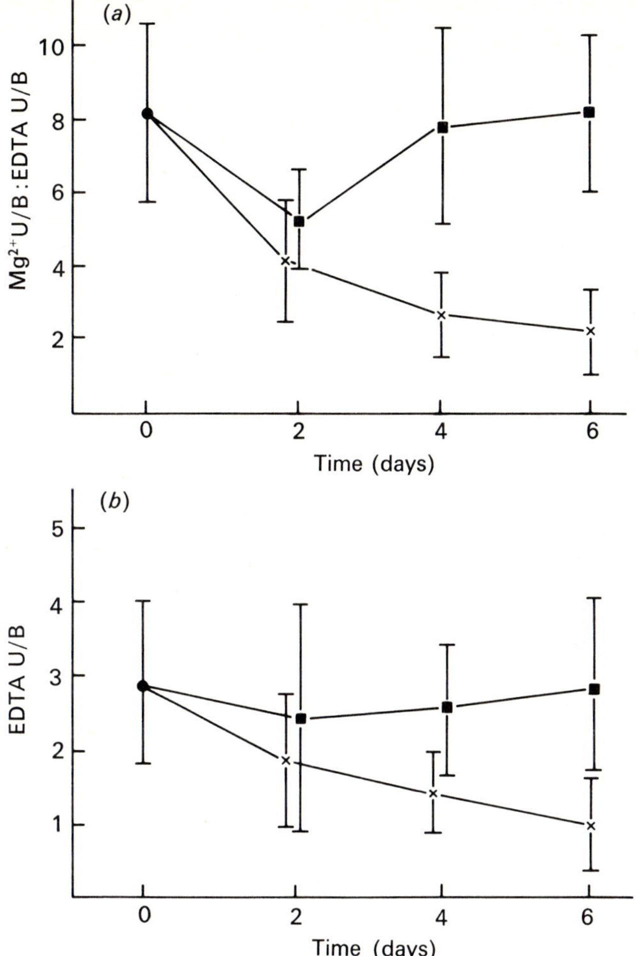

Fig. 6. (*a*) The effect, with time, of repeated injections of ethacrynic acid on the ratio of magnesium U/B to EDTA U/B. (*b*) The effect of repeated injections of ethyacrynic acid on EDTA U/B ratio with time. ●, prior to injection; ×, ethacrynic acid treated; ■, controls. Vertical bars represent standard deviations.

$[^{51}Cr]$EDTA was injected 40 h prior to the first treatment with ethacrynic acid to allow time for equilibration throughout the urinary systems.

Following this treatment, control animals experienced an initial sharp decline in magnesium levels, the U/B ratio falling from 22.8 to 13.4; but recovery then occurred, the ratio rising once more to a level close to the initial value (Fig. 5). By contrast, the magnesium U/B shown by the experimental animals receiving ethacrynic acid not only dropped more sharply initially than that in the controls but continued to fall until the experiment was discontinued. Both rising blood and falling urine magnesium concentrations contribute to the decline in U/B in the treated animals though the transitory fall in U/B in the controls on the second day appeared to be only the result of a decline in urine magnesium concentration.

The decline in [^{51}Cr]EDTA U/B is even more striking in the ethacrynic acid-treated animals than is the decrease in that of magnesium. Ethacrynic acid seems to result in total blockage of net water reabsorption by the sixth day (Fig. 6b). Since some magnesium secretion occurs despite this complete cessation of water movement, it seems unlikely that magnesium transport *per se* is dependent on water movement.

Discussion

Regulation of the magnesium ion level in the haemolymph occurs to a varying extent in all the marine and estuarine crustaceans which have been studied (Robertson, 1949, 1953; Gross, 1964). *Palaemon serratus* maintains the blood magnesium fairly constant, irrespective of the external salinity, at a much lower level that that found in the medium. The urine magnesium level, however, rises markedly with the salinity of the medium. Parry (1954) found a somewhat higher and more variable blood magnesium concentration in the range 50–120% sea water, and a smaller range of urine magnesium concentration than that obtained in this study. The reasons for these differences have not yet been established. It is clear, however, that the prawn has considerable capacity for regulating its blood magnesium level. Ligaturing the eyestalks appears greatly to disturb the mechanism involved in this regulation. Ligaturing sufficiently tightly to preclude blood circulation, as exemplified by rapid darkening of individuals on a pale background (through failure of the supply of chromatophore-controlling hormone), results in a rapid decline of urine magnesium and a concomitant increase in blood magnesium. This is caused by a reduction in the magnesium clearance rate. Ultimately, the urine magnesium concentration of ligatured animals falls to a level equal to, or only marginally greater than, that to be expected from water withdrawal from the primary urine. This indicates that magnesium secretion is under facultative control, and suggests that the eyestalk either is the site of production or storage of a hormone involved in magnesium secretion or is associated with a neural pathway responsible for this. However, Bulleid (1969), working on *Carcinus maenas*, found that ligatured crabs in magnesium-free sea water had a greater magnesium clearance than unligatured crabs. He suggested that a hormonal or neural pathway in the eyestalk caused a reduction in magnesium clearance under these external conditions.

The U/B for the marker [^{51}Cr]EDTA is not greatly affected by eyestalk ligaturing, since, although there is on average a slightly higher U/B at each salinity in experimental animals than in controls, at no salinity is the difference statistically significant. It is improbable, therefore, that eyestalk control of magnesium secretion can be attributed directly to regulation of water reabsorption, although the water movement has an indirect role, as discussed below. It is of course also improbable that the active component involved in concentrating magnesium in the urine could involve solvent drag or bulk flow effects, as has been postulated in vertebrate preparations (Aldor & Moore, 1970; Behar, 1975), since the net water movement and magnesium net transport are in opposite directions. However, it is likely that magnesium already in the excretory system will be passively concentrated by withdrawal of water after formation of the primary urine. The small size (3–9 mV) and incorrect sign (urine positive to blood) of the potential difference obtained when electrodes are inserted into the haemocoel and bladder also make it doubtful that magnesium transport is effected down an electrochemical gradient, unless such gradients are extremely localised. Some form of metabolically driven transport system therefore seems to be indicated.

The effect of the metabolic inhibitor ethacrynic acid appears to support this conclusion. In *Palaemon*, repeated injections of the inhibitor resulted in some inhibition of magnesium transport, as shown by the changes in magnesium U/B and the ratio of magnesium U/B to EDTA U/B. The reduction in the first parameter was caused both by a reduction in urine magnesium concentration and by a concomitant increase in blood concentration of the ion. The decrease in the former results partially from the virtual blockage of water reabsorption, as shown by the substantial decline in the [^{51}Cr]EDTA U/B from 2.95 (prior to injection) to 1.1 (by the sixth day of treatment). The increases in flow rate and volume of urine would serve to reduce the magnesium level by means of a dilution effect and a shorter retention time for the urine in the bladder. However, if no inhibition of magnesium were occurring and if the reduction in urine magnesium were solely the result of a flushing-out effect, the blood magnesium level would not have shown any increase. The decrease to 2.5 in the ratio of magnesium U/B to EDTA U/B shows that magnesium transport itself is substantially inhibited.

The data also show that, as the magnesium U/B increases with increasing salinity, there is a corresponding decrease in the sodium U/B. The latter decrease is caused partly by an increase in blood sodium concentration and partly by a decrease in the urine sodium concentration. Previous authors working with crabs noted that variations in the magnesium concentration in the urine tend to be compensated by inverse changes in the urine sodium level (Robertson, 1949, 1953; Prosser *et al.*, 1955; Green *et al.*, 1959; Riegel & Lockwood, 1961; Gross & Capen, 1966).

Taking into account the sodium reabsorbed in decreasing the urine volume as well as that reabsorbed in order to reduce the sodium concentration of the remaining volume, then more than enough positive charge is moved from urine to blood to compensate for the magnesium being transferred in the opposite direction. The same principle applies in all other salinities, sodium movement (in moles) always being more than twice that of magnesium. It is thus theoretically possible for a forced exchange of magnesium to be occurring, although there is no direct evidence for this. As Gross & Capen (1966) point out, the transfer of the divalent magnesium ion into urine in exchange for two monovalent ions would create an osmotic imbalance resulting in a tendency for water to move in the same direction as sodium, and thus secretion of magnesium should secondarily necessitate water reabsorption. When reabsorption of sodium and water is blocked by ethacrynic acid, the magnesium concentration in the urine falls sharply, although the magnesium secretion is apparently not completely abolished. The converse is not true, however. Sodium and water reabsorption is not dependent on magnesium secretion in the opposite direction; after ligaturing the eyestalks, the [^{51}Cr]EDTA U/B remains normal, or possibly increases slightly, whilst magnesium secretion ceases. If sodium uptake occurs in the absence of magnesium as a counter-ion, then presumably charge imbalance is met by the withdrawal of chloride from the urine and, if this is the case, it may be pre-supposed that the chloride charge is no longer available for exchange with sulphate. Consequently, ligaturing the eyestalks is likely to prevent concentration of sulphate in the urine as well as magnesium, although this hypothesis has yet to be tested.

Theoretically, it is a small step from the withdrawal of sodium and chloride not accompanied by divalent ion secretion in the opposite direction to the production of hypotonic urine by modification of the relative rates of ion and water movement across the epithelium. Whether or not a system originally evolved to transport magnesium into the urine could have been the necessary pre-adaptation for hypotonic urine formation remains an attractive, if highly speculative, concept.

Summary

(1) Measurements have been made of the blood and urine concentration of sodium and magnesium in the prawn *Palaemon serratus* exposed to

salinities in the range 50 –125% sea water. The U/B and clearance of [^{51}Cr]EDTA have also been determined.

(2) The prawn maintains a relatively constant blood magnesium concentration over the range of salinities, urine magnesium concentration increases and sodium U/B decreases with salinity. The mean magnesium U/B in 125% sea water is 33, one of the highest known in the Crustacea.

(3) The U/B for the marker [^{51}Cr]EDTA rises with salinity, implying that water is reabsorbed from the primary urine in the excretory organ, but water reabsorption alone is not sufficient to account for the magnesium U/B values.

(4) Injection of ethacrynic acid blocks water reabsorption, decreasing the [^{51}Cr]EDTA U/B of animals in 100% sea water from the normal average value of 2.7 to 1.1. Magnesium concentration in the urine continues in the presence of ethacrynic acid though at a much reduced rate.

(5) Ligaturing of the eyestalks results in cessation of magnesium secretion, the U/B falling to the level of the [^{51}Cr]EDTA U/B, but without significantly influencing water withdrawal.

We are grateful to Merck, Sharpe & Dohme Ltd for the supply of ethacrynic acid, to Mr M. Lyes for assistance with the potential measurements, to the Natural Environment Research Council for a studentship to S.E.F. and to the College of Fisheries, Kasetsart University, Thailand, for the studentship supporting B.T.

References

Aldor, T. M. & Moore, E. W. (1970). Magnesium absorption by everted sacs of rat intestine and colon. *Gastroenterology*, **59**, 745–53.

Behar, J. (1975). Effect of calcium on magnesium absorption. *American Journal of Physiology*, **229**, 1590–5.

Bentley, P. J. (1969). Actions of vasopressin and aldosterone on the toad bladder: inhibition by ethacrynic acid. *Journal of Endocrinology*, **43**, 347–57.

Binns, R. (1969*a*). The physiology of the antennal gland of *Carcinus maenas* (1). I. The mechanism of urine production. *Journal of Experimental Biology*, **51**, 1–10.

Binns, R. (1969*b*). The physiology of the antennal gland of *Carcinus maenas* (L). II. Urine production rates. *Journal of Experimental Biology*, **51**, 11–16.

Bryan, G. W. & Ward, E. (1962). Potassium metabolism and the accumulation of ^{137}caesium by decapod crustacea. *Journal of the Marine Biological Association, UK*, **42**, 199–241.

Bulleid, N. C. (1969). Investigation of the role of the eyestalks in the excretion of magnesium in *Carcinus maenas*. M.Sc. dissertation, University of Southampton.

Burger, J. W. (1957). The general form of excretion in the lobster *Homarus*.

Biological Bulletin. Marine Biological Laboratory, Woods Hole, Mass., **113**, 207–223.

Duggan, D. E. & Noll, R. M. (1965). Effects of ethacrynic acid and cardiac glycosides upon a membrane adenosine triphosphatase of renal cortex. *Archives of Biochemistry and Biophysics*, **109**, 388–96.

Flemister, L. J. (1958). Salt and water anatomy, constancy and regulation in related crabs from marine and terrestrial habitats. *Biological Bulletin. Marine Biological Laboratory, Woods Hole, Mass.*, **115**, 180–200.

Gordon, E. E. & Hartog, M. de (1971). Localisation and characterisation of the inhibiting action of ethacrynic acid on glycolysis. *Biochemical Pharmacology*, **20**, 2339–48.

Green, J. W., Harsch, M., Barr, L. & Prosser, C. L. (1959). The regulation of water and salt by the fiddler crabs *Uca pugnax* and *Uca pugilator*. *Biological Bulletin. Marine Biological Laboratory, Woods Hole, Mass.*, **116**, 76–8.

Gross, W. J. (1957). An analysis of response to osmotic stress in selected decapod crustacea. *Biological Bulletin. Marine Biological Laboratory, Woods Hole, Mass.*, **112**, 43–62.

Gross, W. J. (1959). The effect of osmotic stress on the ionic exchange of a shore crab. *Biological Bulletin. Marine Biological Laboratory, Woods Hole, Mass.*, **116**, 248–57.

Gross, W. J. (1964). Trends in water and salt regulation among aquatic and amphibious crabs. *Biological Bulletin. Marine Biological Laboratory, Woods Hole, Mass.*, **127**, 447–66.

Gross, W. J. & Capen, R. L. (1966). Some functions of the urinary bladder in a crab. *Biological Bulletin. Marine Biological Laboratory, Woods Hole, Mass.*, **131**, 272–91.

Gross, W. J., Lasiewski, R. C., Dennis, M. & Rudy, P. (1966). Salt and water balance in selected crabs of Madagascar. *Comparative Biochemistry and Physiology*, **17**, 641–60.

Gross, W. J. & Marshall, L. A. (1960). The influence of salinity on the magnesium and water fluxes of a crab. *Biological Bulletin. Marine Biological Laboratory, Woods Hole, Mass.*, **119**, 440–53.

Kamemoto, F. I. & Ono, J. K. (1969). Neuroendocrine regulation of salt and water balance in the crayfish, *Procambarus clarkii*. *Comparative Biochemistry and Physiology*, **29**, 393–401.

Lockwood, A. P. M. & Inman, C. B. E. (1973). Water uptake and loss in relation to salinity of the medium in the amphipod *Gammarus duebeni*. *Journal of Experimental Biology*, **58**, 149–63.

Lockwood, A. P. M. & Riegel, J. A. (1969). The excretion of magnesium by *Carcinus maenas*. *Journal of Experimental Biology*, **51**, 575–89.

Panikkar, N. K. (1941). Osmoregulation in some palaemonid prawns. *Journal of the Marine Biological Association, UK*, **25**, 317–59.

Parry, G. (1954). Ionic regulation in the palaemonid prawn *Palaemon* (= *Leander*) *serratus*. *Journal of Experimental Biology*, **31**, 601–13.

Poat, P. C., Poat, J. A. & Munday, K. A. (1970). The site of action of the diuretic ethacrynic acid on rat kidney and liver tissue. *Comparative and General Pharmacology*, **1**, 400–8.

Prosser, C. L., Green, J. W. & Chow, R. (1955). Ionic and osmotic concentrations in blood and urine of *Pachygrapsus crassipes* acclimated to different salinities. *Biological Bulletin. Marine Biological Laboratory, Woods Hole, Mass.*, **109**, 99–107.

Proverbio, F. R., Robinson, J. W. L. & Whittembury, G. (1969). Sensitivities of Na^+, K^+ ATPase and sodium extrusion mechanism to ouabain and ethacrynic acid in the cortex of the guinea pig kidney. *Biochimica et biophysica Acta*, **211**, 327–36.

Riegel, J. A. & Lockwood, A. P. M. (1961). The role of the antennal gland in the osmotic and ionic regulation of *Carcinus maenas*. *Journal of Experimental Biology*, **38**, 491–9.

Riegel, J. A., Lockwood, A. P. M., Norfolk, J. R. W., Bulleid, N. C. & Taylor, P. A. (1974). Urinary bladder volume and the reabsorption of water from the urine of crabs. *Journal of Experimental Biology*, **60**, 167–81.

Robertson, J. D. (1949). Ionic regulation in some marine invertebrates. *Journal of Experimental Biology*, **16**, 182–200.

Robertson, J. D. (1953). Further studies on ionic regulation in marine invertebrates. *Journal of Experimental Biology*, **30**, 277–96.

Sachs, J. R. & Welt, L. G. (1968). Concentration dependence of active potassium transport in the human red blood cell in the presence of inhibitors. *Journal of Clinical Investigation*, **47**, 949–59.

Teinsongrusmee, B. (1976). Aspects of osmoregulation in the common prawn, *Palaemon serratus* (Pennant, 1777). Ph.D. Thesis, University of Southampton.

Webb, D. A. (1940). Ionic regulation in *Carcinus maenas*. *Proceedings of the Royal Society*, **129**, 107–36.

Wen, S.-F., Evansen, R. L. & Dirks, J. H. (1970). Micropuncture study of renal magnesium transport in proximal and distal tubules of the dog. *American Journal of Physiology*, **219**, 570–6.

Wood, J. L., Jungreis, A. M. & Harvey, W. R. (1975). Active transport of magnesium across the isolated mid gut of *Hyalophora cecropia*. *Journal of Experimental Biology*, **63**, 313–20.

Zieve, P. D. & Solomon, H. M. (1968). Effects of diuretics on the human platelet. *American Journal of Physiology*, **215**, 650–4.

13. Electrolyte transport across the bird coprodeum: role in osmoregulation

By ERIK SKADHAUGE

Birds are, in general, not exposed to an 'unbalanced' external environment. But the lower end of their gut certainly encounters an extremely 'unbalanced' internal environment. The reason for this is the mixing of ureteral urine and faeces (Fig. 1) which takes place in the coprodeum and large intestine (Skadhauge, 1968). The coprodeum is the stage for final salt and water conservation both from renal and intestinal sources. The variable composition of the contents of the coprodeum and large intestine, largely influenced by the variable rate of urine flow and composition of ureteral urine, heavily influences the transmural transport rates in these organs.

The storage of ureteral urine in the lower intestine is, in this survey, assessed from the point of view of its role in osmoregulation. How much of the solutes and water of ureteral urine is absorbed in the intestine under various osmotic situations; and how are the transport parameters of the cloacal epithelium regulated? The discussion is based mainly on observations by the author and co-workers on the cloacal transport parameters in the domestic fowl and in a desert bird, the galah (*Cacatua roseicapilla*). As a necessary background, results on composition of the ureteral urine are included. Quantitative computer calculations have been carried out to understand the interaction of the parameters.

A strong interaction of renal and post-renal functions, modifying the final excretion of solutes and water, is by no means unusual in the lower vertebrates: in birds, reptiles, amphibians, and fish such interactions occur (Skadhauge, 1977). In birds, however, the problem of post-renal modification of ureteral urine – with important implications for the animal's osmoregulation – is amplified by the presence of a concentrating–diluting kidney, which is capable of delivering fluids of widely varying composition to the lower intestine.

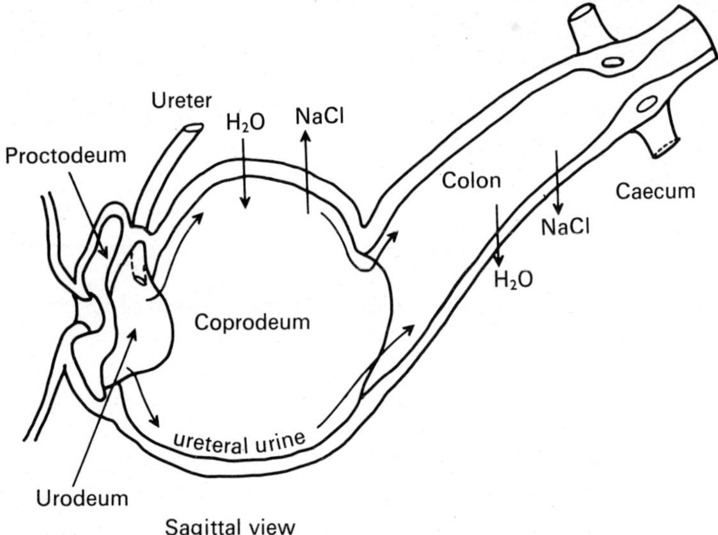

Fig. 1. A schematic drawing of the lower intestine in the domestic fowl. The urine is from the ureteral orificia in the urodeum regurgitated into the coprodeum and large intestine; the uric acid and urates are deposited on the outside of a central core of faeces. The majority of the precipitate is located in the coprodeum. A minor fraction of ureteral urine, up to 20 % of ureterally excreted PEG 4000 (Skadhauge, 1973), is moved as far backwards as the caeca. No absorption is assumed to take place in the urodeum (and proctodeum, where urine/faeces are not stored) because a multilayer squamous epithelium is present. The coprodeum/colon segment has a single layer of tall columnar cells on large irregular villi. This is thus a typical 'resorptive' epithelium. When the hyperosmotic urine in the dehydrated state comes into the coprodeum, water will run into the gut by osmosis. This causes dilution of the fluid, which is further diluted by NaCl absorption. When the fluid reaches the colon, water will be absorbed as solute-linked water flow from the luminal fluid which now has a lower osmolarity than ureteral urine.

Variations in flow rate and composition of ureteral urine

As in mammals, the rate of urine flow is (according to the state of hydration of the bird) roughly inversely proportional to the osmolarity (Skadhauge & Schmidt-Nielsen, 1967). The urine flow is regulated by an neurohypophysial hormone, arginine–vasotocin (Ames, Steven & Skadhauge, 1971). Since, obviously, the relative role of cloacal salt and water absorption/secretion must be quite different depending on whether the bird is hydrated, salt-loaded, or dehydrated, these three osmotic states must be considered separately. The most important problem will be the possible flow of water and salt back to the animal, during hydration and salt loading, respectively, and loss of water (and salt) during dehydration (since dehydrated seed-eaters often lack NaCl). Furthermore, seed-eating birds have, particularly in the dehydrated state, an unusual composition of the osmolarity of ureteral urine. More than 50% of this is constituted by potassium ammonium phosphate (Skadhauge, 1977).

In-vivo measurements of cloacal transport parameters in the domestic fowl

The cloacal absorption rates of NaCl and other solutes and water have been measured both as functions of luminal osmolarity and NaCl concentration and as functions of hydration and NaCl balance of the animal.

The net absorption of sodium was observed to follow saturation kinetics (Bindslev & Skadhauge, 1971a). It can therefore be characterised by a maximal flow rate 'V_{max}' and a luminal sodium concentration for half maximal absorption rate ('K_m'). Dehydration was not observed to change 'V_{max}', but this parameter was apparently halved by sodium-loading the animal (Skadhauge, 1967; Thomas, Skadhauge & Read, 1975). Aldosterone was observed to restore the sodium absorption partly (see later).

Potassium was secreted into the intestine (Skadhauge, 1967; Bindslev & Skadhauge, 1971a, b; Thomas et al., 1975); the flow rate was higher in conjunction with a higher sodium absorption rate, both when this was augmented because of a higher intraluminal sodium concentration and as a result of sodium depletion of the animal. At high luminal potassium concentrations, potassium was absorbed (Skadhauge, 1967).

Chloride largely followed sodium, but at a lower rate. At low luminal chloride concentrations, chloride was even secreted into the intestine. The net result of the flow of the strong electrolytes was, in many experiments, that the net movement of electric charge was close to zero ($Na^+ - K^+ = Cl^-$).

Other abundant ions in solution in dehydrated birds urine are, as noted above, NH_4^+ and phosphate. Phosphate was not subject to any significant absorption in the lower gut, whereas NH_4^+ was absorbed to a minor degree, proportional to the luminal NH_4^+ concentration (Skadhauge & Thomas, in preparation).

Not only did the transmural water flow seem to follow the general osmotic driving force, but a so-called solute-linked water flow was also present. This is water absorption measured in the absence of or even against a transmural osmotic difference. In the domestic fowl this occurred against 65 mOsm (Skadhauge, 1967). In the absence of sodium from the luminal perfusion fluid, water flowed in the direction of the general driving force; osmotic permeability coefficients can therefore be calculated for the cloacal wall (Bindslev & Skadhauge, 1971*b*). The solute-linked water flow was, at osmolarities far from plasma osmolarity, observed to disappear in the general osmotic water flow; only close to plasma osmolarity (± 100 mOsm) was a solute-linked water flow detectable (Bindslev & Skadhauge, 1971*a*). The solute-linked water flow was observed to be augmented in dehydrated birds as compared to hydrated birds.

Quantitative role in osmoregulation of the cloacal storage of ureteral urine in the domestic fowl

A quantitative assessment of the degree of modification of ureteral output of solutes and water can be made, based on the observations decribed in the previous section. Because of the large ureteral flow of water in the hydrated state, and the flow of NaCl in the salt-loaded state, the rate of cloacal absorption of water and NaCl amounted in these two osmotic states, respectively, to only 2% of the ureteral output (Skadhauge, 1973).

In the dehydrated state, however, the situation is quite different: a small amount of hyperosmotic urine runs into the intestine. The problem is whether this inflow leads to a net loss of water across the cloacal wall or a fraction of the urine is diluted so much, as it moves retrogradely into the intestinal tract, that the solute-linked water flow takes over and creates a net water absorption (Fig. 1). Some urine reaches as far, retrogradely, as the top of the caeca (Skadhauge, 1968). The suggested solute-linked water absorption will naturally be at the expense of NaCl absorption as this is the driving force for the water flow. A net water absorption seemed to occur, as judged both by slow infusion experiments with fluids simulating the osmolarity of dehydrated chicken urine (Bindslev & Skadhauge, 1971*a*) and by computer simulations of retrograde flow (Skadhauge & Kristensen, 1972). The cloacal water conservation was calculated to be 15% of ureteral flow and the NaCl absorption to be 70%.

Role in osmoregulation of cloacal storage of ureteral urine in a desert bird

The conclusion mentioned in the previous section was for a bird which is a feeble concentrator of ureteral urine; and the presence of some NaCl in the ureteral urine was assumed. Xerophilic seed-eaters, however, concentrate urine to a higher osmolarity (higher urine to plasma osmotic ratio) and contains less NaCl in the ureteral urine (Skadhauge, 1974*a*). In these birds will the cloacal solute-linked water flow be large enough to result in a net gain of water during the storage of ureteral urine? If, on

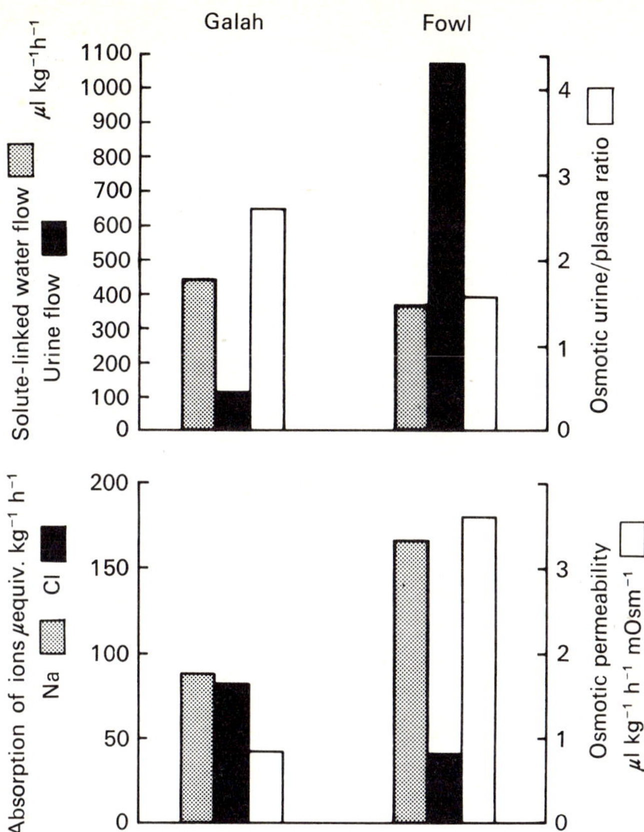

Fig. 2. Cloacal transport parameters in the dehydrated domestic fowl and galah. The parameters rate of sodium and chloride absorption from iso-osmotic NaCl perfusion solutions, solute-linked water flow (measured from the same solutions), and osmotic permeability coefficients in the serosa to mucosa (s–m) direction are shown, together with the crucial renal parameters of the dehydrated birds, rate of urine flow and osmolarity (osmotic urine to plasma ratio). (From Schmidt-Nielsen & Skadhauge, 1967; Bindslev & Skadhauge, 1971*a*, *b*; Skadhauge, 1974*a*.) The galah can, in spite of a high urine osmolarity, resorb a fraction of ureterally excreted H_2O similar to that resorbed by the fowl (Skadhauge, 1974*b*).

a weight basis, the cloacal transport parameters of the chicken were used for the ureteral urine entering the intestine in the budgerigar, this bird would lose too much water to live (Krag & Skadhauge, 1972). It survives very well, however, on dry seeds alone. To solve this problem a larger xerophilic Australian seed-eater, the galah (*Cacatua roseicapilla*), was studied. In Fig. 2 a number of renal and cloacal transport parameters of this bird have been compared with those of the domestic fowl. Sodium absorption in the galah also followed saturation kinetics (Skadhauge, 1974*b*). On a weight basis, a large difference in NaCl handling was not apparent in the relevant concentration range. In spite of the high urine osmolarity, the higher rate of solute-linked water flow, and the lower osmotic permeability coefficient did, however, permit a similar fractional absorption of water and NaCl in the galah, as it did in the chicken. From the point of view of adaptation to arid lands, a reno-intestinal circulation of salt makes sense. Birds, being largely uricotelic, pass large amounts of supersaturated colloid suspensions of uric acid and urates. NaCl helps the urates to stay in suspension (Porter, 1963). The seed-eaters need this sodium, since they receive a low-sodium diet; the ion is taken back in the cloaca by a regulated process (see below). Together with Na^+ (and presumably Cl^-) a small amount of water is saved.

Recent studies on mechanisms of regulation of cloacal epithelial transport

Recent in-vivo perfusion experiments of the coprodeum and large intestine of the domestic fowl (Thomas *et al.*, 1975) have shown a strong dependence of net sodium absorption rate on the sodium content of the diet. The response was graded. Going from a diet of wheat and barley to commercial chicken pellets, which have 0.5 M NaCl included, the net sodium absorption rate decreased to ⅓. A further reduction by 40% occurred after an extra four-fold increase in the NaCl load. In the two salt-loaded groups, chronic aldosterone injections were observed to double the sodium absorption rate. In these groups the potassium secretion rate was also doubled on aldosterone treatment. While perfusion experiments were in progress, acute injections of aldosterone augmented the sodium absorption measurably after 1 h and maximally after 4 h. *In vitro*, however, no effect of aldosterone after addition to the medium could be demonstrated (Thomas, 1973). The effect of this hormone is thus, most likely, the result of an effect through cellular protein synthesis, etc. rather than an effect on simpler passive parameters (see below).

According to the in-vivo findings described above, the total sodium transport of the functional resorptive segment of the gut is influenced by the sodium balance of the bird, but the restoration of the sodium transport rate on aldosterone was only partly to the level observed in sodium-depleted birds.

Recent experiments *in vitro* have, however, shown that the sodium transport of the coprodeum has a much higher sensitivity to the sodium balance of the animal. Choshniak, Munck & Skadhauge (1976) isolated the

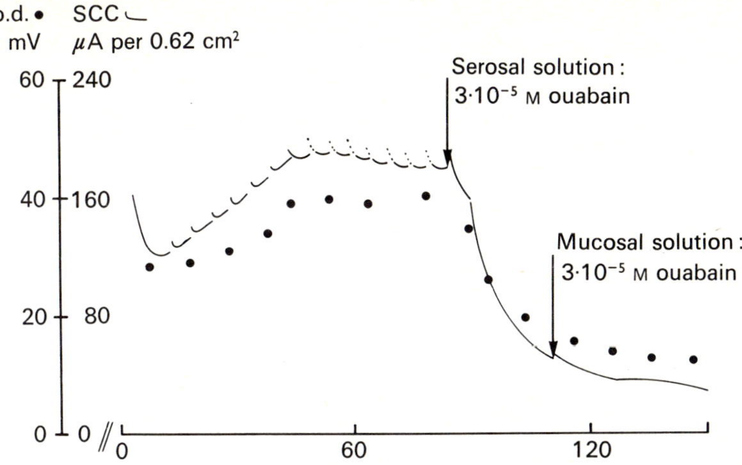

Fig. 3. Short circuit current (SCC) and electric potential difference (p.d.) of the isolated coprodeum of the fowl. These parameters are shown as functions of time after mounting of the tissue in the Ussing-chamber (Lyngdorf-Henriksen, Munck & Skadhauge, in preparation). The p.d. is recorded during unclamping of the preparation. Addition of ouabain to the serosal side inhibits at least 95 % of the sodium transport, whereas no effect can be demonstrated after mucosal application.

mucosa of the coprodeum. After mounting in the Ussing-chamber in a Krebs–Ringer's phosphate buffer with 15 mM glucose, the coprodeum of sodium-depleted birds showed a high transepithelial electric potential difference (p.d.) and a high short-circuit current (SCC), which remained stable for hours (Fig. 3). The sodium depletion was achieved by feeding a low-sodium diet (wheat and barley) for at least a week. The commercial chicken food used as control contains some NaCl. A high NaCl load was given by adding NaCl to the commercial diet. The SCC and p.d. were equally sensitive to the sodium balance (Table 1), both being almost suppressed by commercial chicken food, with little additional effect of further NaCl loading.

The SCC was carried by the net sodium absorption as the SCC was 10.6 ± 0.4 μequiv. cm⁻² h⁻¹ in the birds receiving wheat and barley, and the difference between the two unilateral sodium fluxes was 11.8 μequiv. cm⁻² h⁻¹ (Table 1). The net sodium absorption was reduced to zero in the higher sodium diets. The flux ratio for chloride was unity, and the chloride fluxes were halved in the birds receiving the higher sodium loads.

The sodium inflow across the luminal plasma membrane was measured with the Schultz–Curran technique. The inflow in the sodium-depleted birds was not distinguishable from the isotope flow in the mucosa to serosa

Table 1. *Transport parameters of the isolated coprodeum of the domestic fowl[a]*

	Diet[b]			
	Low Na (wheat + barley)	Medium Na commercial food	High Na commercial food + NaCl	Units
Electrical potential difference	37.3±2.2	4.7±0.6	2.3±0.5	mV
Short-circuit current	10.6±0.4	0.6±0.1	0.4±0.1	μequiv. cm^{-2} h^{-1}
Mucosa to serosa sodium flux	13.2±0.5	1.5±0.2		μequiv. cm^{-2} h^{-1}
Serosa to mucosa sodium flux	1.4±0.2	1.4±0.2		μequiv. cm^{-2} h^{-1}
Sodium inflow across the luminal membrane	11.9±1.0	8.5±0.8	—	μequiv. cm^{-2} h^{-1}

[a] From Choshniak *et al.*, 1976, and Lyngdorf-Henriksen, Munck & Skadhauge, in preparation.
[b] Means ±s.e. are reported.

direction. It was only reduced by 29% in the birds receiving commercial chicken food. It was further observed that addition of ouabain to the serosal side reduced the SCC by at least 95% (Fig. 3). Ouabain was without effect after addition to the mucosal side. Amiloride immediately reduced the SCC to zero, after application from the mucosal side; no effect was observed after addition to the serosal side. Furthermore, the chloride fluxes were unchanged when the chloride of the medium was replaced by bicarbonate showing absence of an obligatory chloride/bicarbonate exchange. The sodium inflow was the same when the chloride of the medium was replaced by sulphate, showing independence of the sodium and chloride inflows.

Voltage-clamp experiments demonstrated that the unilateral sodium flux from serosa to mucosa varied linearly with the electrical driving force, after initially clamping the serosa negative. Furthermore, the flux of sodium in the mucosa to serosa direction showed saturation kinetics as a function of total sodium concentration of the bath, but with a much lower 'K_m' than measured with the in-vivo preparation. Also, in this system the flux of sodium in the serosa to mucosa direction was linearly related to the driving force, the concentration.

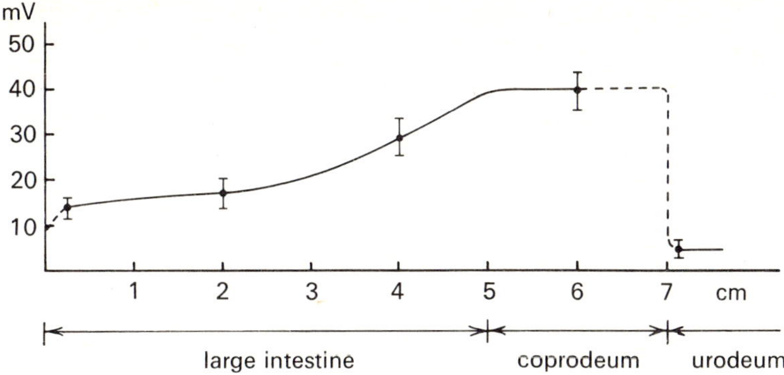

Fig. 4. Transmural electric potential difference (p.d.) of the colon (large intestine) and coprodeum of the domestic fowl *in vivo*. The measurements were carried out in fasted birds which still had the natural semi-solid ureteral urine in the lower gut. The figure is based upon measurements published in less detail by Bindslev & Skadhauge (1971*b*). Note that the p.d. (lumen negative) was much lower in the colon than in the coprodeum.

The picture which emerges from the observations of mechanism of sodium absorption in the coprodeum and its regulation is that there is an active cellular sodium transport. There seems to be a paracellular shunt pathway for chloride and backflow of sodium; the sodium inflow across the luminal membrane is not coupled to chloride. Adaptation to a low sodium diet is, at least partly, mediated through aldosterone, but not primarily via a change of the permeability of the luminal membrane.

The results of the recent experiments *in vitro* are somewhat different from those of the studies *in vivo*. This may be because of the fact that the studies *in vivo* were carried out on both colon and coprodeum. Two lines of evidence now seem to indicate some functional difference between colon and coprodeum. First, the electric potential difference across the gut wall was much lower in the colon than in the coprodeum, as measured *in vivo* (Fig. 4). Although this finding could be explained by a smaller difference in the activity of an electrogenic sodium pump from colon to coprodeum (Skadhauge, 1973), another interpretation would be that the pump is less electrogenic in the colon than in the coprodeum. This might, for example, be the result either of a lower resistance of the epithelium or of a coupled NaCl inflow. Secondly, preliminary Ussing-chamber experiments (Munck & Skadhauge, in preparation) on isolated mucosa from the colon have shown that the resistance of the wall is lower than in the coprodeum. Furthermore, the SCC of the colon was less influenced by the sodium balance of the animal than that of the coprodeum. The sodium transport of the colon was augmented by the presence of amino acids. The colon thus seems to show the properties of small intestine, as known from mammalian studies: low p.d., little influence of mineralocorticoid hormones, low resistance, and augmentation of sodium transport by the presence of amino acids. The coprodeum behaves similarly to mammalian colon having a high p.d., being influenced by mineralocorticoid hormones, little influenced by the presence of amino acids and having a high resistance. The very pronounced sensitivity of sodium transport to the sodium balance of the animal, as observed, is characteristic for such an epithelium.

The demonstrated difference between coprodeum and colon will thus explain the much smaller variation of sodium absorption observed *in vivo* as a function of the sodium balance of the animal, particularly the failure of sodium-loading to suppress net sodium absorption. Since the majority of uric acid and urates are deposited in the coprodeum (Fig. 1), it can be assumed that the fractional difference in absorption pattern from coprodeum to colon is more important than has been previously understood. The colon may well be viewed as an organ largely responsible for the final resorption of salt and water from the chymus. This should function even in the sodium-loaded state. The coprodeum, on the contrary, is the final post-renal 'afterburner' which, through adjustment of sodium transport, precisely regulates the NaCl balance of the animal through absorption from ureteral urine. The great sensitivity of coprodeal sodium absorption revealed in the recent in-vitro studies of the fowl shows that the function of this organ is similar to that of the lizard coprodeum. This also shows a large variation in absorption of sodium from ureteral urine (Bradshaw, 1975).

The work was supported by the Danish Natural Science Research Council and the NOVO foundation.

Expert technical assistance was rendered by Mrs Dorthe Davidsen and Mrs Kirsten Møller.

References

Ames, E., Steven, K. & Skadhauge, E. (1971). Effects of arginine–vasotocin on renal excretion of Na$^+$, K$^+$, Cl$^-$, and urea in the hydrated chicken. *American Journal of Physiology*, **221**, 1223–8.

Bindslev, N. & Skadhauge, E. (1971*a*). Sodium chloride absorption and solute-linked water flow across the epithelium of the coprodeum and large intestine in the normal and dehydrated fowl (*Gallus domesticus*). *In vivo* perfusion studies. *Journal of Physiology*, **216**, 753–68.

Bindslev, N. & Skadhauge, E. (1971*b*). Salt and water permeability of the coprodeum and large intestine in the normal and dehydrated fowl (*Gallus domesticus*). *In vivo* perfusion studies. *Journal of Physiology*, **216**, 735–51.

Bradshaw, S. D. (1975). Osmoregulation and pituitary-adrenal function in desert reptiles. *General and Comparative Endocrinology*, **25**, 230.

Choshniak, I., Munck, B. G. & Skadhauge, E. (1976). Characteristics of electrolyte transport across the isolated coprodeal epithelium of the chicken. *Acta physiologica scandinavica, Supplement* **440**, 76.

Krag, B. & Skadhauge, E. (1972). Renal salt and water excretion in the budgerygah (*Melopsittacus undulatus*). *Comparative Biochemistry and physiology*, **41A**, 667–83.

Porter, P. (1963). Physico-chemical factors involved in urate calculus formation. II. Colloidal flocculation. *Research in Veterinary Science*, **4**, 592–602.

Skadhauge, E. (1967). *In vivo* perfusion studies of the water and electrolyte resorption in the cloaca of the fowl (*Gallus domesticus*). *Comparative Biochemistry and Physiology*, **23**, 483–501.

Skadhauge, E. (1968). Cloacal storage of urine in the rooster. *Comparative Biochemistry and Physiology*, **24**, 7–18.

Skadhauge, E. (1973). Renal and cloacal salt and water transport in the domestic fowl (*Gallus domesticus*). *Danish Medical Bulletin, Supplement* 1, **20**, 1–82.

Skadhauge, E. (1974*a*). Renal concentrating ability in selected West Australian birds. *Journal of Experimental Biology*, **61**, 269–76.

Skadhauge, E. (1974*b*). Cloacal resorption of salt and water in the Galah (*Cacatua roseicapilla*). *Journal of Physiology*, **240**, 763–73.

Skadhauge, E. (1977). Solute composition of the osmotic space of ureteral urine in dehydrated chickens (*Gallus domesticus*). *Comparative Biochemistry and Physiology*, **56A**, 271–4.

Skadhauge, E. (1977). Excretion in lower vertebrates: Function of gut, cloaca and bladder in modifying the composition of the urine. *Federation Proceedings*. (In Press.)

Skadhauge, E. & Kristensen, K. (1972). An analogue computer simulation of cloacal resorption of salt and water from ureteral urine in birds. *Journal of Theoretical Biology*, **35**, 473–87.

Skadhauge, E. & Schmidt-Nielsen, B. (1967). Renal function in domestic fowl. *American Journal of Physiology*, **212**, 793–8.

Thomas, D. H. (1973). Control of active ion and water transfer in the chicken rectum: in-vitro effects of aldosterone. *Journal of Endocrinology*, **59**, xxxiii.

Thomas, D. H., Skadhauge, E. & Read, M. W. (1975). Steroid effects on gut function in birds. *Biochemical Society Transactions*, **3**, 1164–8.

14. Salt glands in marine birds: what triggers secretion and what makes them grow?

By M. PEAKER

As is now well known, the nasal salt glands of marine birds, and of those inhabiting inland salt-lakes, are able to secrete a hypertonic solution consisting mainly of NaCl (see Schmidt-Nielsen, 1960). In this article I shall consider two aspects of salt-gland function on which evidence, more recent than that dealt with in the monograph by Peaker & Linzell (1975), is now available, namely the nature of the coupling process between the gross physiological stimulation of the gland and the cellular mechanisms of secretion, and the process by which growth of the glands is induced when a bird drinks salt water for the first time. Since the salt glands of domestic ducks (*Anas platyrhynchos*) and geese (*Anser anser*) retain the ability to secrete (nasal fluid appearing about 2 min after the administration of an intravenous salt load), these species have been used extensively in studies of salt-gland physiology, including the ones described here.

Excitation–secretion coupling

Following the early work of Fänge, Schmidt-Nielsen & Robinson (1958), it is now well established that secretion is initiated and maintained by a nervous secretory reflex, the efferent limb being cholinergic. From a series of experiments involving infusion, perfusion and cross-circulation studies, Hanwell, Linzell & Peaker (1972) concluded that the receptors are located in the heart or great vessels, with afferent fibres running in the vagus nerves (Fig. 1). It is also now well established that the receptors detect an increase in plasma tonicity rather than changes in the concentrations of specific ions or an alteration in blood volume (Hanwell *et al.*, 1972; Peaker & Linzell, 1975).

The question arises of how acetylcholine, released from nerve terminals on the basal side of the cell, stimulates ion transport for secretion. The scheme for ion transport in the salt gland of marine birds suggested by Peaker (1971) is shown in Fig. 2; the evidence on which this scheme is based is considered in detail by Peaker & Linzell (1975). There, it was argued that neither an increase in the passive permeability of the basal membrane, nor activation of the sodium transport mechanisms, nor an

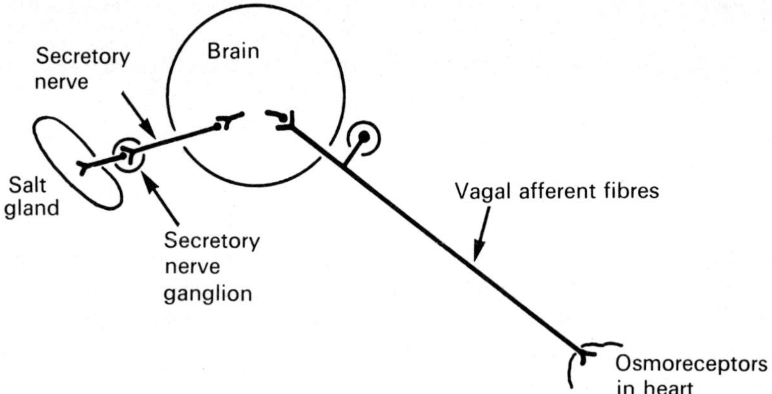

Fig. 1. The proposed reflex arc for stimulation of the salt gland (from Hanwell *et al.*, 1972).

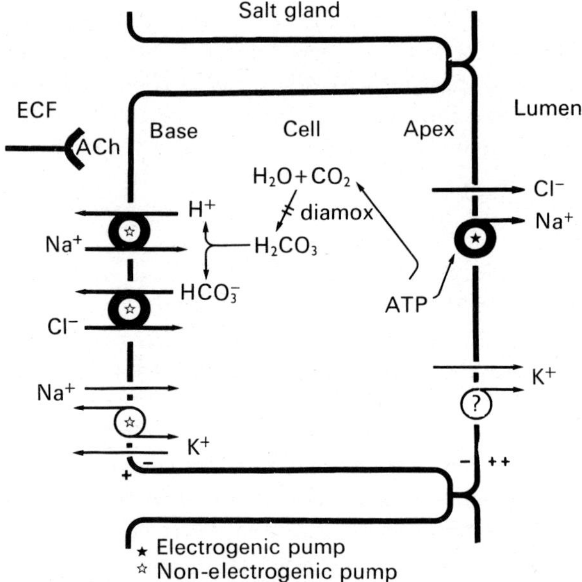

Fig. 2. Suggested scheme for ion transport in the salt gland. Ach = acetylcholine; ECF = extracellular fluid; diamox = the carbonic anhydrase inhibitor acetazoleamide (from Peaker, 1971, with permission).

increase in the intracellular sodium concentration is responsible for stimulating the proposed sodium pump on the luminal membrane. In fact there is some indication that activation of this luminal pump is the primary event associated with the onset of secretion. If this is so, then it can be assumed either that acetylcholine enters the cell (an unlikely possibility) or that some other agent acts as an intracellular 'second messenger'.

Van Rossum (1966) showed that the efflux of ^{24}Na from salt-gland slices previously loaded with ^{24}Na could be stimulated by the cholinomimetic agent methacholine. Therefore this technique has been used to investigate the coupling between cholinergic stimulation and secretion. It should also be pointed out that the efflux of ^{36}Cl was also studied in these experiments and, in all cases, the fluxes of sodium and chloride were parallel; this strengthens the interpretation that the increased efflux of sodium in response to methacholine indicates activity of the luminal sodium pump, with chloride following passively.

The stimulatory effect of metacholine on sodium efflux was almost completely abolished when calcium was omitted from the fluid flowing over the slices (EGTA was added to chelate residual extracellular calcium). In addition the effect of methacholine was mimicked by suddenly raising the external calcium concentration from 2.6 to 9.6 mM, or by treating the slices with a calcium ionophore. Therefore these preliminary studies suggest that an increased intracellular concentration of calcium ions acts to initiate secretion, and that acetylcholine induces an increase in calcium movements into the cell across the basal membrane and possibly also stimulates the release of calcium from the inner side of the cell membrane. Of course, if secretion is mediated by calcium, another question is raised as to how calcium initiates transport at the biochemical level.

The possibility that other putative 'second messengers' could be involved has also been investigated by treating slices with a wide range of concentrations of cyclic-GMP and cyclic-AMP in the monobutyryl and dibutyryl forms. Cyclic-GMP had no detectable effect on sodium efflux, while cyclic-AMP caused a very small increase, in no way comparable in magnitude to that obtained with a cholinomimetic. In agreement with these results, that cyclic nucleotides are not the major mediator of excitation–secretion coupling in the salt gland, is the finding that secretion was not obtained following the infusion of dibutyryl cyclic-AMP at high concentrations into the carotid artery of geese (Peaker, 1971).

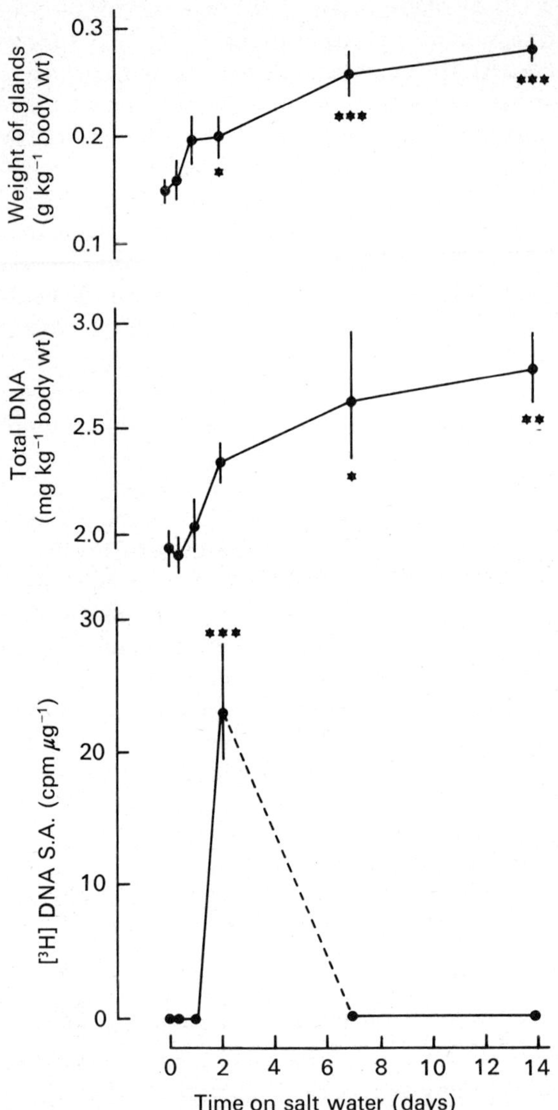

Fig. 3. Changes in salt-gland weight, total DNA content and [³H]DNA specific activity (SA) in ducks given 0.3 M NaCl as their only source of drinking water. [³H]thymidine was given 3 h before the birds were killed. Mean ±s.e.m. *P* values (compared with freshwater group). * < 0.05; ** < 0.01; *** < 0.001.

Table 1. *Summary of events occurring in the salt gland after a bird first drinks salt water (for details see text and Peaker & Linzell, 1975)*

Process	Cells involved	Stimulus to the cell[a]	Intracellular messenger	Time scale
Secretion	Secretory cells	Acetylcholine	Calcium	Seconds–minutes
Vasodilatation	Blood vessels	Acetylcholine and/or local dilators	—	Seconds–minutes
Hypertrophy	Secretory cells	Acetylcholine	?	Hours–days
Hyperplasia (increase in mitotic rate)	Peripheral cells	?	?	Days

[a] Possibility of hormonal modulation also exists.

Adaptive growth

While the salt glands will secrete within minutes of the first administration of salt water, continued ingestion leads to an increase in their size and efficiency such that the concentration of NaCl in the secretion and the output, both in terms of the volume secreted per gland and per unit weight of tissue, increase. A number of workers have studied the changes that take place and their time course. The cells increase in size and there are marked biochemical and cytological changes, including an early increase in the total RNA content of the glands and in the RNA:DNA ratio (see Peaker & Linzell, 1975).

Since the rise in the RNA content of the glands occurs during the first day on salt water, it must be assumed that the 'trigger' for hypertrophy must operate shortly after a bird first ingests salt water. Recent studies involving denervation or the blocking of secretion with atropine have shown that nervous stimulation of the gland is normally responsible for inducing adaptive hypertrophy (Hanwell & Peaker, 1973, 1975; Pittard & Hally, 1973). Therefore it has been suggested that the processes involved in the initiation and maintenance of hypertrophy are closely related to the secretory process (Hanwell & Peaker, 1975). Clearly of interest is whether the same or different intracellular messengers are involved in these two activities of the secretory cell.

Recent evidence indicates that the number of cells in the gland increases during adaptation to salt water – an issue on which there was disagreement (Peaker & Linzell, 1975). The earlier studies were based on DNA measurements or histological findings and so in the more recent experiments the incorporation of [³H]thymidine into DNA has been determined at different stages following the transfer of ducks to salt water. In birds drinking fresh water, the incorporation of ^{3}H was very low, and no change was evident after 6 and 24 h on salt water. However at 48 h there was a marked increase which coincided with the total DNA content beginning to increase. By 7 and 14 days the incorporation had fallen to a level only slightly higher than that of birds on fresh water, although by these times the total DNA had significantly increased (Fig. 3). Therefore there appears to be a burst of mitosis beginning about 2 days after a bird first ingests salt water. The factor (or factors) inducing cell division, which occurs not in the secretory cells but in the peripheral cells at the end of the secretory tubule, is unknown.

A chart showing the time course and control of events occurring in the salt glands when a bird starts to drink salt water is shown in Table 1.

References

Fänge, R., Schmidt-Nielsen, K. & Robinson, M. (1958). Control of secretion from the avian salt gland. *American Journal of Physiology*, **195**, 321–6.

Hanwell, A., Linzell, J. L. & Peaker, M. (1972). The location and nature of the receptors for salt-gland secretion in the goose. *Journal of Physiology* **226**, 453–72.

Hanwell, A. & Peaker, M. (1973). The effect of post-ganglionic denervation on functional hypertrophy in the salt gland of the goose during adaptation to salt-water. *Journal of Physiology*, **234**, 78–80P.

Hanwell, A. & Peaker, M. (1975). The control of adaptive hypertrophy in the salt glands of geese and ducks. *Journal of Physiology*, **248**, 193–205.

Peaker, M. (1971). Avian salt glands. *Philosophical Transactions of the Royal Society*, B, **262**, 289–300.

Peaker, M. & Linzell, J. L. (1975). *Salt Glands in Birds and Reptiles*. London: Cambridge University Press.

Pittard, J. B. & Hally, A. D. (1973). The effect of denervation on genotypic and compensatory growth of the immature avian salt gland. *Journal of Anatomy*, **114**, 303.

Schmidt-Nielsen, K. (1960). Salt-secreting gland of marine birds. *Circulation*, **21**, 955–67.

Van Rossum, G. D. V. (1966). Movements of Na$^+$ and K$^+$ in slices of herring-gull salt gland. *Biochimica et biophysica acta*, **126**, 338–49.

15. Aspects of the adaptation of fish to high external alkalinity: comparison of *Tilapia grahami* and *T. mossambica*

By J. MAETZ & G. DE RENZIS

Tilapia grahami (Boulenger), a small cyclid teleost, inhabits extremely alkaline water in lakes of Kenya's Rift Valley, for example Lake Magadi. Almost the entire lake surface is covered by a solid crust composed of $NaCO_3$ and $NaHCO_3$ derived from alkaline hot volcanic springs, along the edge of the lake. The fish are found in narrow lagoons of open water. In its natural environment, the fish may thrive in temperatures as high as 39 °C (Coe, 1966). The osmolarity may be as high as 600 mOsm and the pH reaches 10.5 (Reite, Maloiy & Aasehaug, 1974). The fish also adjusts to low external salinities during the rainy season.

Physiological investigations of this fish have started only recently. External pH, salinity and temperature tolerance were studied by Reite *et al.* (1974). Regulation of plasma Na^+ and K^+ was compared in two species of fish collected in soda (HCO_3^-) lakes (*T. grahami* and *T. alcalica*) and in three species from freshwater streams or lakes (*T. zillii*, *T. nigra* and *T. leucostica*). As expected, Na^+ plasma values of fish taken from soda lakes are much higher than those taken from fresh water. K^+ values are also higher and the Na^+/K^+ ratio was definitely not related to ambient salinity. Both *T. grahami* and *T. alcalica* showed high mortality under laboratory conditions simulating soda lake water. *T. grahami* resisted little better upon dilution of this medium by fresh water. During adaptation to media of lower salinity, both Na^+ and K^+ plasma levels declined.

Johansen, Maloiy & Lykkeboe (1975) studied blood pH and pCO_2 in *T. grahami* in its natural environment. pH values as high as 8.4 at 35 °C were observed in venous blood drawn by heart puncture within minutes of perturbing the fish. The authors suggested, therefore, that blood CO_2 must be very low in this fish, which is exposed to waters in which free CO_2 can hardly exist. Moreover, a 10–15-fold HCO_3^- concentration gradient occurs across the gill of the fish. Stress, handling or transport is accompanied by a very rapid acidification of the blood. Therefore blood pH measurements were performed in a field laboratory at the lake site. More recently Lykkeboe, Johansen & Maloiy (1975) studied the functional properties of the blood in relation to pH and pCO_2. Its hemoglobin is characterized by a low pH sensitivity or Bohr effect.

T. mossambica is a typical freshwater species also originating from the African continent. It has been exported all over the world for the purpose of fish farming, and is characterized by its osmotic adaptability. For instance, it may be progressively adapted to sea water. Ionic exchanges across the gill in relation to external salinity and its endocrine control have been investigated in our laboratory by Dharmamba and co-workers (Dharmamba & Maetz, 1972; Dharmamba *et al.*, 1973; Dharmamba, Bornancin & Maetz, 1975; Dharmamba & Maetz, 1976). We noted that *T. mossambica* is a species particularly susceptible to stress by handling and confinement in the small aquaria necessary for flux measurements. The fish responds by a considerable increase in the ionic exchanges across the gill.

In the present paper, Na^+ and Cl^- fluxes were compared in *T. grahami* and *T. mossambica* acclimatized to dilute sea water and in the absence of stress. The response of the gill to a sudden increase in external alkalinity by nearly 2 pH units was studied and an attempt was made to acclimatize *T. grahami* to artificial soda lake water and to follow Na^+ and Cl^- effluxes. In a few fish, the gill potential was measured. Data on blood pH, total CO_2 content, as well as plasma Na^+ and Cl^-, are also reported.

Biological material and techniques

T. grahami were collected by Dr D. Johnson from Lake Magadi, transported to Nairobi and kept for several months in tanks at the University; then they were flown to Villefranche. Unfortunately about half of the fish were accidently lost during transport and only a total of 14 fish in good condition (body weight ranging from 15 to 30 g; mean, 20 g) were transferred to a 1 m^3 tank containing sea water diluted 10 times with fresh water ($^1/_{10}$ SW: $[Na^+] = 52$ mM, $[Cl^-] = 61$ mM, $[HCO_3^-] = 1.5$ mM, $[Ca^{2+}] = 1.5$ mM; pH $= 7.6$ to 7.8). The external medium was constantly aerated, filtered and heated at 25 °C, which was also the temperature employed during all the experiments.

Four fish were lost during adaptation to $^1/_{10}$ SW and a further four were lost during attempts to acclimatize the fish to soda lake water. Of the remaining six fish, four were studied in $^1/_{10}$ SW and subjected to a sudden transfer to soda lake water 2.5 to 3.5 months after acclimatization to $^1/_{10}$ SW. Two fish were studied in artificial soda lake water after 1 week adaptation to full-strength alkaline water. About 2 weeks prior to study, fish were transferred to a 40 l tank containing $^1/_{10}$ SW. $NaHCO_3$ (RP Prolabo) was added in four steps of 50 mM final concentration every 2 days. The pH of the external medium increased progressively to 9.55 for one attempt and to 9.75 for the second experiment. Total CO_2 content decreased from 200 mM to 155 or 188 mM. The fish had 3 to 5 months' adaptation to $^1/_{10}$ SW before acclimatization to high alkalinity was attempted. Fish with only 1.5 month adaptation died during acclimatization.

T. mossambica were a gift from the Musée Océanographique of Monaco. They originated from a stock of fish studied 3 years previously by our group. The body weight ranged from 28 to 84 g (mean, 47 g), and the fish were acclimated to $^1/_{10}$ SW in 500 l tanks at 25 °C for at least a month before use. The flux measurements were made first in $^1/_{10}$ SW followed by a period with the external medium replaced by an alkaline solution prepared 2 weeks previously and containing initially 50 mM HCO_3^-. At the time of the flux measurement, the external pH attained 9.6, and the total CO_2 decreased to about 40 mM. The fish survived for not more than 20 h in this medium.

One *T. grahami* and two *T. mossambica* were fitted with indwelling intraperitoneal catheters for the purpose of measuring the gill potential simultaneously with efflux determinations. The technique is described in Maetz & Pic (1975) and discussed in Bentley, Maetz & Payan (1976).

To avoid stress all the fish were confined in the flux-chambers (250 ml) for at least 5 days prior to the flux measurements. The fish without indwelling catheters were handled daily and given a control intraperitoneal injection of saline. This injection was replaced by a mixture of ^{24}NaCl and Na^{36}Cl 1 h before the flux experiment. The indwelling catheter was used for isotope injection in the remaining fish. The Na^+ and Cl^- effluxes were measured during two successive periods of 20 to 30 min separated by a rinsing period of 10 min during which the external medium used during the first period, i.e. $^1/_{10}$ SW, was replaced progressively by the alkaline solution. After this rinsing period, closed-circuit conditions allowing for the measurement of the appearance rate of the isotopes in the external medium were resumed. At the end of the experiment, the fish was removed from the aquarium and blood was collected within 1 min from the caudal vein by means of a heparinized syringe. Blood pH was immediately measured with a micro-glass-electrode connected to a Tacussel pH meter (ISIS 20000). The blood was then centrifuged and plasma CO_2 was measured by means of a Natelson micro-gasometer. In a few cases concerning *T. mossambica*, blood could be collected only after several minutes. No attempt was made to obtain pH and CO_2 values. In all plasma samples Na^+ and Cl^- concentrations were measured with the help of a flame photometer (Eppendorf) and a Buchler chloridometer. The Na^+ and Cl^- effluxes were calculated by dividing the radioactive appearance rate in the external medium (in c.p.m. or d.p.m. h^{-1} per 100 g body weight) by the specific radioactivity of the isotope found in the plasma (in c.p.m. or d.p.m. μequiv.$^{-1}$). The flux expressed in μequiv. h^{-1} per 100 g is the sum of a branchial and a renal component, as no attempt was made to catheterize the urinary bladder. $^{24}Na^+$ efflux was monitored continuously by a γ-flow counter included in the closed circuit (see Tanguy, 1970). At the end of the experiment, a known dilution of plasma was injected in the same counter to obtain the value of plasma radioactivity. $^{36}Cl^-$ efflux was followed by taking 1 ml samples every 3 min from the external medium. After decay of ^{24}Na (one week) the sample was added to 10 ml Bray solution and its radio-activity measured in an Intertechnique liquid scintillation counter with automatic correction of quenching. ^{36}Cl activity was also measured in a 50 μl plasma sample in order to obtain the specific activity of the isotope.

The fluxes being relatively small and the experimental period short, there was no necessity to take into account the changes in specific radioactivity of the internal medium.

Na^+ and Cl^- influxes were measured in *T. mossambica*. For lack of biological material, such measurements could not be performed in *T. grahami*. The total-body counter technique described by Macfarlane & Maetz (1975) for flounders was used. Drinking rates were not measured as the contribution of an intestinal component of ionic influx was considered negligible. The fish were confined to individual 250 ml aquaria 5 days prior to addition of isotopes to the external medium. Fish adapted to $^1/_{10}$ SW were compared to fish transferred to alkaline $^1/_{10}$ SW, 5 min prior to the experiment. After 30 min contact with the radioactive water, fish were transferred to a 'cold' medium prior to total-body counting and blood was collected from the caudal vein for the measurement of plasma ^{24}Na and ^{36}Cl. When the blood sample was of sufficient size, blood pH, plasma CO_2, Na^+ and Cl^- content were determined. For the calculation of HCO_3^- and CO_3^{2-} values in external and internal media, k_2' was taken as 0.5×10^{-9}.

Results
Na^+ and Cl^- effluxes and gill potentials

Figs. 1 and 2 illustrate flux experiments with simultaneous recording of the gill potential. Tables 1 and 2 summarize the results of these experiments.

For three out of the four *Tilapia grahami* tested in $^1/_{10}$ SW, the Na^+ and Cl^- effluxes were found to be in the 90 to 160 μequiv. h^{-1} per 100 g range, while in the fourth, fluxes were found to be 2.5 times higher. Na^+ and Cl^- effluxes were not significantly different. The gill potential measured in one fish was 10 mV (inside positive). Upon transfer to alkaline $^1/_{10}$ SW, the fluxes declined significantly by $21.5 \pm 3.5\%$ for the Na^+ efflux and by $58.0 \pm 12.4\%$ for the Cl^- efflux. The gill potential increased to 17–18 mV.

In *T. mossambica* adapted to $^1/_{10}$ SW, the fluxes were smaller than for *T. grahami*. Fluxes were found to be 2.5 times higher in two out of six

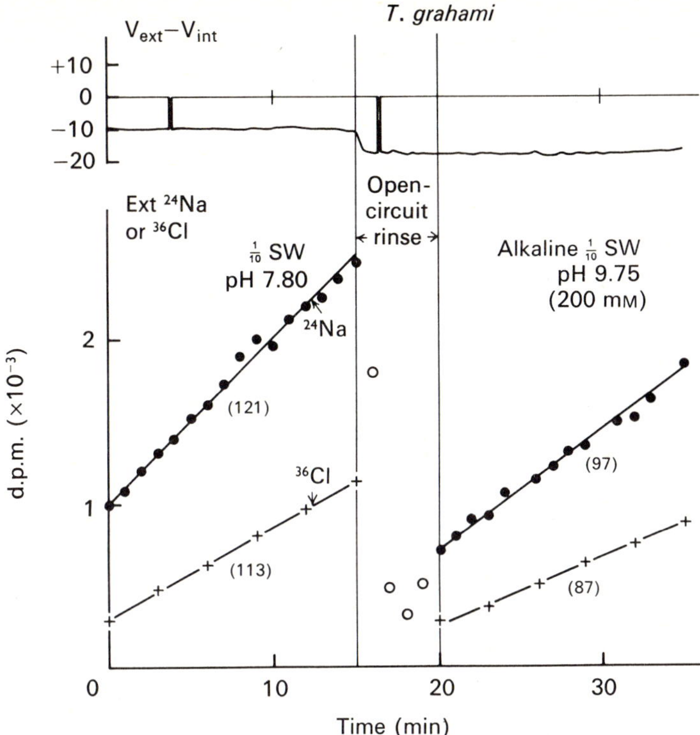

Fig. 1. Na⁺ and Cl⁻ effluxes and gill potential measured simultaneously in *T. grahami*. Effect of alkalinization of the external medium. *Upper trace*: continuous recording of the gill potential in mV. V_{ext} and V_{int} = external and internal voltage, respectively. *Lower traces*: ²⁴Na⁺ and ³⁶Cl⁻ appearances in the external medium in d.p.m. ml⁻¹ as a function of time in minutes. Between the first and second periods the aquarium water (¹/₁₀ SW) was progressively replaced by an alkaline solution. During this rinsing period, the radioactivity accumulated during the first period was removed. The circuit was then closed and the radioactive appearance rate was measured again. Absolute value of fluxes (µequiv. h⁻¹ per 100 g) are given.

fish. Na⁺ effluxes ranged from 30 to 160 µequiv. h⁻¹ per 100 g. Cl⁻ effluxes, ranging from 10 to 100 µequiv. h⁻¹ per 100 g, were consistently smaller. The gill potential varied from 8 to 10 mV (inside positive) in the two fish tested. Upon transfer to alkaline water, both fluxes increased considerably, attaining values 4 to 30 times higher than observed during the control period. The gill potential decreased in both fish by 9 and 5 mV, respectively. In the fish represented in Fig. 2, the potential became negative inside (1 mV).

In the two *T. grahami* adapted for a week to the high alkalinity medium

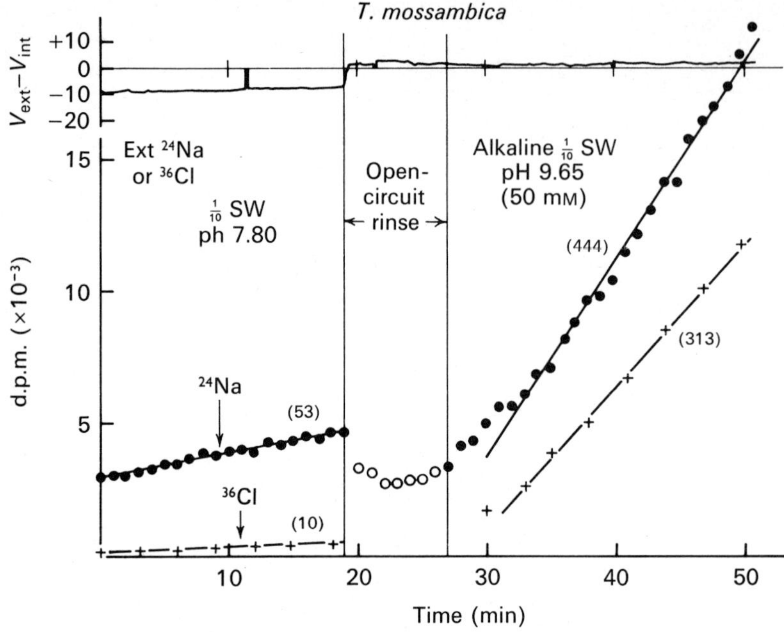

Fig. 2. Na$^+$ and Cl$^-$ effluxes and gill potential measured simultaneously in *T. mossambica*. Effect of alkalinization of the external medium. See caption to Fig. 1.

Table 1. Tilapia grahami: *effect of rising external* pH *on* Na$^+$ *and* Cl$^-$ *effluxes*

	$^1/_{10}$ SW	Alkaline $^1/_{10}$ SW
External Na$^+$	52.0±2.9	259.5±5.5
External Cl$^-$	61.3±3.9	61.3±3.9
External C$_{CO_2}$	2.5−2.7	172±9.2
external pH	7.6−7.8	9.65±0.06
J$_{out}$Na$^+$	176±56.5	140±48
J$_{out}$Cl$^-$	209±70	77±20
Plasma C$_{CO_2}$	—	8.5±1.1
plasma pH	—	7.72±0.04
Plasma Na$^+$	—	174.5±5.3
Plasma Cl$^-$	—	155.5±2.5

Mean ±s.e. for $n = 4$. Body weight: 17.3±1.05 g.
Concentrations in μequiv. ml^{-1} or in mM for CO$_2$ content (C). Fluxes (J) in μequiv. h^{-1} per 100 g.

Table 2. Tilapia mossambica: *effect of rising external* pH *on* Na$^+$ *and* Cl$^-$ *fluxes*

	$^1/_{10}$ SW	Alkaline $^1/_{10}$ SW
External Na$^+$	53.1±3.0 (4)	108.8±4.1 (4)***
External Cl$^-$	58.0±4.3 (4)	65.5±1.9 (4)
External pH	7.6−7.8	9.58±0.12 (4)***
External C$_{CO_2}$	2.5−2.7	40.2±4.4 (4)***
J_{out} Na$^+$	73±21.1 (6)	557±156.5 (6)*
J_{out} Cl$^-$	37±13.6 (6)	340±54.8 (6)***
J_{in} Na$^+$	18±2.6 (6)	155±32.8 (6)**
J_{in} Cl$^-$	10±1.7 (6)	119±48.6 (6)†
Plasma pH	7.63±0.03 (6)	7.94±0.04 (8)***
Plasma C$_{CO_2}$	5.3±0.50 (4)	5.5±0.37 (8)
Plasma Na$^+$	141.4±2.06 (5)	152.6±5.52 (6)
Plasma Cl$^-$	125.7±1.78 (5)	125.4±2.47 (6)

Body weight: 29.3±4.3 g (18). Number of fish or measurements in brackets. Same units as in preceding Table.
Statistics: † $P < 0.05$; *, $P < 0.02$; **, $P < 0.01$; ***, $P < 0.001$.
Comparison of means ±S.E. for all data except effluxes (paired differences).

the Na$^+$ effluxes were, respectively, 226 and 521 μequiv. h^{-1} per 100 g, and the Cl$^-$ effluxes 341 and 300 μequiv. h^{-1} per 100 g, higher therefore than in the fish adapted to $^1/_{10}$ SW. Transfer of these fish back to $^1/_{10}$ SW resulted in contradictory observations: both fluxes declined in one fish and increased in the other.

Na$^+$ and Cl$^-$ influxes in *T. mossambica*

The results are reported in Table 2. Na$^+$ and Cl$^-$ influxes were found to be much lower than the corresponding effluxes both in $^1/_{10}$ SW-adapted fish and in fish transferred to the alkaline $^1/_{10}$ SW.   Both fluxes were found to be more than 10 times higher in the fish subjected to the increase in external pH.

Blood pH, plasma CO$_2$, Na$^+$ and Cl$^-$ content: comparison with external medium

Table 2 contains data for *T. mossambica* plasma values obtained at the end of the influx period in $^1/_{10}$ SW as well as plasma values from fish challenged with high external pH during both influx and efflux experiments. As no significant difference was found between data recovered from influx and efflux experiments, the values were pooled. Following transfer to alkaline $^1/_{10}$ SW, blood pH increased by 0.31 units ($P < 0.001$), while plasma CO_2 and Cl^- contents remained unchanged. Plasma Na^+ increase was not significant. Comparison of the external and internal media at the end of the high alkalinity period reveals an H^+ concentration ratio of 55. In the external medium, HCO_3^- and CO_3^{2-} concentrations may be calculated using the tables given by Harvey (1957) for sea water. Thus, the concentrations were found to be 20.6 and 19.6 mM, respectively. For plasma, pK values were taken from the tables of Baumgardner. HCO_3^- concentration was 5.07 ± 0.51 mM before, and 5.40 ± 0.37 mM after, high alkalinity exposure. pCO_2 changed from 4.15 ± 0.09 to 2.01 ± 0.19 mm Hg, a highly significant decrease ($P < 0.001$). The HCO_3^- concentration ratio across the gill is thus approximately 3.8. Na^+ and Cl^- concentrations in plasma were respectively 40 and 65 μequiv. ml^{-1} higher than in the external medium.

Table 1 reports the plasma values recovered from *T. grahami* after exposure to the high alkalinity medium. Plasma pH is significantly lower than in *T. mossambica* ($P < 0.01$) while CO_2 content is higher ($P < 0.01$). HCO_3^- concentration is 8.25 ± 1.03 mM and pCO_2 5.43 ± 1.14 mm Hg, both values being much higher than for *T. mossambica* subjected to an alkaline challenge ($P < 0.01$). Plasma Na^+ and Cl^- are also much higher in *T. grahami*. Comparison of the external and internal media reveals an H^+ concentration ratio of nearly 100, and the concentrations of HCO_3^- and CO_3^{2-} are, respectively, 53 and 119 mM. Thus the HCO_3^- concentration ratio across the gill is 6.4. In the two *T. grahami* adapted to the high alkalinity medium, blood pH values are surprisingly low (7.33 and 7.69). The CO_2 contents were, respectively, 6.7 and 7.3 mM. The corresponding HCO_3^- and pCO_2 values are 6.3 and 7.1 mM and 10.3 and 4.8 mm Hg. The HCO_3^- gradient reaches 7.9. Plasma Na^+ was 202 and 180 and plasma Cl^- 166 and 144 μequiv. ml^{-1}, values not much higher than those reported in Table 1. Comparison with the Na^+ and Cl^- concentrations in the external medium indicates that plasma Na^+ is lower while plasma Cl^- is higher.

Discussion

Ionic fluxes

Dharmamba *et al.* (1973, 1975) have emphasized the importance of avoiding stress in the assessment of the ionic fluxes across the body surfaces in *T. mossambica*. Na^+ efflux was found to be three to six times higher in handled animals placed in their flux-chambers immediately before flux measurements than in fish adapted for 5 days to these aquaria and trained to receive a daily control injection or fitted with an indwelling catheter which allowed injection of the isotopes without perturbation of the fish. The high Na^+ and Cl^- effluxes recorded in one *T. grahami* and two *T. mossambica* may result from shock despite the precautions taken during the present investigation.

Na^+ and Cl^- effluxes in *T. mossambica*, as in other species of euryhaline fish are related to the salinity of the adaptation medium. The Na^+ and Cl^- effluxes reported by Dharmamba *et al.* (1975) for ⅓ SW-adapted non-stressed fish are, respectively, 125 and 157 and for the freshwater-adapted fish (for both ions) 8 μequiv. h^{-1} per 100 g. The values observed during the present investigation (Table 2) are intermediate, as expected from $^1/_{10}$ SW fish. The gill potential in ⅓ SW fish reported by the above-mentioned authors is about 15 mV (inside positive). In freshwater-adapted fish, unpublished data obtained recently in our laboratory (3.0 ± 2.85 mV, 4 fish, inside negative) indicate a potential not significantly different from 0. The value reported for $^1/_{10}$ SW is intermediate between the ⅓ SW and freshwater values.

The fluxes observed for *T. grahami* adapted to $^1/_{10}$ SW are higher than those found for *T. mossambica*. The discrepancy may partly result from the smaller size of the fish. The gill potential appears to be identical in both species, if one is allowed to generalize from the very few data obtained.

Na^+ and Cl^- influxes which were measured in *T. mossambica* were found to be much smaller than the effluxes and not significantly different from the influxes observed in freshwater-adapted fish (Dharmamba *et al.*, 1975). The discrepancy between influxes and effluxes may result in part from differences in the handling of the fish. Fish destined for efflux measurements were handled daily and received a control injection. Fish destined for influx determination were not handled at all.

The most significant observation concerns the flux changes observed in relation to the sudden alkalinization of the external medium. In *T. grahami*, both Na^+ and Cl^- effluxes decline, while in *T. mossambica*, they increase considerably, even though the HCO_3^- gradient imposed across the gill is much smaller for *T. mossambica*. In the present study, the efflux measured is the total flux, i.e. the sum of renal and branchial components. It is highly probable that the efflux change observed immediately upon modification of the external medium reflects mainly a change in the branchial flux component. The response of the kidney is expected to be delayed by the relatively slow changes of the composition of the internal medium. Moreover, when *T. grahami* and *T. mossambica* are compared, the shifts in gill potential are also very rapid and in opposite directions.

If the Na^+ and Cl^- effluxes result from the passive diffusion of these ions along their chemical gradient across the gill, the observed potential changes should induce a flux change. The expected variation is given by an equation formulated by Maetz & Pic (1975):

$$\frac{J_{\text{out}}(1)}{J_{\text{out}}(2)} = \frac{\Delta V(1)}{\Delta V(2)} \times \frac{\exp\dfrac{zF\Delta V(2)}{RT} - 1}{\exp\dfrac{zF\Delta V(1)}{RT} - 1} \tag{1}$$

where J_{out} is the efflux in medium (1) and (2) in μequiv. h^{-1} per 100 g. $\Delta V = V_{\text{ext}} - V_{\text{int}}$, the gill potential difference is given in mV. z is the charge of the ion considered and R, T and F are the usual thermodynamic coefficients.

Applying this equation for *T. grahami*, it turns out that during the second flux period, the Na^+ efflux should increase by 15 % and the Cl^- efflux decrease by 15 %, while a 21.5 and 58 % decrease are observed for the Na^+ and Cl^- effluxes, respectively. Thus neither of the changes in fluxes may be explained by the potential shift. It is probable that the increase in external pH induces in this species a decrease in the branchial Na^+ and Cl^- permeabilities.

In *T. mossambica*, the potential change is obviously too small to account for the huge increase in the Na^+ and Cl^- effluxes. A considerable increase in the branchial permeability of these ions must occur. While the evolution of the Na^+ efflux remains unexplained, the increase in the Cl^- efflux may be accounted for by a forced HCO_3^-/Cl^- exchange across the gill. In freshwater teleosts, Cl^- is absorbed by the gill in exchange for HCO_3^- of endogenous origin (Maetz & Garcia-Romeu, 1964; de Renzis & Maetz, 1973; de Renzis, 1975). When the Cl^- pump is blocked by SCN^-, a reversed exchange between external HCO_3^- and internal Cl^- is observed in the goldfish (de Renzis, 1975). Addition of HCO_3^- to the external medium was shown to block the Cl^- pump in the goldfish (Maetz & Garcia-Romeu, 1964). It is possible that a reversed HCO_3^-/Cl^- exchange is induced in such conditions with a concomitant increase of the Cl^- efflux. The results of the influx experiments do not confirm this possibility, however. After transfer to the medium with high external HCO_3^-, Cl^- influx does not decrease but increases by more than a factor of 10. Whether the increase in the Na^+ and Cl^- influxes reflects an activation of the Na^+ and Cl^- pumps or results from an enhancement of the passive components of Na^+ and Cl^- entry across the gill remains to be studied. It is probable that the parallel increase of both Na^+ and Cl^- influxes and effluxes reflects a drastic and possibly non-specific augmentation of the gill permeabilities to these ions. *T. grahami* seems more resistant in this respect. Our observations on the effects of alkalinization of the external medium are in contrast with our results in *Carassius* concerning the effects of acidification of aquarium water. At a pH below 4, a considerable increase in both Na^+ and Cl^- effluxes was observed. Na^+ and Cl^- influxes were totally inhibited, however (Maetz, 1972, and unpublished observations).

In the two *T. grahami* adapted for a week to the high alkalinity medium, the Na^+ and Cl^- effluxes were found to be significantly higher than those recorded immediately after transfer to this medium. The increase in Na^+ efflux is to be expected. Motais (1967) has demonstrated in various euryhaline teleosts that a sudden increase in external salinity is not followed by an instantaneous increase in Na^+ efflux. A delayed regulation intervenes, however, which may last several days and which consists of a considerable increase in the Na^+ efflux. The increase in the Cl^- efflux is, however, unexpected because the Cl^- concentration in the high alkaline medium is the same as in $^1/_{10}$ SW. It is possible, in view of the hyperosmoticity of the external medium that the gill may have switched to the Cl^- excretory system which characterizes marine teleosts. It would be of interest to measure the drinking rate of fish kept in the high alkalinity medium. The Cl^- pump would help to get rid of the extra salt-load incurred by the ingestion of the external medium.

Gill potential generation

The gill potential in freshwater and in seawater teleosts results mainly from the free diffusion of Na^+ and Cl^- along their chemical gradients across the branchial epithelium (Potts & Eddy, 1973; House & Maetz, 1974; Kirschner, Greenwald & Sanders, 1974; Maetz & Pic, 1975; Maetz, 1974; Eddy, 1975). Electrogenicity of the ionic pumps may also contribute to the potential: the Cl^- pump in various seawater-adapted teleosts (Pic & Maetz, 1974; Shuttleworth *et al.*, 1974) and possibly the Na^+ pump in some freshwater teleosts (Maetz, 1974). Recent unpublished experiments suggest that *T. mossambica* resembles the goldfish with respect to the origin of the gill potential. Both fish when kept in fresh water (containing in Villefranche 1.5 mM Ca^{2+} and 0.5 Mg^{2+}) exhibit a potential difference not significantly different from $0:3.0\pm2.85$ mV (inside negative) for *Tilapia* and 4.0 ± 2.3 mV (inside positive) for *Carassius*. The gill of both fish becomes cation-selective when external Ca^{2+} and Mg^{2+} is removed by transferring the fish to deionized water containing 2 mM imidazole sulphate as a buffer. The potential reaches about 45 mV (inside negative) in the goldfish and 30 mV (also inside negative) in *Tilapia*. In this experimental situation the gill potential is described by a simplified Hodgkin & Katz (1949) equation:

$$\Delta V = V_{ext} - V_{int} = \frac{RT}{F} \ln \frac{P_{Na}Na_{int}}{P_{Cl}Cl_{int}}$$

where Na_{int} and Cl_{int} are the Na^+ and Cl^- plasma values, and P_{Na} and P_{Cl} are the permeabilities of the gill to Na^+ and Cl^- expressed in

$$\frac{\mu\text{equiv. h}^{-1}\text{ per }100\text{ g}}{\mu\text{equiv. ml}^{-1}}.$$

$Na_{int}/Cl_{int} = 1.14$ in *T. mossambica* (Dharmamba *et al.*, 1975) and 1.23 in *Carassius auratus* (de Renzis & Maetz, 1973). For $P_{Na}/P_{Cl} = 5$, $\Delta V = 40.4$ mV in *Carassius*. For $P_{Na}/P_{Cl} = 3$, $\Delta V = 30.7$ mV in *Tilapia*.

Addition of Ca^{2+} to the external medium is accompanied in both fish by a prompt depolarization of the gill by about 38 mV in *Carassius* (Maetz, 1974) and 22 mV in *T. mossambica*, according to our unpublished observations. In the goldfish, Ca^{2+} acts by reducing P_{Na}, P_{Cl} remaining unchanged. The ratio P_{Na}/P_{Cl} being smaller, it follows from equation (2) that ΔV decreases. For $P_{Na}/P_{Cl} = 1$, $\Delta V = 0$, which is the situation which prevails in fresh water containing Ca^{2+} and Mg^{2+}.

The one catheterized *T. grahami* adapted to $^1/_{10}$ SW survived the second day following the flux measurement. The potential was 10 to 11 mV (inside positive). After transfer to deionized water with imidazole sulphate, the potential reversed and attained 40 mV (inside negative). It then declined and stabilized to 28 mV. Addition of Ca^{2+} (1 mM) produced a rapid depolarization of the gill by about 30 mV. It is probable therefore that *T. grahami* resembles *T. mossambica* and *Carassius* with respect to the origin of its gill potential.

When kept in $^1/_{10}$ SW, the gill potential is 8 to 10 mV positive inside. This potential may result from the diffusion of Na^+ and Cl^- from plasma to external medium with $P_{Na}/P_{Cl} < 1$. A possible contribution of an electrogenic Na^+ pump cannot be excluded. Addition of $NaHCO_3$ to the external medium should, in the light of the diffusion potential hypothesis, produce a shift of the potential which depends on the relative permeabilities of the gill to Na^+ and HCO_3^-. If $P_{Na}/P_{HCO_3} > 1$, the potential should increase. If the ratio is less than 1, the potential should decrease or even reverse. The potential was observed to increase in *T. grahami* and to decrease or reverse in *T. mossambica*. This strongly suggests that the gill of *T. mossambica* is more permeable to HCO_3^- than is the gill of *T. grahami*. A high impermeability of the gill of *T. grahami* to HCO_3^- would explain its peculiar habitat. The gill of *T. mossambica* must be relatively impermeable to HCO_3^-, however, because alkalinization of the external medium does not produce a significant change in plasma HCO_3^- content after 30 min exposure.

Blood data

Plasma Na^+ and Cl^- concentrations reported in Table 2 for *T. mossambica* are significantly lower than previously reported for freshwater fish by Dharmamba *et al.* (1975). Plasma Na^+ values for the two *T. grahami* adapted to the high alkalinity medium are well within the range given by Leatherland, Hyder & Ensor (1974) for fish caught in Lake Magadi, even though the ambient Na^+ level (215 μequiv. ml^{-1}) at the time of capture was definitely lower than that of our artificial solution. For fish adapted to $^1/_{10}$ SW and transferred to the high alkalinity medium, the plasma Na^+ is lower. The values reported by Leatherland *et al.* (1974) for fish adapted to fresh water or to 50% artificial Lake Magadi water with an ambient Na^+ level of 83 μequiv. ml^{-1} are definitely lower than those given in Table 1. Plasma Cl^- levels have not been reported previously. In the absence of control values obtained prior to transfer to the high alkalinity medium, it is impossible to evaluate the increase in plasma Na^+ resulting presumably from the 30 min stay in this medium.

Blood pH in *T. mossambica* adapted to $^1/_{10}$ SW at 25 °C is almost identical to that reported by de Renzis & Maetz (1973) for *Carassius* adapted to fresh water at 16 °C. The CO_2 content is, however, lower than in *Carassius* (9.3±0.3 mM) as would be expected from the difference in temperature of the adaptation medium (Dejours & Armand, 1973). Concerning the changes in acid–base balance of the blood induced by the alkalinization of the external medium, the increase in blood pH reported in Table 2 is not the result of an increased HCO_3^- content of the blood but of a sharp decline in pCO_2.

For *T. grahami*, because of the absense of control values, it is impossible to evaluate the impact of external pH rise on blood pH or CO_2 content. Comparison with the values obtained in *T. mossambica* challenged with a concentration gradient of HCO_3^- and H^+ much smaller than for *T. grahami* strongly suggests that the response of both species is very different. HCO_3^- content and pCO_2 are much higher in *T. grahami*, while pH is significantly lower. Moreover, blood pH in the two fish adapted for more than a week to artificial Lake Magadi water were found to be surprisingly low, almost one unit pH lower than the highest values recently reported by Johansen *et al.* (1975) for fish captured in this lake. According to their report, blood was collected within 10 min of perturbing the fish. During the present investigation, precautions were taken to avoid stress and blood was collected within 1 min of taking the fish out of the aquarium. No explanation is forthcoming to explain this discrepancy. Plasma HCO_3^- (6.7 mM) is very close to the value expected from the nomogram published by Johansen *et al.* (1975), while pCO_2 (7.56 mm Hg) is much higher than that suggested by the nomogram (2 mm Hg).

Summary

Tilapia grahami lives in alkaline lakes of Kenya at pH 10. *T. mossambica* is a typical freshwater teleost.

Na^+ and Cl^- effluxes were measured isotopically in the two species adapted to sea water diluted 10 times with fresh water. Fluxes were found to be higher in *T. grahami*, while gill potential was nearly identical (10 mV inside positive) in both species. Na^+ and Cl^- influxes were also measured in *T. mossambica*.

Both fish were challenged with media of high pH (9.55 to 9.75) obtained by adding to the adaptation medium 50 mM $NaHCO_3$ for *T. mossambica* and 200 mM $NaHCO_3$ for *T. grahami*. Na^+ and Cl^- influxes and effluxes increased drastically in *T. mossambica*, while, in contrast, a decrease in the Na^+ and Cl^- effluxes were observed in *T. grahami*. The gill potential increased in *T. grahami*, while it decreased or even reversed in *T. mossambica*. A few *T. grahami* were studied after a 2 week adaptation in the high alkalinity medium. Both Na^+ and Cl^- effluxes increased as a result.

Data on plasma Na^+ and Cl^- as well as CO_2 content and blood pH are also given. *T. grahami* maintained a surprisingly low blood pH when challenged by high external pH, contrary to *T. mossambica*. These results are discussed in terms of the relative permeability of the gill to HCO_3^- and CO_2.

The authors wish to thank Dr D. Johnson and Mr C. Arnoux for providing the *Tilapia* spp. They are indebted to Dr A. Thomson for correcting the manuscript and to Professor P. Dejours for providing the Baumgardner tables giving the pK values.

References

Bentley, P. J., Maetz, J. & Payan, P. (1976). A study of the unidirectional fluxes of Na and Cl across the gills of the dogfish *Scyliorhinus canicula* (Chondrichthyes). *Journal of Experimental Biology*, **64**, 629–37.

Coe, M. J. (1966). The biology of *Tilapia grahami* Boulenger in Lake Magadi, Kenya. *Acta Tropica*, **23**, 143–77.

Dejours, P. & Armand, J. (1973). L'équilibre acide–base du sang chez la Carpe en fonction de la température. *Journal de Physiologie*, **67**, 264A (abstract).

Dharmamba, M., Bornancin, M. & Maetz, J. (1975). Environmental salinity and sodium chloride exchanges across the gill of *Tilapia mossambica*. *Journal de Physiologie*, **70**, 627–36.

Dharmamba, M. & Maetz, J. (1972). Effects of hypophysectomy and prolactin on the sodium balance of *Tilapia mossambica* in fresh water. *General and Comparative Endocrinology*, **19**, 175–83.

Dharmamba, M. & Maetz, J. (1976). Branchial sodium exchange in seawater-adapted *Tilapia mossambica*: effects of prolactin and hypophysectomy. *Journal of Endocrinology*, **70**, 293–9.

Dharmamba, M., Mayer-Gostan, N., Maetz, J. & Bern, H. A. (1973). Effect of prolactin on sodium movement in *Tilapia mossambica* adapted to sea water. *General and Comparative Endocrinology*, **21**, 179–87.

Eddy, F. B. (1975). The effect of calcium on gill potentials and on sodium and chloride fluxes in the goldfish *Carassius auratus*. *Journal of Comparative Physiology* B, **96**, 131–42.

Harvey, H. W. (1957). *The Chemistry and Fertility of Sea Waters*. London: Cambridge University Press.

Hodgkin, A. L. & Katz, B. (1949). The effect of sodium ions on the electrical activity of the giant axon of the squid. *Journal of Physiology*, **108**, 37–77.

House, C. R. & Maetz, J. (1974). On the electrical gradient across the gill of the seawater-adapted eel. *Comparative Biochemistry and Physiology*, **47A**, 917–24.

Johansen, K., Maloiy, G. M. O. & Lykkeboe, G. (1975). A fish in extreme alkalinity. *Respiration Physiology*, **24**, 159–62.

Kirschner, L. B., Greenwald, L. & Sanders, M. (1974). On the mechanism of sodium extrusion across the irrigated gill of sea water-adapted rainbow trout (*Salmo gairdneri*). *Journal of General Physiology*, **64**, 148–65.

Leatherland, J. F., Hyder, M. & Ensor, D. M. (1974). Regulation of plasma Na^+ and K^+ in five African species of *Tilapia* fishes. *Comparative Biochemistry and Physiology*, **48A**, 699–710.

Lykkeboe, G., Johansen, K. & Maloiy, G. M. O. (1975). Functional properties of hemoglobins in the teleost *Tilapia grahami*. *Journal of Comparative Physiology* B, **104**, 1–11.

Macfarlane, N. A. A. & Maetz, J. (1975). Acute response to a salt load of the NaCl excretion mechanisms of the gill of *Platichthys flesus* in sea water. *Journal of Comparative Physiology* B, **102**, 101–13.

Maetz, J. (1972). Interaction of salt and ammonia transport in aquatic organisms. In *Nitrogen Metabolism and the Environment*, ed. J. W. Campbell & L. Goldstein, pp. 105–54. New York, London: Academic Press.

Maetz, J. (1974). Origine de la différence de potentiel électrique trans-branchiale chez le poisson rouge *Carassius auratus*. Importance de l'ion Ca^{2+}. *Comptes rendus hebdomadaires des Séances de l'Académie des Sciences*, **279**, 1277–80.

Maetz, J. & Garcia-Romeu, F. (1964). The mechanism of sodium and chloride uptake by the gills of a freshwater fish, *Carassius auratus*. II. Evidence for NH_4^+/Na^+ and HCO_3^-/Cl^- exchanges. *Journal of General Physiology*, **47**, 1209–27.

Maetz, J. & Pic, P. (1975). New evidence for a Na/K and a Na/Na exchange carrier linked with the Cl^- pump in the gill of *Mugil capito* in sea water. *Journal of Comparative Physiology* B, **102**, 85–100.

Motais, R. (1967). Les mécanismes d'échanges ioniques branchiaux chez les Téléostéens. *Annales de l'Institut Océanographique de Monaco*, **45**, 1–84.

Pic, P. & Maetz, J. (1974). Différences de potentiel trans-branchial et flux ioniques chez *Mugil capito* adapté à l'eau de mer. Importance de l'ion Ca^{2+}. *Comptes rendus hebdomadaires des Scéances de l'Académie des Sciences*, **280D**, 983–6.

Potts, W. T. W. & Eddy, F. B. (1973). Gill potentials and sodium ionic fluxes in the flounder *Platichthys flesus*. *Journal of Comparative Physiology* B, **87**, 29–48.

Reite, O. B., Maloiy, G. M. O. & Aasehaug, B. (1974). pH, salinity and temperature tolerance of Lake Magadi *Tilapia*. *Nature, London*, **274**, 315.

Renzis, de G. (1975). The branchial chloride pump in the goldfish *Carassius auratus*: relationship between Cl^-/HCO_3^- and Cl^-/Cl^- exchanges and the effect of thiocyanate. *Journal of Experimental Biology*, **63**, 587–602.

Renzis, de G. & Maetz, J. (1973). Studies on the mechanism of chloride absorption by the goldfish gill: relation with acid–base regulation. *Journal of Experimental Biology*, **59**, 339–58.

Shuttleworth, T. J., Potts, W. T. W. & Harris, J. N. (1974). Biolelectric potentials in the gills of the flounder *Platichthys flesus*. *Journal of Comparative Physiology* B, **94**, 321–9.

Tanguy, R. (1970). Ensembles électroniques destinés à l'étude de l'osmorégulation d'animaux aquatiques. *Bulletin d'Information Scientifique et Technique, Saclay*, **144**, 11–15.

16. Osmoregulation in *Tilapia grahami*: a fish in extreme alkalinity

By G. M. O. MALOIY, G. LYKKEBOE, K. JOHANSEN
& O. S. BAMFORD

Teleost fishes have adapted to a wide variety of waters. *Tilapia grahami* must, however, excel among fish in its adaptation to conditions of high salinity, pH and temperature.

T. grahami lives in extreme conditions of temperature, salinity and pH in the alkaline lakes of the Great Rift Valley of Africa. This small cichlid fish inhabits the lagoons and alkaline volcanic springs around the margins of Lake Magadi in Kenya (Leatherland, Hyder & Ensor, 1974; Reite, Maloiy & Aasehaug, 1974). *T. grahami* is also to be found in less alkaline habitats such as Lake Nakuru where the species was introduced early in 1961.

The temperatures of the water in the hot springs of Lake Magadi range between 32 and 40 °C, and the main salts of the springs in this lake are sodium carbonate and bicarbonate. It was, therefore, of interest to find out how *T. grahami* regulates its ionic and osmotic composition in this adverse environment.

Several studies have been carried out on the members of the genus *Tilapia*, especially *T. mossambica* (Potts, 1968; Potts *et al.*, 1967). Recent work on fish (reviewed by Potts, 1968) has clearly indicated that teleosts such as *Tilapia* have a limited ability to adjust to media of high salinities and that the salt flux through the fish is much greater in sea water than in fresh water. Apart from the recent work of Leatherland *et al.* (1974), and Reite *et al.* (1974), little is known about osmoregulation in *T. grahami* or of the water and ionic fluxes in this fish when living in waters of intermediate osmolarities: Lake Magadi water contains only 600 mOsm compared with 1000 mOsm in sea water.

T. grahami can survive in a wide range of habitats. For this reason, in an attempt to understand the osmoregulatory mechanisms which accompany adjustment of the extreme physiochemical conditions prevailing in the habitat of this teleost, we have chosen to investigate the ionic and osmotic composition of its body fluids in relation to Lake Magadi water, bicarbonate/pH relationships of the blood, and the role of the intestine in electrolyte transport. Some preliminary measurements have also been made on the potential between gills and water.

Materials and methods

Animals: Adult *Tilapia grahami*, weighing 1.0–6.0 g, and measuring 3–7 cm in length, were collected in nets in a Lake Magadi spring (Fish Springs). Comparative data were obtained from *T. zillii*, supplied by the Kenya Fisheries Department station at Sagana. The latter fish weighed 21–25 g. From some fish, urine, blood and bile were collected a few minutes after capture. These samples were immediately frozen using 'dry ice' and taken to the laboratory in Nairobi for analysis. Other fish were transported to Nairobi in their native water. These fish were used in controlled laboratory experiments at Nairobi.

Collection of body fluids and chemical analyses

Blood was collected by heart puncture, bile was collected directly from the bile duct and urine from the bladder. Intestinal contents were collected from the anterior portion of the fish intestine. Samples from different fish were pooled together and analyses of Na^+, K^+, Cl^- and osmolarity were made from these pooled samples.

Osmolarities of plasma, urine and bile were measured using either a Ramsay micro-osmometer or a Knauer osmometer. Na^+ and K^+ were determined by emission flame photometry. Cl^- was determined on a Radiometer chloride titrator (CMT10), while Ca^{2+}, and Mg^{2+} were estimated using an atomic absorption spectrophotometer.

The muscle water content was estimated by taking the difference between wet weight and the weight of the same tissue after 24 h of desiccation at 110 °C. Na^+ and K^+ concentrations were measured by dissolving weighed fish in 5 ml of concentrated nitric acid and then diluting to 50 ml with distilled water.

The potential difference between the fish gill and the ambient media was measured using a high impedance millivoltmeter with calomel electrodes. The pH of the venous blood was determined using a microelectrode assembly and an acid–base analyzer PHM 71 (Radiometer Copenhagen) and buffer capacities were measured using a Radiometer BMS 2 and Wösthoff mixing pump.

pH, salinity and temperature tolerance

pH tolerance was studied by transferring fish from Lake Magadi water to tap-water solutions of pH between 3 and 12, at 22–23 °C. The pH of the solutions was checked regularly and maintained when necessary by adding NaOH or HCl. The effects of the change to tap-water *per se* were negligible, mortality being very low.

Tolerance to salinity changes was examined at 22–23 °C by keeping fish in solutions of NaCl in tap-water ranging between 2 and 6%. Temperature tolerance was studied in Lake Magadi water at temperatures between 10 and 40 °C.

Measurement of drinking rates

Drinking rates were measured in fish kept in Magadi water at 570 mOsm, pH 9.7, 22 °C, using the polyvinylpyrrolidone (PVP) method of Evans (1968). Fish were placed in 200 ml of water containing 1.6 ml of labelled PVP for 1 h then into tap water for 30 min before killing with an overdose of MS222. The gut was dissected out and activity counted.

Results

Composition of body fluid

Table 1 shows the ionic composition of water from Lakes Magadi and Sagana together with values for body fluids in *T. grahami* and *T. zillii* from these lakes. The Sagana water, where *T. zillii* was studied, had very low concentrations of Na^+ and K^+ (0.4 and < 0.1 mequiv. l^-), respectively, and the osmolarity was about 5 mOsm. In the laboratory, *T. grahami* was also acclimated to a mixture of Magadi water and fresh tap-water. This water had an osmolarity of about 125 mOsm.

The ionic and osmotic composition of *T. grahami* urine is presented in Table 1. Compared with Lake Magadi water the urine had a lower pH, and lower osmolarity, Na^+, and Cl^- concentrations.

Interesting changes were observed in the plasma composition of *T. grahami* acclimated for 1 week to the diluted Lake Magadi water of 125 mOsm. Under these conditions, plasma osmolarity for the fish averaged 252 mOsm and the Na^+ and K^+ concentrations were 118 and 13 mequiv. l^{-1}, respectively. These latter values are lower than those obtained from *T. grahami* in Lake Magadi water. Differences are apparent in the ionic and osmotic concentrations of the plasma between *T. grahami* and *T. zillii* (Table 1).

Table 1 also shows the Na^+ and K^+ concentrations in the muscles of both *T. grahami* and *T. zillii*. Differences are apparent in the muscle electrolyte concentration of the two species.

Significant changes were also observed in muscle electrolyte concentration of *T. grahami* acclimated for a week to water with an osmolarity of 125 mOsm. The muscle Na^+ and K^+ concentrations under these conditions were 16 and 140 mequiv. l^{-1}, respectively. The compositions of the bile of the two fishes are shown in Table 1.

Table 1. *Ionic composition of Lake Magadi water and body fluids of* Tilapia *species*

Sample	Na^+	K^+	Cl^-	Ca^{2+}	Mg^{2+}	HCO_3^-	CO_3^{2-}	pH[e]	Osmolarity (mOsm)
Lake Magadi water[a]	314[c]	2.5	91	0.4	—	87	116	9.8	539
Tilapia grahami[d]									
Plasma	217	8	110	0.8	3	6–9	—	7.1	426
Urine	184	15	42	4	1.3	—	—	4.8	423
Bile	304	11	22	22	4	1.6	—	7.9	536
Muscle	117	456	—	—	—	—	—	—	—
Anterior intestine	291	6	136	—	—	—	—	—	544
Sagana water	0.4	< 0.1	—	—	—	—	—	—	5
Tilapia zillii									
Plasma	137	4.2	108	1.8	0.16	—	—	—	303
Bile	180	6	—	—	—	—	—	—	—
Muscle	12	262	—	—	—	—	—	—	—

[a] Water samples were taken from different sites in Lake Magadi where fish were collected.
[b] Ionic concentrations are all given in mequiv. l^{-1}.
[c] Values are means.
[d] Values represent samples pooled from several fish.
[e] pH at 35 °C, $pCO^2 = 0$.

pH, salinity and temperature tolerance

Fish transferred from Lake Magadi water to Nairobi tap-water survived well at 22 °C. In tap-water made acid or alkaline by addition of HCl or NaOH, fish were alive and healthy after 24 h at pH between 5 and 11. At pH 3–4 or pH 12, fish died within 2–6 h. Most of these experiments were performed at 22 °C but some experiments at the habitat ambient temperature of 35 °C gave similar results.

Fish kept in solutions of NaCl in tap-water survived over 24 h in 2–3 % NaCl but for only 8–10 h in 4 % NaCl. At higher concentrations death was fairly rapid.

Fish kept in Lake Magadi water survived for over 24 h at all temperatures between 16 and 40 °C, though sudden transfer from low to high temperatures without acclimation was usually lethal. Below 10–12 °C the fish died within 1–2 h. Upper temperature limit was hard to determine, but for a survival time of 24 h it was about 40 °C.

Intestinal contents and structure

Figures for composition of anterior intestinal contents are given in Table 1. Na^+ concentration was slightly lower than in the water, while Cl^- concentration was considerably higher, suggesting some active extrusion of Cl^- into the gut. Concentrations of Na^+ and Cl^- in the anterior intestine were higher than in the plasma, though K^+ occurred at the same concentration in both.

A study of the fine structure of the intestinal epithelium in *T. grahami* indicated that the cranial intestine is involved in water and ion transport as well as lipid absorption (Kayanja, Maloiy & Reite, 1975).

Drinking rate

Drinking rates measured in six fish ranged from 0.05 to 0.007 ml H_2O g^{-1} h^{-1}. The mean value was 0.024 ml g^{-1} h^{-1} and the mean body weight 2.63 g. There was a slight but statistically insignificant tendency for large fish to have lower drinking rates per unit body weight.

pH/bicarbonate relationship

Venous blood pH was found to be remarkably high though variable. The highest value recorded was pH 8.4, but it is likely that the effects of stress during capture and handling caused a drop in pH so the true values may well be higher. Bicarbonate concentrations in the blood, although

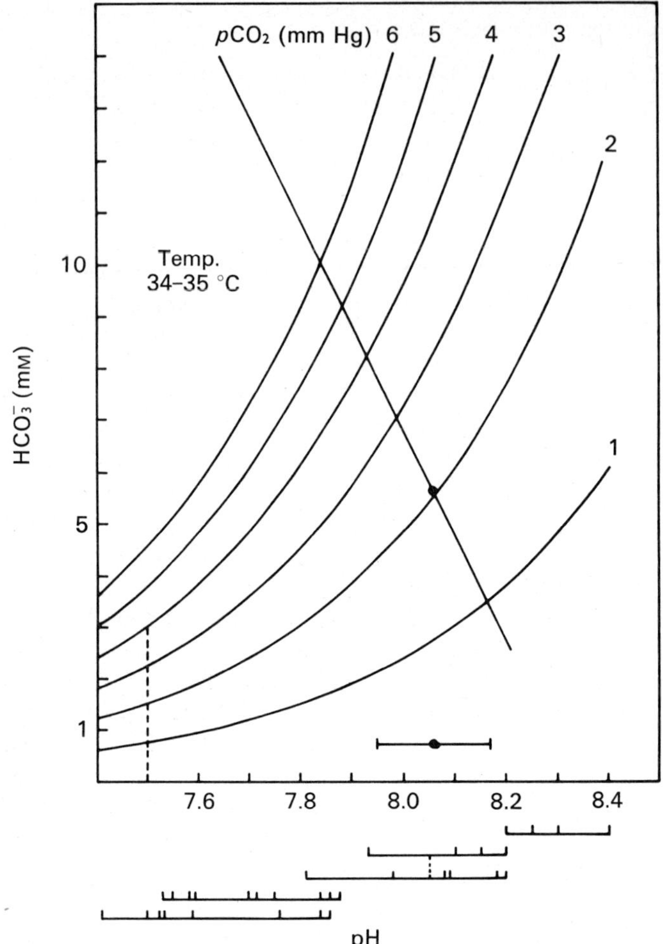

Fig. 1. Acid-base nomogram for *Tilapia grahami*. In-vivo blood pH values indicated below the nomogram pH axis as short vertical lines (Johansen *et al.*, 1975). The broken vertical line at pH 7.5 indicates values of pCO_2, HCO_3^- and pH predictable for fish at 35 °C. (Rahn & Baumgardner, 1972).

unusually high for fish, were well below that of lake water, and a bicarbonate concentration gradient of 10–15-fold existed across the gills.

An acid–base nomogram was constructed for *Tilapia* blood (Fig. 1) using standard values for pK and α (Johansen, Maloiy & Lykkeboe, 1975). It was found that observed values for blood pH differed markedly from those predicted using standard acid–base nomograms, probably because of the virtual absence of free CO_2 in the lake water.

Electrical potentials

Only a few measurements have so far been made of gill potentials in *T. grahami* and the values are somewhat variable. Mean values recorded under three different conditions were as follows: lake water, 35 °C, $+1.4$ mV; lake water, 22 °C, $+1.0$ mV; lake water made up to 300 mequiv. l^{-1} Cl^- with NaCl, 22 °C, $+4.8$ mV. The fish were consistently positive with respect to the water, and the increase in potential in increased external Cl^- concentrations also occurred with all fish tested.

Discussion

This study has provided new information on ionic and osmotic regulation in a fish naturally adapted to conditions of extreme alkalinity. For comparison some tests were carried out on *T. zillii*, which cannot tolerate extreme conditions.

The only other report on ionic concentrations in body fluids of East African species of *Tilapia* appears to be that of Leatherland *et al.* (1974). Our determinations of plasma Na^+ concentrations and Na^+/K^+ ratios in *T. grahami* and *T. zillii* agree with the results of these authors. We found an Na^+/K^+ ratio of 28 for *T. grahami* under Lake Magadi conditions, and for *T. zillii* from Sagana River the ratio was 34. The ratios reported for these fish by Leatherland *et al.* (1974) were 24 and 31, respectively. The plasma ionic concentrations in *T. grahami* were found to vary markedly with the ionic content of the medium; and determinations of osmolarity of the plasma varied between 426 mOsm, at an external osmolarity of 540 mOsm, to 252 mOsm, at an external osmolarity of 125 mOsm. Thus plasma osmolarity can be varied over an enormous range and may be below or above ambient. Further work is needed on the limits of toleration for plasma osmolarity, but the preliminary work on survival in water at different salinities suggests that the limiting external osmolarity is about 1000 mOsm, approximately double that of Lake Magadi water at the habitat of these fish. Thus, even under these conditions, *T. grahami* is nowhere near its tolerance limit for ambient osmolarity. To date, no measurements have been made on body fluids at the limiting ambient concentrations.

Tolerance to changing pH and temperature similarly shows *T. grahami* to be very well adapted to Lake Magadi conditions (Reite *et al.*, 1974). The pH tolerance extended from pH 5 to 11 and temperature tolerance from 16–40 °C. Lake conditions are highly variable, but pH at the springs where the fish were caught is normally in the range 8.5–10 and water temperature seldom rises much above 35 °C.

Measurements on ionic composition of anterior intestinal contents show a reduced Na^+ and increased Cl^- and K^+ contents, compared with water. The increase in Cl^- may be because of gastric secretion, but the slight reduction in Na^+ concentration is of interest, if confirmed by further measurements. The very slight increase in K^+ can probably be accounted for by the presence of plant material in the gut, and this may also be the source of the slight fall in sodium compared with lake water. Although the fine structure of the epithelium of an anterior intestine appears to be adapted to water and ion transport (Kayanja *et al.*, 1975), no great differences were found between ionic compositions of water and gut contents. Either the role of intestinal ion transport is limited or flow of food and water through the gut is sufficiently high to maintain near-ambient conditions in the anterior intestine. Further data are needed to decide between these possibilities.

The mean drinking rate measured was 0.055 ml g^{-1} h^{-1}. This can be compared with rates of 0.0076 ml g^{-1} h^{-1} in the euryhaline *Aphanius dispar* in fresh water, rising to 0.020 ml g^{-1} h^{-1} in 100 % sea water and 0.027 ml g^{-1} h^{-1} in 200 % sea water (Lotan, 1969).

Maetz & Skadhauge (1968) and Skadhauge & Lotan (1974), working on the eel and *Aphanius*, respectively, report values similar to those obtained by Lotan (1969). Drinking rate in *T. grahami* appears to be remarkably high for a fish living in an osmolarity which is the same as that of 50 % sea water, and presumably ion fluxes are correspondingly high. Measurements of Na^+ flux are currently being undertaken to investigate this.

Blood pH and bicarbonate concentrations were unusually high, though bicarbonate was far lower than in the lake water (Johansen, Maloiy & Lykkeboe, 1975). Since plasma Cl^- was found to be a little above ambient, this may be evidence for a Cl^-/HCO_3^- exchange, as described by Maetz (1971), though the stoichiometry is not convincing. Blood pCO_2 is very low, presumably because of the high external pH which creates an effective CO_2 vacuum. The maintenance of normal bicarbonate concentrations would be energetically very costly under Lake Magadi conditions, and the values for pH and bicarbonate found presumably represent a compromise between an optimal internal environment and what is energetically feasible. *T. grahami* is known to grow faster in the more dilute water of Lake Nakuru, so osmotic and ionic pumping in Lake Magadi must impose considerable loads on the energy economy of the fish.

The gill potentials measured are rather unexpected in view of the ionic compositions of plasma and lake water (Table 1). The potential was found to increase when external Cl^- concentration was trebled, and this would appear to rule out any active electrogenic Cl^- transfer. As the Na^+ concentration was also increased in this experiment this result may simply indicate a preferential Na$^+$ influx. The critical experiment of raising external Cl$^-$ concentration by adding choline chloride has not yet been performed. As normal sodium levels can be maintained in a dilute medium there are probably inwardly directed sodium pumps that do not operate under Magadi conditions.

Much work remains to be done on problems of regulation in *T. grahami*. In particular, ion fluxes need to be measured and the measurement of electrical potentials extended to different ambient concentrations. The picture that emerges from work so far is of a fish using much of its metabolic energy in pumping water and ions. Oxygen consumption measurements in different ambient osmolarities are needed, as are studies on enzyme systems that can operate at the high pH values found in this fish.

Summary

Tilapia grahami, a small cichlid fish, inhabits hot alkaline springs of some of the Lakes of the Great African Rift Valley system. This fish is exposed to adverse environmental conditions of temperature, salinity and pH.

A study has been made of the osmotic regulation and ionic composition of the body fluids and muscles of *T. grahami*, as well as pH/HCO$_3^-$ relationships and the potential difference between the fish and the ambient media. In Lake Magadi, the water had Na$^+$ concentrations of 314 mequiv. l^{-1}, Cl$^-$ was 91 mequiv. l^{-1}, osmolarity was about 600 mOsm, pH ranged between 9.6 and 10.5 and the water temperature rose as high as 43 °C.

The osmolarity of both urine and plasma, as well as the Na$^+$ concentration in these fluids was always lower than that of the media. The urine and plasma osmolarities were 423 and 426 mOsm, respectively, in fish kept in Lake Magadi water. The urinary and plasma Na$^+$ concentrations under these conditions were 184 mequiv. l^{-1} and 217 mequiv. l^{-1}, respectively. The plasma Cl$^-$ concentration was higher (110 mequiv. l^{-1}) than that of Lake Magadi water.

The ionic composition of the bile of *T. grahami* is also reported. These data are compared with those for *T. zillii*, a fish found in less extreme habitats. A potential difference of up to +4.8 mV was measured between the fish gills and the ambient media. Altering the external Cl$^-$ concentration from 100 mm to 300 mm increased the diffusion potential measured.

This study received support from the Danish International Development Agency, The National Geographic Society and the Leverhulme Trust Funds. The authors would like to thank Professor E. Skadhauge for help in carrying out a chemical analysis of body fluids. The able technical assistance of Mr J. K. Kanja is greatly appreciated.

References

Evans, D. H. (1968). Measurement of drinking rates in fish. *Comparative Biochemistry and Physiology*, **25**, 751–3.

Johansen, K., Maloiy, G. M. O. & Lykkeboe, G. (1975). A fish in extreme alkalinity. *Respiration Physiology*, **24**, 159–62.

Kayanja, F. I. B., Maloiy, G. M. O. & Reite, O. B. (1975). The fine structure of the intestinal epithelium of *Tilapia grahami*. *Anatomischer Anzeiger*, **138**, 451–62.

Leatherland, J. F., Hyder, M. & Ensor, D. M. (1974). Regulation of plasma Na^+ and K^+ concentrations in five African species of *Tilapia* fishes. *Comparative Biochemistry and Physiology*, **48A**, 699–710.

Lotan, R. (1969). Sodium, chloride and water balance in the euryhaline teleost *Aphanius dispar* (Ruppell) (Cyprinodontidae). *Zeitschrift für vergleichende Physiologie*, **65**, 455–62.

Maetz, J. (1971). Fish gills: mechanisms of salt transfer in fresh water and sea water. *Philosophical Transactions of the Royal Society*, B, **262**, 209–49.

Maetz, J. & Skadhauge, E. (1968). Drinking rates and gill ionic turnover in relation to external salinities in the eel. *Nature, London*, **217**, 371–3.

Potts, W. T. W. (1968). Osmotic and ionic regulation. *Annual Review of Physiology*, **30**, 73–104.

Potts, W. T. W., Foster, M. A., Rudy, P. P. & Parry Howells, G. (1967). Sodium and water balance in the cichlid teleost, *Tilapia mossambica*. *Journal of Experimental Biology*, **47**, 461–70.

Rahn, H. & Baumgardner, F. W. (1972). Temperature and acid–base regulation in fish. *Respiration Physiology*, **41**, 171–82.

Reite, O. B., Maloiy, G. M. O. & Aasehaug, B. (1974). pH, salinity and temperature tolerance of Lake Magadi *Tilapia*. *Nature, London*, **247**, 315.

Skadhauge, E. & Lotan, R. (1974). Drinking rate and oxygen consumption in the euryhaline teleost *Aphanius dispar* in waters of high salinity. *Journal of Experimental Biology*, **60**, 547–56.

17. The internal environment of marine fish

By RAGNAR FÄNGE

Fish have a large capacity to keep a constant composition of the fluids that bathe their cells. However, the mode of regulation of electrolyte concentrations differs between the major groups of fishes. Large inter-species variations may be found in the concentrations of many organic components. The composition of the blood plasma of fishes, including cyclostomes, has been surveyed many times (Urist & Van de Putte, 1967; Holmes & Donaldson, 1969; Love, 1970; Robertson, 1976). But our knowledge is incomplete, because data are based on studies of a relatively limited number of species, and the fish world comprises many thousands of different species.

I and my colleagues have studied the composition of the blood and a few other body fluids of marine fish. The material consisted of more than 200 specimens belonging to 30 different species. The animals were caught from 100–500 m depth in the Scagerac Sea, part of the North Atlantic. The external environment was more or less the same for all fishes investigated: dark, cold, well-oxygenated sea water with a salinity of 34–36‰. During hauling of the trawl the fishes were enormously stressed and probably asphyxiated. Samples of body fluids were drawn as quickly as possible on board the investigation ship. Clinical-chemical methods, when necessary modified, were used for the analyses (Larsson, Johansson & Fänge, 1977; Fänge, Lidman & Larsson, 1976). The parameters measured were haematocrit, haemoglobin, inorganic ions, glucose, lactate, free fatty acids, cholesterol and protein.

Results

Inorganic ions in blood plasma

The ions measured were sodium, potassium, calcium, magnesium and chloride (Table 1).

For Na^+ and Cl^- a stepwise decrease in plasma levels was found when passing from cyclostomes (*Myxine glutinosa*), holocephalans (*Chimaera monstrosa*) and elasmobranchs to teleosts. Within the teleosts the average concentration of Na^+ in the plasma varied from 162 ± 8 mM (mean and

Table 1. *Inorganic ions in the blood plasma of marine fish*[a]

Species	Mean concentration (mM)				
	Na$^+$	K$^+$	Ca^{2+}	Mg^{2+}	Cl$^-$
Cyclostomes					
Myxine glutinosa	462	9.0	6.3	10.1	453
Holocephalans					
Chimaera monstrosa	317	7.9	4.3	3.8	319
Elasmobranchs					
Raja radiata	272	5.0	4.1	1.4	269
Teleosts					
Trachurus trachurus	198	—	—	—	159
Scomber scombrus	191	8.0	4.1	2.4	168
Clupea harengus	191	5.5	4.0	5.3	201
Cyclopterus lumpus	184	5.3	4.4	—	169
Gadus morhua	176	6.1	3.8	1.0	164
Coryphaenoides rupestris	174	4.4	2.7	3.9	181
Lophius piscatorius	168	4.3	2.7	—	161
Melanogrammus aeglefinus	162	6.2	3.9	2.8	166

[a] From Fänge *et al.*, 1976.

standard deviation) in the haddock (*Melanogrammus aeglefinus*) to 198±12
mM in the horse mackerel (*Trachurus trachurus*). In most species the Na$^+$
level was around 170–180 mM. The values for Cl$^-$ usually followed those
of Na$^+$ rather closely. Apparently, in marine teleosts the concentration of
NaCl in the plasma is very efficiently controlled, being kept at a level of
37–40% of the external environment (the sea water).

The concentrations of K$^+$, Mg^{2+} and Ca^{2+} were found to vary slightly
more than those of Na$^+$. Because K$^+$ and Mg^{2+} dominate intracellularly,
leakage from asphyxiated cells might have influenced the results. On the
whole our results concerning K$^+$, Mg^{2+} and Ca^{2+} in the plasma of cyclostomes, holocephalans and elasmobranchs are in good agreement with
results obtained by Urist & Van de Putte (1967: cyclostomes, holocephalans, elasmobranchs), Read (1971: the holocephalan *Hydrolagus*) and
Robertson (1976: *Myxine*, *Chimaera*). We found Ca^{2+} to vary between 1.8
and 4.5 mM. The lowest value was measured in *Gadus virens*, a good
swimmer, and relatively low values (2.7 mM) were also found in two
bottom-living less active fishes (*Coryphaenoides*, *Lophius*). No distinct
correlation could be found between the plasma concentration of Ca^{2+} and
the swimming capacity. Probably calcium varies according to seasons
(spawning periods), which may explain relatively large interspecies
variations.

Table 2. *Haematocrit and haemoglobin in blood of marine teleostean fish[a]*

	Haematocrit (%)	Haemoglobin (g per 100 ml)
Group I		
Lophius piscatorius	17.2	3.2
Cyclopterus lumpus	19.3	3.3
Coryphaenoides rupestris	24.1	6.4
Group II		
Trachurus trachurus	49.9	12.5
Clupea harengus	51.3	14.0
Gadus virens	52.1	11.4
Scomber scombrus	52.5	12.7

[a] From Larsson *et al.*, 1977.

Haematocrit and haemoglobin

A positive correlation was found, in all species, between the haematocrit value and the concentration of haemoglobin of blood samples. Low values of both these parameters were found in the non-teleostean fish. In these the haematocrit varied between 15 and 20%, and the haemoglobin concentration was less tha 4 g per 100 ml.

Among teleosts two groups could be distinguished which differed markedly. Low haematocrit and haemoglobin values were measured in *Lophius piscatorius* (angler), *Cyclopterus lumpus* (lumpsucker), and *Coryphaenoides rupestris* (rat-tail or grenadier), which are bottom-living poor swimmers (group I, Table 2). In another group of teleosts, comprising actively swimming species, the haematocrit varied around 50%, and the haemoglobin concentration exceeded 19 g per 100 ml. This group included *Trachurus trachurus* (horse mackerel), *Clupea harengus* (herring), *Gadus virens* (saithe) and *Scomber scombrus* (mackerel). Undoubtedly there is a good correlation between the degree of swimming activity and the oxygen-carrying capacity of the blood. High haemoglobin content of the blood may be an adaptation to a high oxygen requirement.

Glucose, lactate and protein in blood plasma

When the concentrations of glucose and lactate in the blood plasma of different species of teleosts were compared, a similar pattern of interspecies variation was observed as for the haematocrit and the haemoglobin concentration. Bottom-living, less active fish (group I, Table 3) showed lower values than active swimmers (group II, Table 3). Several authors have found that fish subjected to stress for a few hours develop Hyperglycemia

Table 3. *Some organic compounds in the blood plasma of marine teleostean fish*[a]

	Glucose (mg %)[c]	Lactate (mg %)	Cho-lesterol (mg %)	FFA[b] (μM)	Protein (g %)
Group I					
Lophius piscatorius	1.1	9.0	190	212	3.8
Coryphaenoides rupestris	21.9	32.9	237	405	2.0
Cyclopterus lumpus	30.5	14.2	695	474	2.5
Group II					
Gadus virens	33.1	52.7	575	1078	4.8
Scomber scombrus	43.3	113.7	358	430	4.3
Gadus morhua	49.5	54.1	840	1260	5.5
Melanogrammus aeglefinus	57.7	59.6	460	1542	4.2
Clupea harengus	66.9	111.3	649	849	3.3

[a] From Larsson *et al.*, 1977.
[b] Free fatty acid.
[c] mg % = mg per 100 ml plasma.

(Fletcher, 1975). In our material the stress period might have been too short to produce an increase in blood glucose concentrations. Instead blood glucose might have been low because of consumption of glucose by the muscles.

Lophius (angler) was remarkable in showing extremely low glucose values. Probably *Lophius*, in similarity to the eel (*Anguilla anguilla*: Lewander *et al.*, 1976), tolerates marked hypoglycaemia without reacting with coma or convulsions.

The total plasma protein was, on average, higher in the group of active fish than in the less active teleosts (Table 3). The finding of a low protein content in the plasma of *Lophius* is in agreement with results of other authors (Palacios, Rubió & Planas, 1972). As a consequence of their low plasma protein concentration, some teleosts supposedly have a low colloid osmotic pressure of the blood.

Cholesterol and free fatty acids (FFA) in the blood plasma

In non-teleosts the cholesterol concentrations in the blood plasma varied between 86 and 474 mg per 100 ml and the concentration of free fatty acid between 97 and 290 μM (Larsson & Fänge, 1977). Values of similar magnitudes were found in the blood plasma of less active teleosts (group I, Table 3), whereas, on average, considerably higher values were measured in the blood plasma of more active teleosts (group II, Table 3). Remarkably high cholesterol values were found in gadids (codfish), which, when compared to mammals, showed a marked hypercholesterolaemia. Evidence was obtained that in *Gadus* species the highest values of blood cholesterol are found in the winter and early spring, probably in connection with spawning.

The lymph system

Lymph-like fluids constitute a large although often neglected part of the internal environment of vertebrates. In some fishes like the elasmobranchs, lymph vessels seem to be poorly developed, but teleosts as a rule possess a voluminous lymph system. In the plaice (*Pleuronectes platessa*) the volume of the lymph contained in the muscles and the skin exceeds the total blood volume many times (Wardle, 1971). An especially high lymph content is found in mesopelagic fish without a swimbladder. In these the lymph, which is poor in electrolytes in comparison with the sea water, apparently helps the fish to establish buoyancy (Blaxter, Wardle & Roberts, 1971).

A special case is found in the deep water fish *Coryphaenoides rupestris* (rat-tail or grenadier fish) and other macruorids. In these the head contains wide canals filled with a lymph-like fluid. This canal endolymph resembles blood plasma in its electrolyte composition (with the possible exception of K^+ which is low), but concentrations of protein and glucose are much lower than in the blood plasma (Fänge, Larsson & Lidman, 1972).

In several species of teleosts lymph may be sampled either from the longitudinal lymph trunk dorsal to the spinal cord (Kampmeier, 1969) or from lymph sinuses in the orbit. Our preliminary results from analyses of lymph from the marine teleosts *Anarrhichas lupus* and the cod (*Gadus morhua*) show that the lymph has a considerably lower protein concentration than the blood plasma. Supposedly, as in mammals, lymph is formed by ultrafiltration from small blood vessels. In spite of the relatively low arterial blood pressure in teleostean fish, apparently large quantities of lymph are formed. To reach an understanding of the mechanism of lymph formation in fish it is desirable to get data on the colloid osmotic pressure of lymph in relation to blood plasma.

Conclusions

The internal environment of marine fish, the plasma and the lymph, is characterised by an efficient electrolyte homeostasis. The total level of inorganic ionic concentrations differs between the major groups, being highest in cyclostomes (myxinids) and lowest in teleosts.

Large variations were found in the concentrations of some organic substances in the blood plasma. To a considerable extent the concentrations of organic compounds are correlated to the degree of muscular activity or metabolism of the fish. Remarkably high levels of cholesterol were observed in gadids (codfish).

In teleostean fish the lymph, in addition to the plasma, is an important part of the internal environment.

The article is based on work done in co-operation with Å. Larsson, U. Lidman and M.-L. Sjöbeck.

References

Blaxter, J. H. S., Wardle, C. S. & Roberts, B. L. (1971). Aspects of the circulatory physiology and muscle systems of deep-sea fish. *Journal of the Marine Biological Association*, **51**, 991–1006.

Fänge, R., Larsson, Å. & Lidman, U. (1972). Fluids and jellies of the acustico-lateralis system in relation to body fluids in *Coryphaenoides rupestris* and other fishes. *Marine Biology*, **17**, 180–5.

Fänge, R., Lidman, U. & Larsson, Å. (1976). Comparative studies of inorganic substances in the blood of fishes from the Scagerac Sea. *Journal of Fish Biology*, **8**, 441–8.

Fletcher, G. L. (1975). The effects of capture, 'stress', and storage of whole blood on the red blood cells, plasma proteins, glucose and electrolytes of the winter flounder (*Pseudopleuronectes americanus*). *Canadian Journal of Zoology*, **53**, 197–206.

Holmes, W. N. & Donaldson, E. M. (1969). The body compartments and the distribution of electrolytes. In *Fish Physiology*, ed. W. S. Hoar & D. J. Randall Jr, vol. 1, pp. 1–89. New York, London: Academic Press.

Kampmeier, O. F. (1969). *Evolution and Comparative Morphology of the Lymphatic System*. Springfield, Ill.: C. C. Thomas.

Larsson, Å. & Fänge, R. (1977). Cholesterol and free fatty acids (FFA) in the blood of marine fish. *Comparative Biochemistry and Physiology*, **57B**, 191–6.

Larsson, Å., Johansson-Sjöbeck, M.-L. & Fänge, R. (1977). Comparative study of some haematological and biochemical blood parameters in fishes from the Skagerrak. *Journal of Fish Biology*, **9**. (In Press.)

Lewander, K., Dave, G., Johansson-Sjöbeck, M.-L., Larsson, Å. & Lidman, U. (1976). Metabolic effects of insulin in the European eel, *Anguilla anguilla* L. *General Comparative Endocrinology*, **29**, 455–67.

Love, R. M. (1970). *The Chemical Biology of Fishes*. New York: Academic Press.

Palacios, L., Rubió, M. & Planas, J. (1972). Plasma chemical composition in the goosefish (*Lophius piscatorius*, L.). *Journal of Fish Biology*, **4**, 99–102.

Read, L. J. (1971). Chemical constituents of body fluids and urine of the holocephalan *Hydrolagus colliei*. *Comparative Biochemistry and Physiology*, **39A**, 185–92.

Robertson, J. D. (1976). Chemical composition of the body fluids and muscle of the hagfish *Myxine glutinosa* and the rabbit-fish *Chimaera monstrosa*. *Journal of Zoology*, **178**, 261–77.

Urist, M. R. & Van de Putte, K. A. (1967). Comparative biochemistry of the blood of fishes: identification of fishes by the chemical composition of serum. In *Sharks, Skates and Rays*, ed. P. W. Gilbert, R. F. Mathewson & D. P. Rall, pp. 271–85. Baltimore: Johns Hopkins Press.

Wardle, C. S. (1971). New observations on the lymph system of the plaice *Pleuronectes platessa* and other teleosts. *Journal of the Marine Biological Association, UK*, **51**, 977–90.

18. Control of catecholamine release from chromaffin tissue in a teleost fish

By T. ABRAHAMSSON & S. NILSSON

Chromaffin cells, similar to those found in the mammalian adrenal medulla and containing large quantities of catecholamines (adrenaline and noradrenaline), are also present in the lower vertebrates. Such cells, which contain intracellular granules resembling the adrenergic storage granules of mammalian adrenal medullary cells, are found in the heart of cyclostomes (Bloom *et al.*, 1961). Stress induces a release of amines in the sea lamprey (*Petromyzon marinus*), where the concentration of noradrenaline in the blood plasma increases from 0.21 to 4.39 μg per 100 ml plasma (Mazeaud, 1969*a*).

In elasmobranchs, the chromaffin tissue is arranged segmentally in chromaffin bodies, sometimes associated with sympathetic ganglia (Young, 1933; Nicol, 1952; Mazeaud, 1969*b*; Gannon, Campbell & Satchell, 1972; Nilsson, Holmgren & Grove, 1975). These fish also have the ability to increase their plasma levels of catecholamines, noradrenaline dominating, during stress. Normal plasma concentrations in *Scyliorhinus canicula* of 2.4 μg adrenaline per 100 ml plasma and 4.8 μg noradrenaline per 100 ml plasma are elevated to 8.5 and 13.8 μg per 100 ml plasma, respectively, after keeping the animal in air for 30 min (Mazeaud, 1969*b*).

The gills of elasmobranchs are dilated by catecholamines, and Davies & Rankin (1973) concluded from their experiments that the plasma of *Scyliorhinus canicula* contained a concentration of catecholamines high enough to play an important role in controlling the vascular resistance of the gills and catecholamines released from the anterior chromaffin bodies in *Heterodontus portusjacksoni* are thought to control the heart of this species (Gannon *et al.*, 1972). Similar conclusions were reached in a study of the spleen and arteries of *Scyliorhinus* (Nilsson *et al.*, 1975), where the concentration producing 50% of maximal contraction (EC_{50}) of spleen and artery strips *in vitro* was found to be 11.8 and 42.3 μg per 100 ml (spleen) and 3.6 and 16.9 μg per 100 ml (arteries) for adrenaline and noradrenaline, respectively. These values, compared with the plasma concentrations estimated by Mazeaud (1969*b*), show that circulating catecholamines can contribute to the control of these organs. The sympathetic nervous control of the spleen of *Scyliorhinus* appears to be little developed, and thus circulating amines may be the major way of splenic smooth muscle control in this species. In another elasmobranch, *Squalus acanthias*, splenic innervation is better developed, but reinforcement of this innvervation by circulating catecholamines is probable (Nilsson *et al.*, 1975).

In teleosts, the chromaffin cells line the walls of the posterior cardinal veins usually inside the head kidney. In the cod (*Gadus morhua*), catecholamine analysis shows that adrenaline is the dominating catecholamine in most tissues, and especially high levels of adrenaline and noradrenaline could be found in the chromaffin cells in the walls of the posterior cardinal vein (Abrahamsson & Nilsson, 1976). The dominating plasma catecholamine is adrenaline in most species, including the cod, although a dominance of noradrenaline has been demonstrated in a few species (Euler & Fänge, 1961; Mazeaud, 1971; Grove *et al.*, 1972; Abrahamsson & Nilsson, 1976). The adrenaline of the chromaffin cells is synthesized from noradrenaline by the action of the methylating enzyme phenylethanol-*N*-methyltransferase (PNMT), which is present in the chromaffin tissue (Mazeaud, 1972; Abrahamsson & Nilsson, 1976). This enzyme is also responsible for intraneuronal synthesis of adrenaline in sympathetic neurons of the cod (Abrahamsson & Nilsson, 1976). Experiments performed on the spleen of the cod show that sympathetic nerve sectioning or injection of reserpine of 6-hydroxydopamine significantly decreases the content of both adrenaline and noradrenaline, which suggests that both amines are stored in the neurones and thus may act as transmitter substances (Abrahamsson & Nilsson, 1975).

The rate of turnover of catecholamines in fish plasma is similar to that measured in other vertebrates, and the main metabolites of the amines are conjugated *o*-methylated catechols in both elasmobranchs and teleosts (Mazeaud & Mazeaud, 1973*a, b*).

Recent work on the cod, *Gadus morhua*, has led us to believe that, in this teleost also, circulating catecholamines contribute to the over-all sympathetic control of the circulatory system (Wahlqvist & Nilsson, 1977), and Holmgren (1977) demonstrated a marked effect on heart rate by catecholamines released *in situ* from the chromaffin tissue in this species. In addition, the contraction force of the heart is probably influenced by fluctuations in the concentration of circulating catecholamines *in vivo* (Holmgren, 1977).

Innervation of the chromaffin tissue in the cod

The chromaffin cells line the walls of the posterior cardinal veins inside the head kidney. The arrangement of blood vessel and nerves in this area is diagrammatically outlined in Fig. 1.

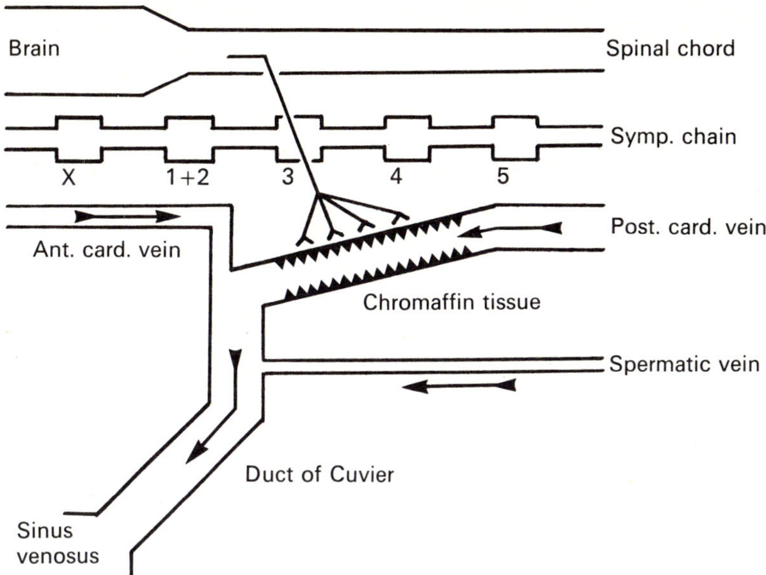

Fig. 1. Diagrammatic representation of the main features of the anatomy in the head kidney region on the left side in the cod, *Gadus morhua*. Nerve fibres to the chromaffin tissue may leave the central nervous system in rami communicantes of spinal nervess other than the third, and run in the sympathetic chain to leave in the distinct medullated nerve to the chromaffin cells. Arrows indicate the direction of blood flow in the veins. Numbers at the sympathetic (Symp.) ganglia refer to the corresponding cranial (X) or spinal (1–5) nerve. Post = posterior; Ant = anterior; Card. = cardinal.

A detailed anatomical and histochemical investigation of the sympathetic nervous system of the cod revealed medullated fibres supplying the walls of the cardinal veins, leaving the sympathetic chains at the level of the sympathetic ganglion corresponding to the third spinal nerve (Nilsson, 1976). These fibres form a distinct nerve on the left side, and it proved possible to reach the fibres for electrical stimulation *in situ* without damaging the head kidney. A study was undertaken to find out whether or not these fibres were responsible for the control of catecholamine release from the chromaffin tissue in the cod (Nilsson, Abrahamsson & Grove, 1977).

In these experiments, the catecholamine concentration was measured fluorimetrically (Häggendal, 1963) in fractions of perfusate during perfusion of the head kidney (inflow through the left posterior cardinal vein, outflow collected from the left duct of Cuvier; see Fig. 1). Alternatively, experiments where the chromaffin tissue was pre-loaded with tritium-labelled adrenaline were performed. The tritium level in fractions of perfusate was then used as indication of catecholamine release.

Both injection of acetylcholine and electrical stimulation of the nervous supply to the chromaffin tissue proved effective in releasing both adrenaline and noradrenaline, adrenaline dominating. Furthermore, the release of tritium evoked in both ways could be inhibited by the ganglion blocking agent hexamethonium, suggesting that the medullated fibres stimulated are preganglionic and cholinergic. Thus, the nervous control of the chromaffin tissue in the cod closely resembles the arrangement in the adrenal medulla of mammals.

Effects of stress

In an attempt to see whether the sympathetic control of the chromaffin tissue is of importance *in vivo*, a series of experiments was performed where plasma levels of catecholamines were determined fluorimetrically (Häggendal, 1963) in 'unstressed' control animals and animals 'stressed' by being kept out of water for 15 min.

Three groups of fish were used: controls, sham-operated fish where only incisions in the dorsal skin and muscles were made, and operated fish where the first to the fourth spinal nerves with their preganglionic sympathetic fibres were sectioned bilaterally close to their exit from the skull/spinal column, 4 days prior to the experiment. In the operated animals no significant increase in plasma concentration of catecholamines was induced by the stress (Table 1), which favours the view that the sympathetic outflow at these levels is responsible for control of catecholamine release *in vivo*.

Table 1. *Effect of stress on the plasma levels (in μg per 100 ml plasma) of catecholamines in the cod*, Gadus morhua

Each value represents the mean of 8–10 animals. Levels of significance (Student's *t* test) between unstressed and stressed fish in each vertical group are indicated as *** = $P < 0.001$, ** = $P < 0.01$. (For details see Nilsson *et al.*, 1977.)

	Untreated		Sham-operated		Operated	
	A	NA	A	NA	A	NA
Unstressed	0.73	0.08	0.62	0.10	0.36	0.07
Stressed	4.10***	1.05***	3.92**	0.51***	0.60	0.10

A = adrenaline; NA = noradrenaline.

Conclusions

In the cod, *Gadus morhua*, circulating catecholamines may contribute to the overall sympathetic control of the circulatory system, reinforcing the adrenergic innervation of the heart and blood vessel. The circulating amines possibly also affect other factors such as chromatophores, alimentary canal and other smooth muscle systems, and metabolism of carbohydrates and fat (Nakano & Tomlinson, 1967; Larsson, 1973). The catecholamines, mainly adrenaline, are released from stores in the chromaffin tissue by the action of a sympathetic nervous system mainly at the levels of the first to the fourth spinal nerves and run as medullated, presumably preganglionic cholinergic fibres to the chromaffin tissue.

The circulating catecholamines of elasmobranchs and teleosts may represent a primitive slower and more general mechanism of sympathetic control, which has been taken over to a larger extent by adrenergic nerve fibres, with their rapid, more direct and restricted action in higher vertebrates.

References

Abrahamsson, T. & Nilsson, S. (1975). Effects of nerve sectioning and drugs on the catecholamine content in the spleen of the cod, *Gadus morhua*. *Comparative Biochemistry and Physiology*, **51C**, 231–3.

Abrahamsson, T. & Nilsson, S. (1976). Phenylethanolamine-*N*-methyl transferase (PNMT) activity and catecholamine content in chromaffin tissue and sympathetic neurons in the cod, *Gadus morhua*. *Acta Physiologia Scandinavia*, **96**, 94–9.

Bloom, G., Östlund, E., Euler, U. S. v., Lishajko, F., Ritzén, M. & Adams-Ray, J. (1961). Studies on catecholamine-containing granules of specific cells in cyclostome hearts. *Acta Physiologica Scandinavica*, **53**, Supplement 185, 1–35.

Davies, D. T. & Rankin, J. C. (1973). Adrenergic receptors and vascular responses to catecholamines of perfused dogfish gills. *Comparative Pharmacology*, **4**, 139–48.

Euler, U. S. v. & Fänge, R. (1961). Catecholamines in nerves and organs of *Myxine glutinosa*, *Squalus acanthias* and *Gadus callarias*. *General and Comparative Endocrinology*, **1**, 191–4.

Gannon, B. J., Campbell, G. D. & Satchell, G. H. (1972). Monoamine storage in relation to cardiac regulation in the Port Jackson shark, *Heterodontus portusjacksoni*. *Zeitschrift für Zellforschung und mikroskopische Anatomie*, **131**, 437–50.

Grove, D. J., Starr, C. R., Allard, D. R. & Davies, W. (1972). Adrenaline storage in the pronephros of the plaice, *Pleuronectes platessa* L. *Comparative and General Pharmacology*, **3**, 205–12.

Häggendal, J. (1963). An improved method for fluorimetric determination of small amounts of adrenaline and noradrenaline in plasma and tissue. *Acta physiologica Scandanavica*, **59**, 242–54.

Holmgren, S. (1977). Regulation of the heart of a teleost, *Gadus morhua*, by autonomic nerves and circulating catecholamines. *Acta physiologica scandinavica*, **99**, 62–74.

Larsson, Å. (1973). Metabolic effects of epinephrine and norepinephrine in the eel, *Anguilla anguilla* L. *General and Comparative Endocrinology*, **20**, 155–67.

Mazeaud, M. (1969*a*). Andrénalinémie et noradrénalinémie chez la lamproie marine, *Patromyzon marinus* L. *Comptes rendus des Séacnes de la Société de biologie*, **163**, 349–52.

Mazeaud, M. (1969*b*). Influence de stress sur les teneurs en catecholamines du plasma et des corps axillaires chez un Selacian, la Rousette (*Scyliorhinus canicula* L.). *Comptes rendus des Séances de la Société de biologie*, **163**, 2262–6.

Mazeaud, M. (1971). Recherches sur la biosynthèse, la sécrétion et le catabolisme de l'adrénaline et de noradrénaline chez quelques espèces de cyclostomes et de poisson. Thesis, Faculty of Science, Paris.

Mazeaud, M. (1972). Epinephrine biosynthesis in *Petromyzon marinus* (cyclostomata) and *Salmo gairdneri* (teleost). *Comparative and General Pharmacology*, **3**, 457–68.

Mazeaud, M. & Mazeaud, F. (1973*a*). Excretion and catabolism of cat catecholamines in fish. Part I: Excretion rates. *Comparative and General Pharmacology*, **4**, 183–7.

Mazeaud, M. & Mazeaud, F. (1973*b*). Excretion and catabolism of catecholamines in fish. Part II: Catabolites. *Comparative and General Pharmacology*, **4**, 209–17.

Nakano, T. & Tomlinson, N. (1967). Catecholamine and carbohydrate concentration in the rainbow trout (*Salmo gairdneri*) in relation to physical disturbance. *Journal of the Fisheries Research Board of Canada*, **24**, 1701–5.

Nicol, J. A. C. (1952). Autonomic nervous systems in lower chordates. *Biological Reviews*, **27**, 1–49.

Nilsson, S. (1976). Fluorescent histochemistry and cholinesterase staining of sympathetic ganglia in a teleost, *Gadus morhua*. *Acta zoologica, Stockholm*, **57**, 69–77.

Nilsson, S., Abrahamsson, T. & Grove, D. J. (1977). Sympathetic nervous control of adrenaline release from the head kidney of the cod, *Gadus morhua*. *Comparative Biochemistry and Physiology*, **55C**, 123–8.

Nilsson, S., Holmgren, S. & Grove, D. J. (1975). Effects of drugs and nerve stimulation on the spleen and arteries of two species of dogfish, *Scyliorhinus canicula* and *Squalus acanthias*. *Acta physiologica scandanavica*, **95**, 219–30.

Wahlqvist, I. & Nilsson, S. (1977). The role of sympathetic fibres and circulating catecholamines in controlling the blood pressure and heart rate in the cod, *Gadus morhua*. *Comparative Biochemistry and Physiology*, **57**, 65–8.

Young, J. Z. (1933). The autonomic nervous system of selachians. *Quarterly Journal of Microscopical Science*, **75**, 571–624.

Part 3

Fluid mechanics in biology

19. Fluid propulsion by cilia and flagella

By M. A. SLEIGH

The normal function of cilia and flagella is to propel fluids, and organisms of almost all groups, from bacteria to mammals, use them in the performance of one or more physiological functions. The cilia and flagella have been adapted in a number of ways to increase their effectiveness in various situations. This paper will consider some of the principles of fluid propulsion by cilia and flagella and discuss their adaptations in a review of the propulsive functions performed by these organelles.

Some general principles

Cilia and flagella have an identical internal structure (Warner, 1974) and a uniform diameter of about 0.25 μm; a few flagella are somewhat thickened by additional structures and some types of cilia form compound structures by aggregation of many units, but both of these are adaptive features that will be considered later. Most cilia and flagella fall in the length range between 5 and 100 μm, although a few shorter organelles are known and sperm tail flagella and compound cilia may reach lengths of over 1 mm. The difference between cilia and flagella is a functional rather than a structural one, as illustrated in Fig. 1, but as a result of the different form of movement of flagella some structural adaptations have been developed that would not be appropriate for cilia (see below). Typical flagella propagate planar or three-dimensional undulations along their length and propel water parallel to the long axis of the organelle, whilst typical cilia perform an eccentric beat towards one side and propel water parallel to the surface bearing the organelle; intermediate patterns of movement are common, so that there is no sharp distinction between the two types of organelle. The 'flagella' of bacteria are thinner (around 20 nm in diameter) and are believed to be more or less rigid helical structures that are rotated by forces applied at the base where the 'flagellum' penetrates the cell membrane (Berg, 1975). By contrast, it is quite clear that the undulations of cilia and true (eucaryote) flagella are generated within the organelles themselves, presumably through forces generated by active sliding between the component axial fibrils that are enclosed within an extension of the cell membrane (Goldstein, 1974; Satir, 1974).

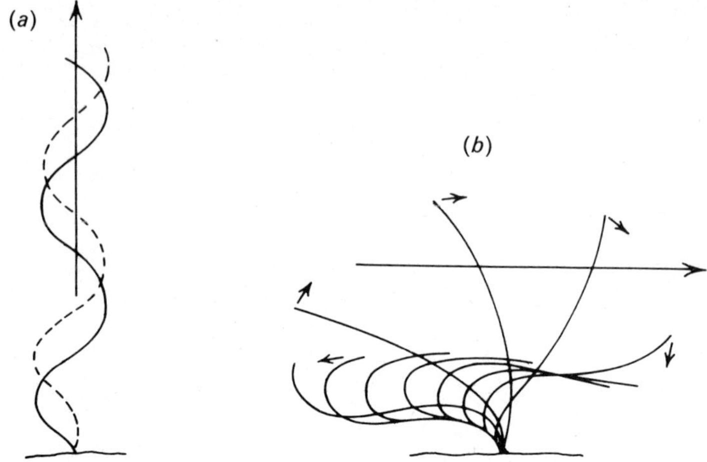

Fig. 1. A comparison of the motion of a flagellum (*a*) with that of a cilium (*b*); the direction of the main flow of water is indicated by the large arrows.

Two types of forces are important in the propulsion of fluid by a moving body: inertial forces (dependent upon mass and acceleration) and viscous forces (dependent upon surface area and viscous shear stress). The ratio of these two forces (inertial forces/viscous forces) is known as the Reynolds number (Re), which is also given by

$$\text{(fluid density} \times \text{speed} \times \text{length)/fluid viscosity.}$$

In the cases of cilia and flagella both the speed and length are very small, so that Re is small (frequently about 10^{-2} or 10^{-3}) and intertial forces are negligible in comparison with viscous forces. By contrast, a fish swimming in the same medium might have an Re of more than 10^3, and viscous forces are then of little importance in comparison with inertial forces.

The fluid immediately adjacent to a surface cannot move relative to that surface, i.e. we have a no-slip zone in the fluid. If a body like a cilium moves in water, the water imediately around it is carried along at the same speed as the body. Because of viscous stresses in the fluid, the water further from the body is also swept along, and at progressively slower speeds as the distance from the body increases. The extent of this zone of shear depends upon the size of the moving body, its speed of motion, the viscosity of the fluid and (if the body is non-spherical) the orientation of the body relative to the direction of motion – a cylinder moving long-itudinally through a fluid carries less fluid than the same cylinder moved at the same speed through the same fluid in a direction perpendicular to its long axis. Similar no-slip zones and regions of decreasing shear occur around cell surfaces, and a cilium attached to a cell will be surrounded by a conical zone of fluid in which its influence is dominant over that of the cell surface, the diameter of the cone at any point along the cilium being approximately equal to the distance of that point from the cell surface. These features were considered in some detail by Blake & Sleigh (1974).

The beat cycle of a cilium is divided into effective and recovery strokes. In the effective stroke the cilium moves relatively quickly, maintains an orientation approximately perpendicular to the direction of motion and is extended at full length above the cell surface; all of these features are associated with maximal fluid propulsion. In the recovery stroke the cilium moves more slowly as it is drawn along its longitudinal axis at a level rather close to the cell surface, a combination of features that gives minimal fluid transport. A large volume of fluid is carried in one direction in the effective stroke, but a much smaller volume is carried back in the recovery stroke, so that there is a considerable net transport in the direction of the effective stroke. The net transport is greatest near the tip of the cilium in the area that only feels the effective stroke, but is reduced to nothing or may even be negative near the ciliary base where the influence of the recovery stroke is prolonged. Clearly the net transport can be made greater if the cilium is longer or moves faster in its effective stroke, and if the recovery stroke takes place as slowly and as close to the surface as possible. The functional advantages of grouping cilia into compound structures of greater length and stiffness, and of metachronal coordination that makes maximum use of collaboration between cilia and minimises interciliary interference, also contribute significantly to the net transport of fluid in various ciliary systems. These aspects of the flow of water around cilia have been discussed by Sleigh (1976).

Measurements of the average velocity of particles at the centre of the water stream that is propelled by the cilia, i.e. at the level of the ciliary tips, gave average flow rates of about ¼ of the ciliary tip velocities (Sleigh & Aiello, 1972). In the same studies it was observed that the rate of flow decreased sharply beyond the tips of the cilia, and the influence of the cilia did not extend to more than two or three ciliary lengths from the cell surface, a feature that influences the disposition of cilia that produce mass flows for such functions as feeding. Since cilia are commonly only 5–20 μm long, fluid propulsion is largely limited to within 20–50 μm of the cell surface, but because this fluid comes from one side and is pushed towards the other, extensive flows can be generated in the fluid, especially where many ciliary tracts work in parallel. Ciliary lengths of 12 μm are about average, and at a common frequency of about 30 Hz a ciliary tip velocity of about 4 mm s^{-1} would be expected; this would result in a flow of fluid at an average rate of about 1 mm s^{-1} at the level of the ciliary tips.

Comparable propulsion by flagella, details of which can be found in reviews by Holwill (1966, 1974), could give a water flow of about 100 μm s^{-1} by propagating waves of length of about 25 μm at a frequency of about 40 Hz along the flagellum at a speed of about 1 mm s^{-1}.

Locomotion by cilia and flagella

Body size and methods of propulsion

Small organisms tend to swim by flagella or cilia, while large organisms use muscles for swimming. This is seen in Fig. 2, where it is also evident

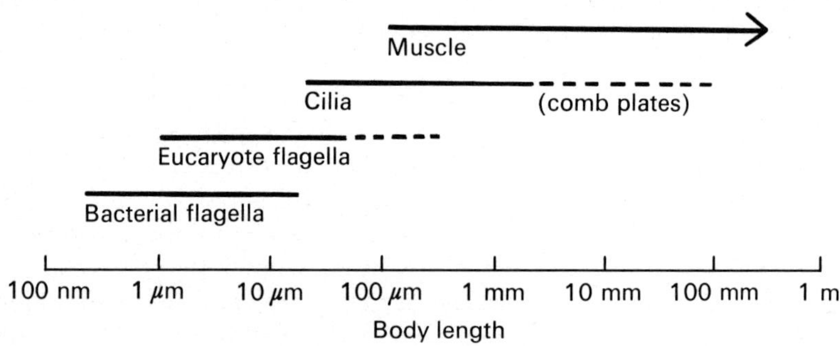

Fig. 2. The range of body lengths of organisms that swim by different methods.

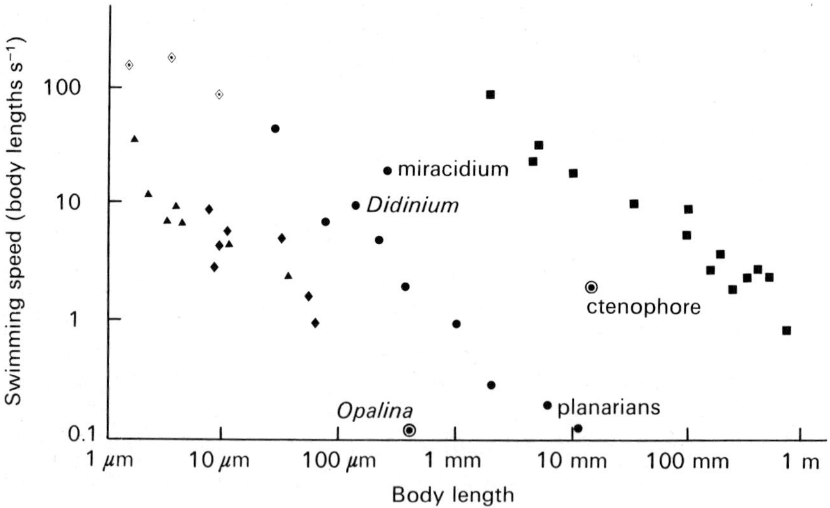

Fig. 3. The relation between maximal swimming speed and body length for a variety of organisms that use various methods of propulsion. ▲ Bacterial flagella; ◆ eucaryote flagella; ◊ animal sperm flagella; ● cilia; ◉ 'special cilia'; ■ muscle. Data from Vaituzis & Doetsch (1969), Holwill (1966), Kearn (1967), Pearl (1903) and papers in Pedley (1977).

that there is substantial overlap, since some organisms propelled by muscles are smaller than the largest organisms propelled by flagella, whilst those propelled by cilia occupy an intermediate position (Sleigh & Blake, 1977). In each of these categories of propulsion there is a tendency for the speed of swimming to fall within a narrow range, so that for each category one can plot values of body length and swimming speed (in body lengths s^{-1}) on logarithmic scales (Fig. 3) and obtain a scatter of points about a line with a gradient of about -1. Each of the series of lines for propulsion by bacterial flagella, eucaryote flagella, cilia and muscles is in general quite distinct, although the position of the ctenophores is rather intermediate. Propulsion by bacterial flagella and by muscles is outside the scope of this paper, but selections of recent discussions will be found in the books edited by Wu, Brokaw & Brennen (1975) and by Pedley (1977).

Swimming by flagella

Male gametes of many phyla of animals and plants as well as flagellated unicellular organisms and the zoospores of lower plants swim by means of flagella. Frequently the gamete cells propelled by flagella are very small, consisting of little more than a nucleus, an acrosome and a small mitochondrial component; such sperm tend to swim at rather high speeds, propelled by a posteriorly directed flagellum which propagates more or less planar waves from base to tip, and occupy a rather anomalous position in Fig. 3 because it is difficult to assign an appropriate 'body length' to the cell. Other sperm cells of animals may have much more complex head and tail structures (see e.g. Baccetti, 1970; Phillips, 1974) and may move in a diversity of ways with planar or helical distally propagated waves of greater or lesser amplitude; the many individual variants of this general type are too diverse to describe here, but often the motion is rather feeble, possibly because the animals concerned are fertilised internally and the sperm may only be required to swim a short distance under their own power. The male gametes of a number of plants, particularly such higher plants as ferns and cycads, are relatively complex and multiflagellate and their motion is not well known.

The full diversity of functional adaptations of flagella for swimming is probably seen among the flagellate algae and protozoa, and the major variants are illustrated in Fig. 4a–e (Sleigh, 1973). The cell may be propelled by planar flagellar waves or by helical undulations of the flagellum, and the waves may pass from base to tip so that the cell is pushed along by the flagellum, or the waves may be propagated from tip to base along the flagellum so that the cell is drawn through the water by a flagellum held in front. The hydrodynamic properties of the flagellum may

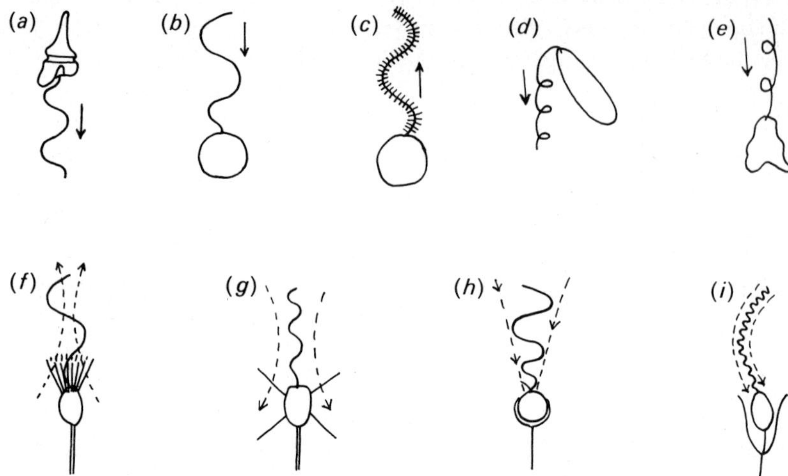

Fig. 4. Various ways in which flagella are used to propel water: (1) in swimming, *a* to *e* (where all organisms progress towards the top of the page and the arrows indicate the direction of propagation of the flagellar waves); and (2) in feeding, *f* to *i* (where the arrows indicate water currents).
(*a*) *Ceratium*, (*b*) *Crithidia*, (*c*) *Ochromonas*, (*d*) *Euglena*, (*e*) *Mastigamoeba*, (*f*) *Codonosiga*, (*g*) *Actinomonas*, (*h*) *Ochromonas*, and (*i*) *Bicoeca* (after Sleigh, 1973).

be modified by the presence of lateral hairs on the axis; if these are long and flexible and wrap around the flagellum they merely serve to increase its effective diameter (Holwill, 1974), but if they are stiff and project laterally in two rows in the plane of undulation of the flagellum then distally directed waves on the flagellum generate a proximally directed water flow, as in the case of *Ochromonas* (Holwill & Sleigh, 1967). Functional adaptations involving changes in the amplitude of the wave as it passes along the flagellum, or the generation of waves of restected amplitude, can alter the form of the water flow around the naked or hispid flagella, as can be seen in the feeding currents of attached phagotrophic cells, Fig. 4*f–i* (Sleigh, 1964); similar flow patterns presumably occur around the flagella of detached cells when they swim, but these are less easy to study. Where large numbers of flagella occur on a single cell they lack appendages and show metachronal coordination with propulsive forces provided by the wave fronts that pass across the body surface; a comparable situation is seen in *Opalina* (see Fig. 3), whose ciliary organelles form metachronal waves with a continuous outline (Machemer, 1974).

Swimming by cilia

Ciliated organisms that swim through the water seem to burrow through the fluid, dragging a narrow zone of water backwards over their surface as they progress. It is interesting that the volume of fluid influenced by the passage of a swimming ciliated protozoon is less than that influenced if the same body were to fall through the same distance passively under the influence of gravity (Blake & Sleigh, 1974). Such a swimming action may depend upon a complete covering of cilia or upon bands or rows of cilia restricted to parts of the body. Where the coverage of cilia is complete, the cilia tend to be simple, short (around 10 μm long) and evenly spaced over the surface. The effective stroke is directed backwards, sometimes a little obliquely so as to produce some rotation of the body as it moves along, and lines of cilia in the plane of motion of the effective stroke move synchronously. Waves of metachronal coordination pass to one side or the other at right angles to the lines of synchrony in a direction that is correlated with the path taken by the ciliary tip during the recovery stroke (Aiello & Sleigh, 1972; Machemer, 1974). In their effective strokes the cilia are extended to their full length, but in the recovery strokes they are bent low and to one side, close to the cell surface, so that the reverse flow in this phase of beat is minimised; if the cilia move to the right in the recovery stroke the waves move to the right, and vice versa. This form of swimming is found in ciliated protozoa, larvae and small worms ranging in length from about 25 μm to several mm, and is associated with swimming speeds in the range of about 0.5–1 mm s^{-1}.

The restriction of cilia to one or a few bands of cilia seems to be associated with somewhat faster swimming at rates between 1 and 5 mm s^{-1}, exemplified in Fig. 3 by *Didinium* and a miracidium larva. These bands may be transverse rings around a spherical or cylindrical body (e.g. trochophore larvae, *Didinium*), or may be arranged around the margin of special lobes of the body (e.g. rotifers, veliger larvae), or around an extended lip (e.g. *Stentor*, *Vorticella*) (Sleigh & Barlow, 1976; Sleigh & Blake, 1977). In these cases the cilia are densely aggregated within the band and tend to be longer than on continuously ciliated surfaces; the cilia may remain separate but closely bunched within the metachronal waves, as in *Didinium*, or they may move as compound organelles with greater or lesser cohesion between the component cilia, as in *Stentor* or molluscan veliger larvae. The ciliated band is normally orientated at right angles to the direction of locomotion, the effective stroke is perpendicular to the band, and the waves travel along the band to the observer's right (e.g. *Stentor*) or to his left (e.g. molluscan veliger). The compound cilia attain lengths of some 30–50 μm, permitting greater tip speeds and the possibility of faster propulsion. The compound cilia of ctenophores are discussed later as a special case.

The functioning of these cilia as organelles for locomotion is complicated by the fact that they are often also used for collection of particulate food from the water; so that both the arrangement of the cilia and the way in which they move are dependent upon a compromise between swimming and feeding. One way of avoiding this is illustrated by *Vorticella*, in which the feeding cilia can be used for swimming if the cell is accidentally broken from its stalk, but migratory cells normally develop a second independent ciliary girdle for swimming and tuck in the feeding cilia during migration.

The activity of most cilia and flagella appears to be under the control of the cell in that their activity can be stopped for a shorter or longer period by appropriate stimulation. In the case of the multicellular forms like veliger larvae, this control is mediated by nerves (Carter, 1926). Many ciliated protozoa can also reverse the direction of swimming in response to stimulation by reversing the direction of the effective stroke of the ciliary beat following a depolarisation of the ciliated cell and a rise in the intracellular calcium concentration (Naitoh & Eckert, 1974). Such a reversal is rare in flagellates, but *Crithidia* provides an example in which reversal of the propulsive wave can follow stimulation (Holwill, 1965).

Crawling by cilia

The smaller flatworms and a number of other small invertebrates progress over surfaces by ciliary action. This form of motion occurs in aquatic and terrestrial situations, though in the latter case the ciliary propulsion is accompanied by the secretion of much mucus. It is not clear how far this can be regarded as swimming on a very thin layer of fluid. The cilia involved are generally rather short, often only 6 or 7 μm in length, and tend to form a complete cover over the surface of the sole, upon which the animal moves. Flatworms show wave patterns similar to those of uniformly ciliated swimming animals, with longitudinal lines of synchrony and wave propagation to right or left, e.g. *Convoluta* (Dorey, 1965). Planarians with lengths of up to at least 11 mm can crawl by cilia and maintain a speed of about 1.5 mm s^{-1} (Pearl, 1903).

Swimming of ctenophores by comb plates

The largest known ciliary structures are the comb plates of ctenophores in which several hundreds of thousands of cilia form a rectangular paddle, with cross-linking between the cilia in the plane of the flat surface of the plate. These plates may attain lengths of from several hundred μm up to

1 mm or more and have widths of 0.5 mm or more. Normally the plates are disposed in rows with the flat surface of the plate perpendicular to the midline of the row, and the plates move in sequence with typical ciliary beat pattern and an antiplectic metachronism (Sleigh, 1974). By this means ctenophores of up to 10 cm or more in length are able to swim, but larger, strap-shaped, ctenophores like *Cestus*, that may attain lengths of 0.5 m or more, move by muscular undulations and only use the comb plates for maintaining their orientation in the water. The Reynolds number of comb plates approaches 1, and this means that inertial forces cannot be neglected; indeed observations of particle flow around moving comb plates show a persistence of flow after the adjacent plates have completed their effective stroke (Sleigh & Aiello, 1972), and the water is expelled from between the plates as a jet. The effect of motion of comb plates is comparable in size and rate with the beating of the limbs of small crustaceans, so that it is not too surprising that in Fig. 3 the ctenophore occupies a position intermediate between the line for animals propelled by cilia and the band of data for animals propelled by muscular activity.

Propulsion of fluids over fixed surfaces
Water transport by gills and similar structures

Ciliated gills and tentacles of invertebrate animals serve two principal functions, the circulation of water for respiratory exchange and the maintenance of water currents from which food can be filtered out. Those gills which serve only the former purpose may occur over the whole body and may have only rather simple ciliary bands that produce a local flow of fluid around the gills, but do not normally maintain an extensive current unless the animal lives in a tube. Where the gills or tentacles are also used in feeding they tend to be organised in a fairly regular pattern of bars and slits, with the water-propelling cilia arranged along the walls of the slits and filter systems on the bars. Since the main zone of water propulsion is within one or two ciliary lengths from the cell surface, it is not too surprising that the width of the slits is related to the length of the cilia, so that the distance between two adjacent ciliated bars is only two or three times the ciliary length; it is also interesting that the metachronal waves of the bands of cilia at either side of the slits pass in opposite directions to enhance the propulsive effect (Sleigh, 1974). This pattern is found in the gills of lamellibranch molluscs (Aiello & Sleigh, 1972), in ascidians (Mackie *et al.*, 1974), in the tentacles of polychaete worms and in the lophophores of polyzoans and related animals where the feeding structure often has the form of a conical funnel of ciliated bars (Atkins, 1932). Such an arrangement provides an extensive surface for filtration and allows the maintenance of substantial flows of water from one side of the gill or tentacle fan to the other, e.g. a *Mytilus* 6 cm long may pump 2.5 l of water through its gills in an hour (Dral, 1967).

Propulsion by cilia in fluid-filled tubes

In addition to the circulation of fluids in more open cavities like coelomic spaces and the ventricles of the vertebrate central nervous system, cilia are used to move fluids through tubes, most commonly in excretory and reproductive systems. Such a system can be studied in the nephridia of annelids, where the cilia line the walls of the tube and project about half-way to the centre of the tube. Blake & Sleigh (1974) showed theoretically that a maximal constant velocity across the central region of the tube could be maintained if the ratio of ciliary length to tube radius was about 0.3–0.5. The propulsion of fluid in the flame cells of some lower invertebrates depends upon the activity of flagella.

Propulsion of mucus

Mucus glands are often associated with ciliated epithelia that serve a function of transporting particles either for cleansing or for feeding. The epithelium may be functioning in an aquatic environment or may be covered with air and have only a thin aqueous layer surrounding the cilia. Packets of the long-chain mucopolysaccharide molecules are extruded from goblet cells and disperse a little as they become diluted with water and spread into small rafts of mucus that are carried along near the level of the ciliary tips (Iravani & van As, 1972). The rafts tend to adhere together to form viscoelastic strands or even sheets that are drawn across the surface of the epithelium above a more fluid layer that surrounds the ciliary shafts. There seems to be agreement that the recovery strokes of the cilia take place beneath the mucus, whilst in the effective strokes the tips of the cilia penetrate the mucus and claw it forward. The cilia involved in the transport of mucus in tetrapod vertebrate respiratory systems are only 5–7 μm long and achieve maximal tip speeds of rather less than 1 mm s^{-1} (Aiello & Sleigh, in preparation), but are associated with mucus transport rates of the order of 0.2–0.5 mm s^{-1}, the relatively high coupling of cilia to mucus presumably resulting from the high viscosity of the mucus. Although the mucus is moved forward fairly quickly, the resultant motion of the watery fluid within the ciliary layer is supposed to be very small because forward and backward motions of the cilia are more or less balanced. The rafts and normal strands of mucus are usually not more than 5 μm or so thick, but mucus of thicknesses of

20 μm and more seems to be successfully transported by the cilia. The strands of mucus bearing food particles that are carried along the gill bars of *Mytilus* by the bands of frontal cilia appear to behave in very much the same way as the strands on epithelia exposed to the air.

References

Aiello, E. & Sleigh, M. A. (1972). The metachronal wave of lateral cilia of *Mytilus edulis*. *Journal of Cell Biology*, **54**, 493–506.

Atkins, D. (1932). The ciliary feeding mechanism of the Entoproct Polyzoa, and a comparison with that of the Ectoproct Polyzoa. *Quarterly Journal of Microscopical Science*, **75**, 393–423.

Baccetti, B. (1970). *Comparative Spermatology*. Rome: Accademia Nationale dei Lincei.

Berg, H. (1975). How bacteria swim. *Scientific American*, **232** (2), 36–44.

Blake, J. R. & Sleigh, M. A. (1974). Mechanics of ciliary locomotion. *Biological Reviews*, **49**, 85–125.

Carter, G. S. (1926). On the nervous control of the velar cilia of the nudibranch veliger. *Journal of Experimental Biology*, **4**, 1–26.

Dorey, A. E. (1965). The organisation and replacement of the epidermis in acoelous turbellarians. *Quarterly Journal of Microscopical Science*, **106**, 147–72.

Dral, A. G. (1967). The movements of the laterofrontal cilia and the mechanism of particle retention in the mussel (*Mytilus edulis* L.). *Netherlands Journal of Sea Research*, **3**, 391–422.

Goldstein, S. F. (1974). Isolated, reactivated and laser-irradiated cilia and flagella. In *Cilia and Flagella*, ed. M. A. Sleigh, pp. 111–30. London: Academic Press.

Holwill, M. E. J. (1965). The motion of *Strigomonas oncopelti*. *Journal of Experimental Biology*, **42**, 125–37.

Holwill, M. E. J. (1966). Physical aspects of flagellar movement. *Physiological Reviews*, **46**, 696–785.

Holwill, M. E. J. (1974). Hydrodynamic aspects of ciliary and flagellar movement. In *Cilia and Flagella*, ed. M. A. Sleigh, pp. 143–74. London: Academic Press.

Holwill, M. E. J. & Sleigh, M. A. (1967). Propulsion by hispid flagella. *Journal of Experimental Biology*, **47**, 267–76.

Iravani, J. & van As, A. (1972). Mucus transport in the tracheobronchial tree of normal and bronchitic rats. *Journal of Pathology*, **106**, 81–93.

Kearn, G. C. (1967). Experiments on host-finding and host-specificity in the monogenean skin parasite *Entobdella solea*. *Parasitology*, **57**, 585–605.

Machemer, H. (1974). Ciliary activity and metachronism in Protozoa. In *Cilia and Flagella*, ed. M. A. Sleigh, pp. 199–286. London: Academic Press.

Mackie, G. O., Paul, D. H., Singla, C. M., Sleigh, M. A. & Williams, D. E. (1974). Branchial innervation and ciliary control in the ascidian *Corella*. *Proceedings of the Royal Society*, B **187**, 1–35.

Naitoh, Y. & Eckert, R. (1974). The control of ciliary activity in Protozoa. In *Cilia and Flagella*, ed. M. A. Sleigh, pp. 305–52. London: Academic Press.

Pearl, R. (1903). Movements and reactions of freshwater planarians, a study in animal behaviour. *Quarterly Journal of Microscopical Science*, **46**, 509–714.

Pedley, T. (ed.) (1977). *Scale Effects of Animal Locomotion*. London: Academic Press.

Phillips, D. M. (1974). Structural variants in invertebrate sperm flagella and their relationship to motility. In *Cilia and Flagella*, ed. M. A. Sleigh, pp. 379–402. London: Academic Press.

Satir, P. (1974). The present status of the sliding microtubule model of ciliary motion. In *Cilia and Flagella*, ed. M. A. Sleigh, pp. 131–42. London: Academic Press.

Sleigh, M. A. (1964). Flagellar movement of the sessile flagellates *Actinomonas*, *Codonosiga*, *Monas* and *Poteriodendron*. *Quarterly Journal of Microscopical Science*, **105**, 405–14.

Sleigh, M. A. (1973). *The Biology of Protozoa*. London: E. Arnold.

Sleigh, M. A. (1974). Metachronism of cilia of Metazoa. In *Cilia and Flagella*, ed. M. A. Sleigh, pp. 287–304. London: Academic Press.

Sleigh, M. A. (1976). Fluid propulsion by cilia and the physiology of ciliary systems. In *Perspectives in Experimental Biology*, ed. P. Spencer Davies, vol. 1, pp. 125–34. Oxford: Pergamon Press.

Sleigh. M. A. & Aiello, E. (1972). The movement of water by cilia. *Acta Protozoologica*, **11**, 265–77.

Sleigh, M. A. & Barlow, D. (1976). Collection of food by *Vorticella*. *Transactions of the American Microscopical Society*, **95**, 489–93.

Sleigh, M. A. & Blake, J. R. (1977). Methods of ciliary propulsion and their size limitations. In *Scale Effects of Animal Locomotion*, ed. T. J. Pedley, pp. 243–56. London: Academic Press.

Vaituzis, Z. & Doetsch, R. N. (1969). Motility tracks: technique for quantitative study of bacterial movement. *Applied Microbiology*, **17**, 584–8.

Warner, F. D. (1974). The fine structure of the ciliary and flagellar axoneme. In *Cilia and Flagella*, ed. M. A. Sleigh, pp. 11–37. London: Academic Press.

Wu, T. Y., Brokaw, C. J. & Brennen, C. (1975). *Swimming and Flying in Nature*, 2 vols. New York: Plenum Press.

20. Fluid skeletons in aquatic and terrestrial animals

By H. D. JONES

Animal life was primarily aquatic and there are many physiological problems associated with the adoption of a terrestrial habitat. One of these is that of support for the body and locomotion, since air is very much less dense than water. It is the function of the skeletal system in association with the muscular system to provide support and posture for the body and to allow muscle antagonism, and skeletal systems are thus stressed compressively.

Two groups of animals are generally regarded as being successful in the terrestrial habitat, the Insecta and the Vertebrata. Both rely on rigid skeletal elements in the form of levers which apply forces at a distance from the muscle and allow for an increase in speed. The success of these two groups is in no small way the result of the presence of a rigid skeleton.

Many different types of invertebrates rely on a hydrostatic or fluid skeleton with varying degrees of 'success', examples being found in: Turbellarian Platyhelminthes; Nemertini; Nematoda; Oligochaete Annelida; Pulmonate Mollusca; Onychophora; Geophilomorph centipedes; Insecta (larvae); Collembola (Insecta). General descriptions of the principles of hydrostatic skeletons are given elsewhere (Trueman, 1975); it is my intention to review the problems of support and locomotion in terrestrial hydrostatic animals and to see if there are any differences in the skeletal system of terrestrial animals when compared with aquatic ones.

Principles and problems

Water, air and density

The density of water at 293 °K is 998 kg m^{-3}, that of sea water is 1025 kg m^{-3} and that of air is 1.293 kg m^{-3}. The density of proteins is about 1330 kg m^{-3}, so that a muscle with about 20 % by weight of protein content is about 5 % denser than sea water (Denton, 1961).

Consider an animal with a body wall of muscle and a liquid-filled body cavity, the liquid having about the same density as sea water. By weight, the liquid weighs about half of the total body weight in air. The animal on land will have to bear the entire weight of the liquid contents, whereas in sea water the weight of the contents is nil and the weight in sea water of the animal is half of $5\% = 2.5\%$ of its weight in air.

Obviously in considering real animals the proportion of flesh to contents and the densities of the tissue, contents and media (fresh water or sea water) will all vary from case to case, but the essential point is that a terrestrial animal has to support the weight of its liquid contents as well as its tissues. When the system of support and locomotion is itself a liquid, which of course flows, it can be appreciated that problems arise, such as the provision of an increased pressure in the liquid and its containment, and the maintenance of the external form of the animal at the same time as allowing change of shape for locomotion to occur.

Pressure, shape and rigidity

Apart from a sphere, the most economical shape for containing fluids under pressure is a cylinder. Since hydrostatic skeletons function through the agency of liquids under pressure it is not without coincidence that most soft-bodied animals are basically cylindrical.

Flexible tubes subjected to internal pressure will become cylindrical and the higher the pressure then the more round will become the cross-sectional shape.

The containment of high pressure by flexible tubes, and thus the ultimate strength in compression of a hydrostatic skeleton, is dependent upon the *tensile* properties of the tube wall rather than on the properties of the contents. Many hydrostatic animals possess a basement membrane in the body wall which acts as reinforcement. This is made of fibrous collagen, and the fibres are orientated spirally or geodetically around the body wall. They are inelastic and of constant length. Clark (1964) and Trueman (1975) give convenient descriptions of the system. Geodetic fibres have been found in the body wall of turbellarians, nemertines, nematodes, a polychaete (Elder, 1972), earthworms (Laverack, 1963), possibly molluscs (Jones, 1975) and arthropod cuticle also contains spiral fibres (Wainwright *et al.*, 1976). Thin-walled tubes under internal pressure are most effectively reinforced against bursting by such geodetic fibres (Wainwright *et al.*, 1976). Man uses spiral reinforcing in hose-pipes and in radial-ply car tyres.

In addition to reinforcement, these fibres provide a means whereby the length:circumference ratio of the body may be controlled, but this is only possible if the animals are not maximally inflated.

When a thin-walled cylinder (engineers talk of 'thin shells') collapses under load it does so by buckling or kinking. The precise conditions under which it buckles depending on the load, the length:circumference ratio,

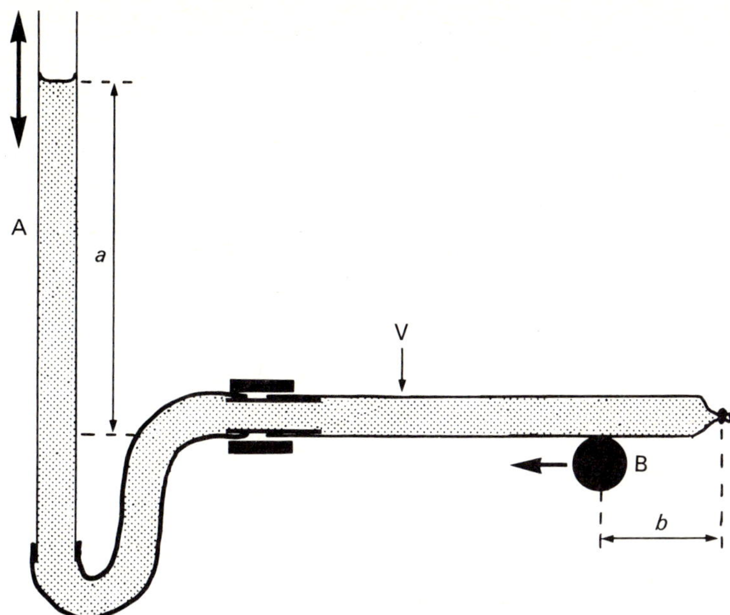

Fig. 1. Apparatus for investigating the effect of internal pressure on the rigidity of dialysis tubing. See text. A = manometer; a = head of pressure; B = moveable bar; b = length of unsupported tube.

the rigidity of the tube wall and the contained pressure. An inflated tube, if bent sufficiently, suddenly buckles, the buckling being an inpushing of the tube wall on the inner surface of the bend. An internal pressure will oppose the inpushing and thus a higher pressure might be expected to reduce buckling. A search of texts of pneumatics and hydraulics failed to reveal any details of this effect and so some simple experiments were conducted.

Lengths of water-filled 8/32 dialysis tubing (Fig. 1, V) were used as test material, tied off at one end and the other end attached to a simple manometer (A), which could be raised or lowered thus altering the head of pressure (a). The tubing was positioned horizontally and a moveable bar (B) supported the tube. By moving the bar, the length of unsupported tube (b) could be increased until buckling occurred under the weight of the tube and contents.

The results (Fig. 2) show that even at no pressure the tube is rigid for some length, presumably as a result of the stiffness of the tube material itself. As the pressure increased, the supportable length increased but not linearly. Above a pressure of about 1176 Pa the rate of increase decreases.

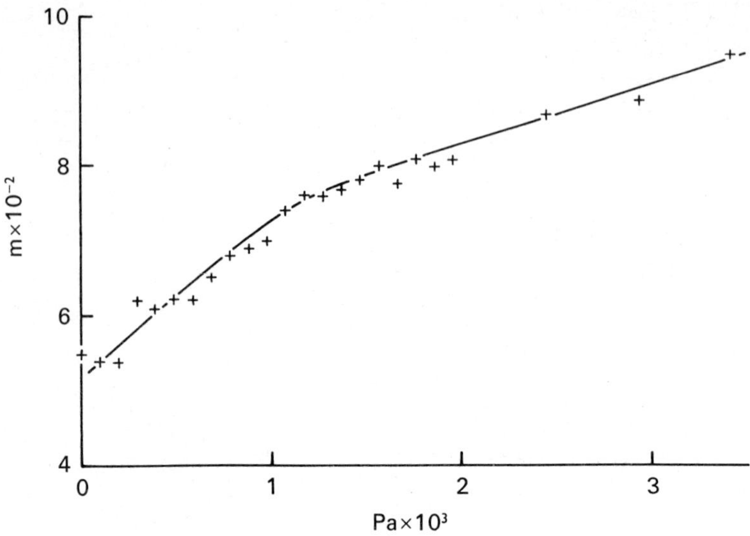

Fig. 2. Results of test on horizontal dialysis tubing. Length at which tubing collapsed ($m \times 10^{-2}$) against internal pressure ($Pa \times 10^{3}$).

I subsequently came across confirmation, from tests on steel pipe-bends under internal pressure, that the pressure effect on rigidity is non-linear (Kafka & Dunn, 1957).

There is then a rapid increase in stiffness as pressure first increases but after a certain value the rigidity increase with increasing pressure is less. The precise value will depend on the properties of the tube wall.

The stability of a vertically orientated tube was also briefly investigated to compare with tentacles of snails. A length of dialysis tubing was knocked sideways by a pendulum, the tubing being under varying pressure. The results (Table 1) show that the tube is stable and capable of remaining erect when it is under pressure in excess of 1471 Pa, a value similar to the value mentioned above in connection with rigidity.

It must be remembered that, to endow rigidity and stability to a tube, values for pressure will depend on the properties of the tube material and on the dimensions of the tube. However, the results of the tests on tubing do indicate the sorts of values for pressure that might occur in hydrostatic terrestrial animals in order that they maintain body shape and posture.

Pressures involved in locomotory forces may be higher than pressure required for postural purposes, though not necessarily so in all cases.

It should be added that lengths of dialysis tubing held under water at

Table 1. *Results of knock-down tests on vertical dialysis tubing*

The results are expressed as number remained erect/number collapsed. A slight hysteresis is apparent.

Pressure (Pa)	Decreasing pressure	Increasing pressure	Total
1961	5/0	5/0	10/0
1863	5/0	5/0	10/0
1754	5/0	5/0	10/0
1667	5/0	5/0	10/0
1569	5/0	5/0	10/0
1471	5/0	5/0	10/0
1373	5/0	4/1	9/1
1274	5/0	2/3	7/3
1176	5/0	2/3	7/3
1078	4/1	2/3	6/4
980	1/4	0/5	1/9
882	0/5	0/5	0/10
784	0/5	0/5	0/10
686	0/5	0/5	0/10
588	0/5	0/5	0/10

no pressure (open-ended) will more or less inflate themselves and lengths many times greater than in air will 'float-out' with no difficulty, clearly indicating that pressure requirements are minimal under water.

Plathyhelminths and nemertines

The presumed primitive form of locomotion in both these groups of animals is ciliary (see Jones & Trueman, 1977). At some indeterminate size as these animals get larger the comparatively weak forces produced by cilia are assisted or replaced by the larger forces produced by muscle. These muscular forces are transmitted by the parenchymatous mesodermal tissue, and it is this material that provides the hydrostatic medium in both groups of animals.

Terrestrial platyhelminths are no exception, using a mixture of ciliary and muscular forces for locomotion (Pantin, 1950). The cilia are always restricted to a narrow, ventral, longitudinal strip; as the animals are cylindrical this is the only part of the body in contact with the substratum during surface creeping. What happens during burrowing is not clear. *Microplana britannicus* Percival, a species Pantin (1950) was not able to study, is apparently much more extensible than other terrestrial species, that is it is able to change body length by elongation to about seven times its contracted length. *M. terrestris* (O. F. Müller) has an extensibility factor of about three (see Clark, 1964). Forward locomotion of *M. britannicus* is by ventral cilia, but it is able to reverse by the use of retrograde peristaltic waves in the manner of an earthworm. The circular and longitudinal muscles presumably act on the parenchyma. *M. terrestris* and *Rhynchodemus bilineatus* (Mecznikow) both have a ventral ciliated sole but locomotion may be assisted by muscular waves passing backwards, either along the ventral surface as in *Rhynchodemus*, or peristaltically involving the whole body as in *M. terrestris*. In either case, locomotion is not strictly comparable with that of earthworms as the muscular waves remain stationary relative to the substratum and the animal leaves a dotted slime track, much the same as 'loping' in terrestrial snails (see Jones, 1975). In these cases the propulsion of the planarians is probably still by cilia, the muscular waves simply lifting regions of the body off the substratum and reducing sliding friction and allowing the animals to move faster. This needs experimental verification.

Nemertines are also able to produce peristaltic waves along the body, but in the terrestrial *Geonemertes dendyi* Dakin normal locomotion is ciliary. The animal effectively swims through a mucous tube of its own secretion, the whole surface of the body being ciliated (Pantin, 1950). The escape reaction of *Geonemertes* is, however, muscular and hydrostatic. The proboscis of nemertines is contained in a liquid-filled cavity, the rhynchocoel, and Pantin (1950) describes how the proboscis may be suddenly everted by contraction of the circular muscles of the body wall. The tip is then attached to the substratum and the rest of the body is drawn forwards.

The values of hydrostatic pressures involved in muscular locomotion of platyhelminthes and nemertines are unknown and direct measurement in any case would not be easy, though it may be possible to calculate them from external forces. Since terrestrial species are much more rounded than aquatic species, the pressures are probably relatively higher in the former.

Nematodes

Nematodes have a remarkable uniformity of body form and structure, having a body wall made up of a stiff outer cuticle and longitudinal muscle fibres. There are no circular fibres, nor cilia. The body cavity is pseudocoelomic and the fluid is under constant very high pressure, giving the worms their typical very rounded cross-section. Pressures have been recorded in *Ascaris* by Harris & Crofton (1957), resting pressure typically being over 9000 Pa and highest pressures exceeding 30 000 Pa. The pressures are apparently produced by the action of the longitudinal muscle on the contents of the pseudocoel, antagonism being provided for by the elastic properties of the body wall in general and the behaviour of the geodetic fibres in the cuticle. It might be that osmotic pressure forces keep the animals fully inflated.

All terrestrial nematodes are very small and restricted to movements in the water-film in soil or on plants (see Lee, 1965).

A few nematodes are able to perform peristaltic types of locomotion or even 'looping', but these are exceptional (Lee, 1965).

Annelids

Terrestrial annelids are typified by earthworms of the genus *Lumbricus*, and their locomotion has been studied in some detail. The body wall has an outer cuticle and a layer each of longitudinal and circular muscle fibres. The hydrostatic medium is the coelomic fluid, and the coelom in earthworms is divided into numerous more or less watertight compartments by transverse septa. This is in clear contrast to other annelids where septa, if present, are incomplete.

Coelomic cavities, unlike parenchymatous tissue, pseudocoels or haemocoels, have openings to the exterior in the form of coelomic ducts and loss of fluid may be a problem. However, Chapman (1950) has calculated probable water losses and shown them to be minimal.

Each segment in an earthworm is to some degree isolated from its neighbours, but the pressure difference bearable by a single septum is not very large, up to 323 Pa (Seymour, 1976). Different regions of the body may of course be under very different pressures.

Chapman (1950) has calculated that an earthworm can produce ten times the coelomic pressure during longitudinal muscle contraction (shortening and fattening) than that which is produced by circular muscle contraction (elongating and thinning). Pressure measurements confirm this, about 20 000 Pa occurring during shortening and up to 2000 Pa during elongation (Seymour, 1969). This latter value represents the sort of pressure, and it is a maximum, required to inflate and protrude the animal. Pressures greater than the circular muscle can bear can only be produced in a worm in soil where the resultant lateral force is used in enlarging the burrow laterally after the segments have entered the pre-existing crevices that occur in terrestrial soils. The transverse septa themselves prevent the worms flattening during this process and enable the worm to concentrate lateral forces onto one or two circumferential points rather than round the whole circumference (Seymour, 1970).

In a septate worm, each segment is capable of *reciprocal* elongation and thickening and the resultant locomotion must be by retrograde peristaltic waves (Jones & Trueman, 1977). In non-septate worms a segment may simultaneously elongate and thicken and direct peristaltic waves are possible in addition to retrograde waves. The type of wave present will depend on neuromuscular, rather than mechanical, factors.

Burrowing in most aquatic worms, and here we can include such as sipunculids and priapulids (Trueman, 1975), involves the penetration of homogeneous soils like sand and mud, and the lack of transverse septa allows the whole of the body musculature to be brought to bear on the proboscis apparatus. Consolidation of the burrow in a homogeneous soil requires forces round the entire circumference of the worm and again the presence of septa is unnecessary.

The presence of septa in earthworms is then an adaptation to the type of soil found on land. Problems of support are almost non-existent, since worms lie in or on soil. It would be interesting to know the length of worm that is able to support itself in air and the pressure necessary in the coelom.

Molluscs

Jones (1975) has recently reviewed the problems of support and locomotion in terrestrial gastropods and only the more salient points need be given here.

Gastropods can be considered as inflated sacs with various hollow appendages, one of which consists of the top-heavy mass of the viscera and shell. Slugs of course lack the shell, but tentacles are present.

A simple hollow aquatic animal that seems appropriate for consideration at this point is the polyp of the sea-anemone (Coelenterata; Anthozoa). These are hollow sacs with hollow appendages, the tentacles. Batham & Pantin (1950) have measured pressures in the coelenteron and these are really minute (30 Pa), illustrating the sorts of pressures necessary in simple aquatic animals. Sea-anemones are quite unable to support themselves in air.

Gastropods are of course rather more complex animals than sea-anemones, the cavity involved in hydrostatics is the haemocoel and the fluid is the blood. Therefore there are likely to be pressures present because of the circulatory gradients in the blood system, and pressure measurements show that the *residual* pressure in the haemocoel is about 300–500 Pa in aquatic or terrestrial animals. Pressures above this value occur in most of the haemocoel, the highest figure being in the ventricle during systole.

In terms of support it is relevant to know what sorts of pressure are present at the base of tentacles and at the base of the visceral mass. Table

Table 2. *Blood pressure (in Pa) and ventricle weight of some aquatic and terrestrial gastropods, the latter marked* *****

Species	Pressure in the head	Pressure in the ventricle at systole	Dry ventricle weight as % dry tissue weight of animal
Viviparus viviparus	805	—	—
Lymnaea stagnalis	681	956	0.10
L. pereger	1142	—	—
Planorbarius corneus	649	—	—
*Helix aspersa**	1681	2481	0.22
*H. pomatia**	1725	2284	—

After Jones, 1975.

2 gives pressures at the base of the tentacles in snails. It is clear that the two terrestrial species (*Helix*) have greater pressures than any of the aquatic ones. *Lymnaea pereger* is prone to make excursions out of water and has pressures intermediate to those of the terrestrial species and of the other aquatic species.

These pressure measurements make an interesting comparison with the tests on dialysis tubing reported earlier.

The source of the haemocoelic pressure is the ventricle. Since terrestrial snails require higher blood pressures, the ventricle of land snails is hypertrophied relative to the aquatic ones (Table 2).

It is difficult to measure pressures beneath the visceral mass of snails but the haemocoel here is in close contact with that under the tentacles and pressures are similar. In *Helix* the shell and viscera seem to require a pressure of about 784 Pa to support it (Dale, 1973).

In terrestrial molluscs therefore we have some idea of pressures actually in the hydrostatic skeleton, and these related very nicely to the expected pressures.

Snails of the family Achatinidae must surely be amongst the largest terrestrial hydrostatic animals, but we have no idea as yet of any size-limitations imposed on fluid skeletons.

Ciliary locomotion is presumed to be the primitive form of locomotion in the molluscs, several aquatic and some terrestrial gastropods still rely on ciliary forces for locomotion. Most gastropods, however, use pedal muscular waves for locomotion. There is a marked contrast between terrestrial pulmonates on the one hand and most other muscular locomotory gastropods on the other. Terrestrial snails and slugs use only direct monotaxic pedal waves with a large number – up to nineteen – occurring on the sole at any one time, the number depending on the size of the animal (Jones, 1975). Aquatic snails on the other hand use a variety of different types of pedal wave but the number of waves is always much less, three being a high number. It may be that terrestrial animals have to adopt a lower gear – taking more short steps – than aquatic animals because of their increased weight.

The pedal muscle involved in locomotion of gastropods, either aquatic or terrestrial, is able to work against the pressure in the pedal haemocoel, this pressure being present as a result of the circulation of the blood. The locomotory musculature thus becomes localised in gastropods, a feat not accomplished in other hydrostatic animals where the whole body is affected by locomotion.

Terrestrial snails provide an example of animals which use a compressible fluid for some body movements. The mantle cavity, which contains air for respiratory purposes, is involved in the pumping movements which occur when the animals protrude from their shell (Dale, 1974).

Onychophora

Peripatus is considered to have a hydrostatic skeleton (Manton, 1973), the haemocoel being the skeletal element. The only pressure measurements available were done with simple equipment (Picken, 1936) and are not likely to be precise. However, the pressure (up to 1961 Pa) is high enough to account for the basically cylindrical body form and the turgid limbs. A comparison with the hydrostatics of parapodial locomotion of polychaete Annelida may prove illuminating.

Geophilomorph centipedes

Certain centipedes such as *Orya barbarica* are able to actively enlarge burrows in soil by exerting strong dorso-ventral forces (Manton, 1965). These forces are produced by shortening the body and thus increasing the pressure in the haemocoel. From measurement of lateral force and its area of application, pressures were calculated to be as high as 35000 Pa. This pressure is larger than in earthworms, though Seymour (1970) has shown that calculated pressures in annelids exceed measured ones because of the stiffness of the body wall, so actual pressures may not be so high (Trueman, 1975).

It is only larger species that need to enlarge burrows in this way, others being small enough to live interstitially.

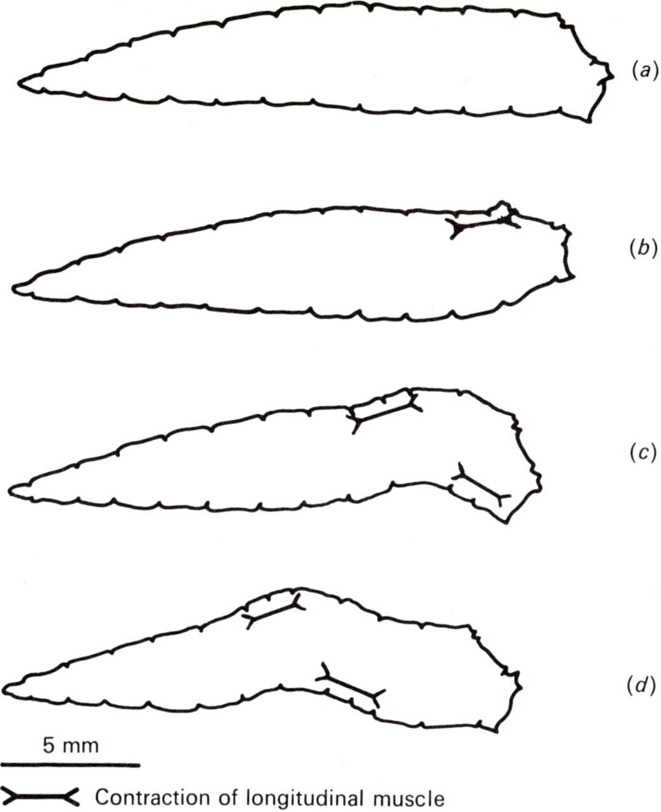

Fig. 3. (a)–(d). Successive stages in the locomotion of the larva of *Calliphora vomitoria*. (After Roberts, 1971.)

Insect larvae

Larvae of insects, particularly of Diptera (maggots) and of Lepidoptera (caterpillars) are basically soft-walled liquid-filled cylinders, and the skeletal system unlike most insects is principally hydrostatic. However, details of the functioning of the system are lacking. The fact that the body is more or less round in cross-section again suggests relatively high pressures.

The locomotion of maggots (Fig. 3) is basically similar to that of caterpillars, waves of muscular contraction starting posteriorly and passing forwards in both cases. Only one wave of contraction passes along the body at once. The muscles involved are the longitudinal ones of the dorsal and ventral body wall, other muscles may be involved simply to oppose the

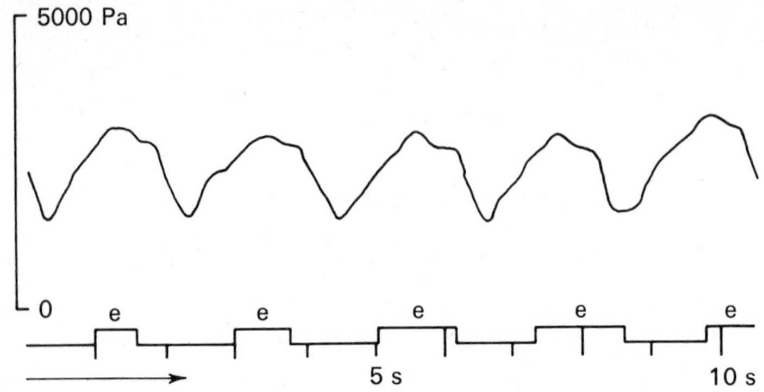

Fig. 4. Pressure recording from the haemocoel of *Calliphora* during surface crawling. Five locomotory cycles are shown. The event marks (e) were manually coordinated with posterior contraction of the body, i.e. Fig. 3 (*b*).

increased internal pressure. In most larvae the wave of contraction is shorter than the body, but in the looping caterpillars (Geometridae) the wave is longer than the body. Geometrid caterpillars also spend long periods resting attached only by their pro-legs, the anterior portion of the body being held out rigidly. It would be interesting to know the haemocoelic pressure during this behaviour.

Jones (unpublished) has measured and recorded haemocoelic pressure in *Calliphora* larvae during surface crawling (Fig. 4) and pressure cycles between about 1471 Pa and 3432 Pa during locomotion. The highest pressure occurs when the tail is drawn forwards, and the pressure then decreases to its lowest value as the wave of contraction passes off the anterior.

There is no obvious comparison with insect larvae among aquatic animals, except perhaps between leeches and looping caterpillars, but a study of the hydrostatic skeleton of insects would doubtless be rewarding.

Collembolans

Collembolans or spring-tails are able to jump by the action of their peculiar jumping organ on the abdomen. Manton (1972) has described the mechanism and the anatomy of the jumping mechanism and concluded that the organ was propelled by hydrostatic pressure in the haemocoel, the pressure being produced by the slight shortening of the body as a result of muscular contraction in the thorax and abdomen. The thoracic limbs are specially reinforced against the high pressures needed during jumping. The value of the pressure should be calculable from details of the jump height, of the mass of the animal and of the mechanics of the jumping organ.

Discussion and conclusions

We do then have some idea of the physiological details of the hydrostatic skeleton, but only in the annelids and to a lesser extent the molluscs can our knowledge be said to be comprehensive enough to draw any conclusions.

From the experiments on dialysis tubing and from pressure measurements from various animals it seems that pressures of the order of 1000–2000 Pa are necessary to inflate and give rigidity to terrestrial hydrostatic animals, lower pressures being insufficient, but higher pressures may be produced to overcome resistance during locomotion.

Earthworms, like many annelids are burrowing animals and the modifications that we have seen – lack of proboscis and presence of septa – seem to be related to the type of soils occurring on land and forces required to progress through them.

In terrestrial molluscs, which are essentially surface crawling animals, increased pressures are necessary, relative to aquatic animals, measurements and calculation being in good agreement. Locomotion has been modified in terrestrial molluscs, but the comparative mechanics of molluscan locomotion are not yet worked out (see Jones & Trueman, 1977). The shape of the shell and the contained viscera, discoidal and narrow conical being the two extremes, is of importance in terms of the postural stability of the shell and the habits of the animals (Cain, 1977). Slugs lack the demarcated viscera and shell of the snails and have a greater tendency to burrow than snails, though the mechanism of burrowing is not understood (Jones, 1975).

The locomotion of insect larvae is a potentially large field of comparative research as yet untapped.

One of the major problems associated with terrestrial life that we have not so far considered is that of water loss. Most of the animals mentioned are restricted to moist habitats, many of them secreting greater or lesser quantities of mucus, which is principally water. Terrestrial turbellarians are all more or less rounded in cross-section, unlike their aquatic relatives (flat-worms). Therefore the surface area to volume ratio is minimal, reducing the potential for evaporation, but copious quantities of mucus are secreted and they are restricted to very humid habitats. Likewise the terrestrial molluscs only inhabit damp environments, the few species that have colonised more or less arid habitats being active only when local humidity is high and otherwise retreating into their shells. Only some of the insect larvae, particularly of the Lepidoptera, have developed a more or less impermeable outer layer to the skin and this enables them to live in relatively dry air. Such animals must be among the more numerous of the animals with hydrostatic skeletons and certainly more successful in adapting to an aerial existence than any other group of hydrostatic animals. The relative lack of success of the other groups results more from problems of water loss than any limitations imposed by a hydrostatic skeleton.

It is a pleasure to record my thanks to Professor E. R. Trueman for his continuing encouragement and many helpful discussions.

References

Batham, E. J. & Pantin, C. F. A. (1950). Muscular and hydrostatic action in the sea-anemone *Metridium senile* (L.). *Journal of Experimental Biology*, **27**, 264–89.

Cain, A. J. (1977). Variation in the spire index of some coiled gastropod shells, and its evolutionary significance. *Philosophical Transactions of the Royal Society, B.* (In Press.)

Chapman, G. (1950). Of the movement of worms. *Journal of Experimental Biology*, **27**, 29–39.

Clark, R. B. (1964). *Dynamics in Metazoan Evolution.* Oxford: Clarendon Press.

Dale, B. (1973). Blood pressure and its hydraulic functions in *Helix pomatia* L. *Journal of Experimental Biology*, **59**, 477–90.

Dale, B. (1974). Extrusion, retraction and respiratory movements in *Helix pomatia* in relation to distribution and circulation of the blood. *Journal of Zoology, London*, **173**, 427–39.

Denton, E. J. (1961). The buoyancy of fish and cephalopods. *Progress in Biophysics*, **11**, 177–234.

Elder, H. Y. (1972). Connective tissues and body wall structure of the polychaete *Polyphysia crassa* and their significance. *Journal of the Marine Biological Association, UK*, **52**, 747–64.

Harris, J. E. & Crofton, H. D. (1957). Structure and function in the nematodes: internal pressure and cuticular pressure in *Ascaris*. *Journal of Experimental Biology*, **34**, 116–30.

Jones, H. D. (1975). Locomotion. In *Pulmonates*, ed. V. Fretter & J. Peake, vol. 1, pp. 1–32. London: Academic Press.

Jones, H. D. & Trueman, E. R. (1977). Crawling and burrowing. In *The Mechanics and Energetics of Animal Locomotion*, ed. R. McN. Alexander & G. Goldspink. London: Chapman & Hall.

Kafka, P. G. & Dunn, M. B. (1957). Stiffness of curved circular tubes with internal presssure. *Journal of Applied Mechanics*, **24**, 153–5.

Laverack, M. S. (1963). *The Physiology of Earthworms.* Oxford: Pergamon.

Lee, D. L. (1965). *The Physiology of Nematodes.* Edinburgh: Oliver & Boyd.

Manton, S. M. (1965). The evolution of arthropodan locomotory mechanisms. Part 8. Functional requirements and body design in Chilopoda, together with a comparative account of their skeleto-muscular systems and an Appendix on a comparison between burrowing forces of annelids and chilopods and its bearing upon the evolution of the arthropodan haemocoel. *Journal of the Linnean Society of London (Zoology)*, **46**, 278–483.

Manton, S. M. (1972). The evolution of arthropodan locomotory mechanisms. Part 10. Locomotory habits, morphology and evolution of the hexapod classes. *Journal of the Linnean Society of London (Zoology)*, **51**, 203–400.

Manton, S. M. (1973). The evolution of arthropodan locomotory mechanisms. Part 11. Habits, morphology and evolution of the Uniramia (Onychophora, Myriapoda, Hexapoda) and comparisons with the Arachnida, together with a functional review of uniramian musculature. *Journal of the Linnean Society of London (Zoology)*, **53**, 257–375.

Pantin, C. F. A. (1950). Locomotion in British terrestrial nemertines and planarians: with a discussion on the identity of *Rhynchodemus bilineatus* (Mecznikow) in Britain and on the name *Fasciola terrestris* O. F. Müller. *Proceedings of the Linnean Society of London*, **162**, 23–37.

Picken, L. E. R. (1936). The mechanism of urine formation in invertebrates. I. The excretory mechanism in certain Arthropoda. *Journal of Experimental Biology*, **13**, 309–28.

Roberts, M. J. (1971). On the locomotion of cyclorrhaphan maggots (Diptera). *Journal of Natural History*, **5**, 47–58.

Seymour, M. K. (1969). Locomotion and coelomic pressure in *Lumbricus terrestris* L. *Journal of Experimental Biology*, **51**, 47–58.

Seymour, M. K. (1970). Skeletons of *Lumbricus terrestris* L. and *Arenicola marina* (L.). *Nature, London*, **228**, 383–4.

Seymour, M. K. (1976). Pressure difference on adjacent segments and movement of septa in earthworm locomotion. *Journal of Experimental Biology*, **64**, 743–50.

Trueman, E. R. (1975). *The Locomotion of Soft-bodied Animals*. London: Arnold.

Wainwright, S. A., Biggs, W. D., Currey, J. D. & Gosline, J. M. (1976). *Mechanical Design in Organisms*. London: Arnold.

21. The fluid mechanics of circulatory systems

By T. J. PEDLEY

The main function of an animal's circulatory system is to transport respiratory gases, nutrients, metabolites and heat between the tissues and the respiratory, digestive, excretory and other organs. In most vertebrates the system operates in a roughly similar manner; that is, the blood is pumped by the heart into one or more large arteries, which distribute it to the tissues via a diverging network of branched vessels. The diameters of the individual vessels decrease with distance from the heart until, in the capillaries, the diameters are close to red cell diameter; the total cross-sectional area of all the vessels in parallel increases with distance from the heart, so the average blood velocity decreases. Most of the mass transport between the blood and the tissues takes place across the capillary walls. Having passed through the capillaries, the blood returns to the heart via the venous system, which superficially resembles the arterial system except that the blood flows from the smaller vessels to the larger. In fishes the respiratory organs (gills) are in series with the systemic organs, whereas in most other vertebrates they (the lungs) are in parallel. I know very little about the circulation of invertebrates, except what I have read in Professor Schmidt-Nielsen's book (Schmidt-Nielsen, 1975). In many invertebrates, the circulation is 'open', in that in the periphery the blood (which often does not have a respiratory function) leaves the blood vessels and passes directly through the interstitium, filtering back slowly into the return vessels. The pumping of the blood is often not confined to a single heart, and may be driven by peristaltic muscular contractions of the walls of several vessels.

In this paper I shall describe the dynamic principles governing the mammalian circulation, because that is what I know most about. However, despite the many differences in detail, the same basic physics underlies the circulation of all vertebrates, and most of what I say can be applied directly. I shall refer in passing to just a few of the differences. Most circulatory systems have not been studied in detail from the fluid mechanical point of view, and I regard it as one of the main functions of this paper to stimulate interdisciplinary research into them by both zoologists and fluid dynamicists. I shall conclude the paper by discussing the scaling of circulatory systems, in order to see if any important dynamic changes can be expected as animal size changes.

There are several medical, physiological and engineering text-books in which aspects of the fluid mechanics of the mammalian circulation are described (for example Burton, 1965; Whitmore, 1968; Bergel, 1972; Charm & Kurland, 1974; Lightfoot, 1974; McDonald, 1974; Lighthill, 1975). They are usually either too qualitative (as if biologists cannot cope with quantitative physical ideas) or take too much mathematical or fluid mechanical knowledge for granted. There is one book, written by R. C. Schroter and myself with C. G. Caro and W. A. Seed (Caro *et al.*, 1977), which sets out to bridge this gap, providing full physical explanations of the dynamic phenomena occurring in the circulation, but not requiring previous knowledge of fluid mechanics or advanced mathematics. If we have succeeded in our aims, then I think this is the most appropriate further reading to recommend.

The problems to which most research on the fluid mechanics of the circulation has been addressed have largely had a medical motivation. For example, one wishes to be able to interpret measurements of the wave-forms of pressure and of flow-rate everywhere in the arteries of man, in order both to understand their normal function and to diagnose the cause of abnormalities. Another line of research has been to investigate the influence of the flow on the structure of the artery wall, because the level of the shear stress exerted on the wall by the flowing blood is known to affect its permeability to large molecules such as lipoproteins (Caro, 1973) and there is strong evidence that this is a factor in the initiation of atherosclerosis (Porter & Knight, 1973). So we have to know what the shear stresses on the wall are, and these are determined by the detailed pattern of flow in the artery. There are also, presumably, other more purely biological questions to be answered, related perhaps to the energy required to maintain the circulation, or to the 'design' of the cardiovascular system. I hope the approach outlined in this paper will shed some light on the answers.

Basic physical principles

The motion of continuous fluids, like that of discrete particles, is governed by Newton's laws of motion, which can effectively be summarized in the statement that the acceleration experienced by a particle or fluid element is directly proportional to the net force acting on it, the constant of proportionality being the reciprocal of the element's mass. The motion of fluid along a tube can be driven only by a *pressure gradient*, since the pressure on one side of an element must be different from that on the

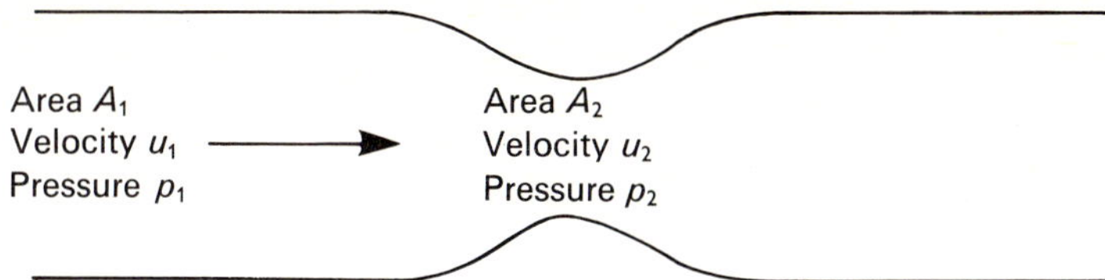

Fig. 1. Steady flow through a constriction; area A_1 upstream, A_2 at constriction. Conservation of mass requires that the velocity at the constriction, u_2, exceeds the velocity upstream according to $u_1A_1 = u_2A_2$. This convective acceleration of the fluid requires the upstream pressure (p_1) to exceed that at the constriction (p_2) according to equation (2).

opposite side if a net force is to be exerted (this neglects gravity, as justified below). An element also experiences viscous retarding forces, so Newton's law can be recast in the following way:

$$\text{Pressure gradient force} \qquad \text{(term I)}$$
$$= \text{Mass} \times \text{acceleration or inertia} \qquad \text{(term II)} \qquad (1)$$
$$+ \text{Viscous retarding force} \qquad \text{(term III)}$$

Several conclusions can be drawn immediately from this equation. (i) Fluid at rest (like blood in the aorta at the end of diastole) requires a pressure gradient to give it (*local*) acceleration. (ii) When fluid encounters a constriction in a tube (Fig. 1), it has to speed up in order to conserve the volume flow-rate. This emphasizes that fluid elements can experience accelerations (*convective* accelerations) even when the fluid velocity at every point remains constant in time. This acceleration must be associated, from equation (1), with a pressure gradient: if viscous forces are small, the result can be expressed in the form known as Bernoulli's equation,

$$p_1 + \tfrac{1}{2}\rho u_1^2 = p_2 + \tfrac{1}{2}\rho u_2^2 \qquad (2)$$

where p_1 and u_1 are the pressure and average velocity, respectively, at station 1 (Fig. 1), and p_2, u_2 those at station 2; ρ is the fluid density. (iii) Even fluid flowing steadily along a straight pipe, in which there are no accelerations, needs a favorable pressure gradient to overcome the viscous resistance.

In examining the fluid dynamics of circulatory systems there are two separate questions to be answered:

(i) What is the pressure gradient driving the fluid motion at any location?

(ii) What is the fluid motion resulting from that pressure gradient?

In answering the second question, we want to find out both the overall flow-rate through a blood vessel, and the detailed fow pattern within it. In answering either question, it will be instructive to consider the relative importance of the inertia (term II in equation 1) and the viscous forces (term III) in balancing the pressure gradient (term I).

Blood pressure

Before discussing the events of the circulation in the light of equation (1), it is important to define what we mean by blood pressure. The actual pressure in a blood vessel can be thought of as having three components: (i) Atmospheric pressure, experienced equally by every tissue, and approximately equal to the mean pressure in the left (or right) atrium when the air pressure in the lungs is the same as that outside. (ii) Hydrostatic pressure, equal to ρgh, where ρ is the density of blood, g is the gravitational acceleration, and h is the vertical distance below the heart (h is negative above the heart). Because the system is closed this has no direct effect on the motion of the blood, but does of course influence the transmural pressure of blood vessels and hence, because they are elastic, their cross-sectional area (e.g. veins in the foot of a standing man are distended, despite venous tone, while those in a raised arm are collapsed). (iii) The pressure generated by the heart, which is needed to drive the blood flow. It is this last component, varying between 10.7 and 16.0 kN m^{-2} (80 and 120 mm Hg) in the aorta of a normal man, and between 0 and 0.7 kN m^{-2} (0 and 5 mm Hg) in his vena cava, which is commonly referred to as blood pressure. The word 'pressure' in this article refers to this component, unless otherwise stated.

Heart and valves

When the muscle of the left ventricle has contracted sufficiently for left ventricle pressure to exceed aortic pressure, the aortic valve opens, and blood is ejected (i.e. accelerated out of the heart) until ventricular pressure starts to fall below aortic pressure, when the valve starts to close. Valve closure occurs smoothly because the pressure in the sinuses of Valsalva, behind the cusps (Fig. 2), is roughly the same as that at the downstream end of the cusps in the aorta, while the average pressure on the aortic side of the cusps is less, on account of the reversed pressure gradient. There is negligible backflow, because it takes a significant time for the aortic flow to reverse after the pressure gradient has reversed as a result of its inertia: the reversed pressure gradient has to decelerate the blood before accelerating it backwards again, by which time the valve has closed. The valve between the right ventricle and the pulmonary artery operates in the

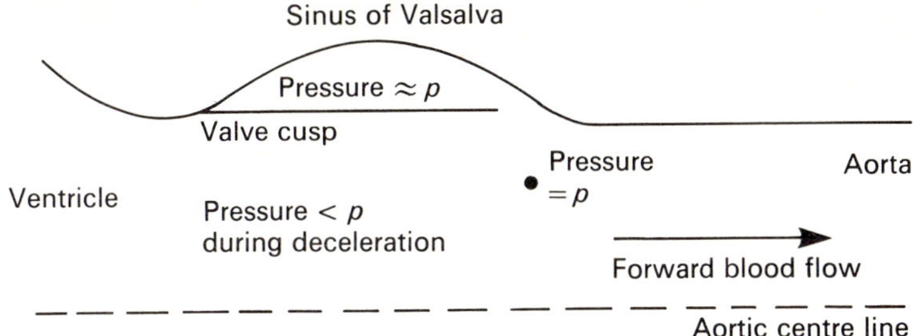

Fig. 2. The pressure distribution on the cusp of the aortic value, when the aortic pressure gradient has reversed, is such as to cause smooth closure of the valve before the blood flow reverses.

same way, and similar principles apply to the atrio-ventricular valves, which are also cusped. Bellhouse (1972) gives a fluid mechanical description of mitral and aortic valve action. It should be emphasized that all cusped valves operate entirely passively, involving no active muscular contraction. The right atrio-ventricular valve in a *bird's* heart, however, is of the sphincter type, opening and closing actively (Jones & Johansen, 1972). I do not know what functional advantage this might have (it may merely be associated with the shape of the right ventricle, which is a thin chamber, wrapped round the left ventricle), but it seems an interesting problem.

We should also note that in amphibia there are two valved atria, but only one ventricle; nevertheless in many species the systemic arteries are supplied with oxygenated blood, and the pulmocutaneous arteries with deoxygenated blood. This means that the two streams must pass through the ventricle with negligible mixing, aided by a long spiral ridge or baffle in the bulbus arteriosus, which separates the ventricle from the aorta and pulmocutaneous arteries (Hughes, 1963; Robb, 1965). How this works is a fluid mechanical problem of great interest.

Aorta and large arteries

Blood vessels have elastic walls, so when blood enters the aorta and aortic pressure rises the cross-sectional area of the first segment increases at the same time as the blood is accelerated forwards. The inertia of the blood, once it is set in motion, carries it on to the next segment where the pressure also rises and distention begins. Meanwhile elastic forces both aid this distension and cause the area of the first segment to return towards its undisturbed value; in fact it overshoots because of the fluid inertia. The process is repeated continuously throughout the arterial system, and a *wave* is transmitted along. This is the pulse wave, in which the important dynamic features are the elastic restoring force of the walls and the inertia of the blood. The propagation of this wave determines the instantaneous pressures throughout the systemic arteries.

A very simple theory leads to the prediction that the speed of propagation of the pulse wave is equal to c, where

$$c^2 = (\rho D)^{-1} \tag{3}$$

and D is the *distensibility* of the artery, defined as

$$D = \frac{1}{A}\frac{\mathrm{d}A}{\mathrm{d}p_{\mathrm{tm}}}, \tag{4}$$

where A is the cross-sectional area and p_{tm} is the transmural pressure. If the artery is *assumed* to be a thin-walled, linearly elastic, circular tube of diameter d, wall-thickness h and Young's modulus E, equation (3) reduces to the well-known Moens–Korteweg equation:

$$c^2 = Eh/\rho d. \tag{5}$$

However, the measurements from which an effective value of E is inferred are simply those relating area to transmural pressure, so equation (3) is more appropriate.

The predicted wave speeds in the aorta and large arteries of a dog (and a man) are about 5 m s^{-1} in the proximal aorta, rising to about 8 m s^{-1} in the peripheral large arteries, and these values agree very well with measured wave speeds. They mean that the wavelength of the wave, for a heart-rate of 2 Hz, ranges from 2.5 m to 4.0 m, which is considerably longer than a single arterial pathway in a dog. This is the basis for the *Windkessel* model of the circulation, proposed by Otto Frank (1899), in which the large arteries are treated as a single capacitance, in series with a peripheral resistance. Examination of the aortic wave-form shows that this model is actually quite inadequate to predict its detailed shape, but it has been a helpful concept. The bulbus (or conus) arteriosus in fishes is a very compliant chamber, which can reasonably be regarded as a single capacitance, so a modified *Windkessel* model may well be appropriate for it. It has been suggested that the function of this compliant chamber is to reduce the pressure pulse transmitted to the fragile blood vessels of the gills, in order to protect them (Chapter 6 in Gordon, 1968).

The above theory to predict wave speed is very successful, despite the fact that the viscous properties of blood are completely ignored, at least from the point of view of the bulk motion of the blood. This means that the viscous forces (term III in equation (1)), which are mathematically represented by the expression $\mu \partial^2 u/\partial r^2$, where μ is the blood viscosity, u

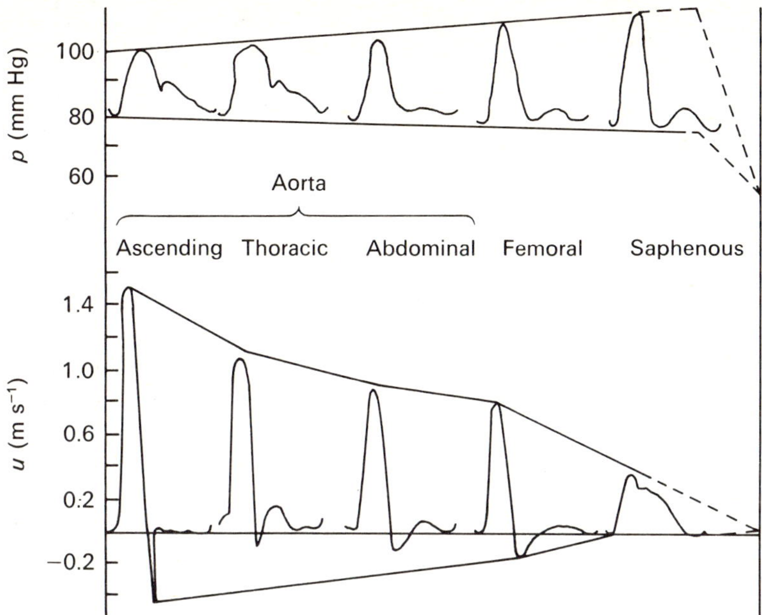

Fig. 3. Variation of the pressure and velocity wave-forms with distance along a canine arterial pathway. 1 mm Hg = 0.133 kN m^{-2}. (From McDonald, 1974.)

is the longitudinal velocity, and r is the radial coordinate, are negligibly small compared with the mass-acceleration terms (term II), represented by $\rho \partial u/\partial t$, where t is the time. A rough scaling of these expressions indicates that, for a wave of angular frequency ω ($= 2\pi f$ where f is the frequency in Hz, i.e. the heart-rate), the viscous forces should be negligible as long as the quantity

$$\alpha^2 = \left(\frac{d}{2}\right)^2 \frac{\rho \omega}{\mu} \tag{6}$$

is much greater than 1. α is often called the Womersley parameter, after the man who first recognized its importance in a cardiovascular context (Womersley, 1955). In a dog, with heart-rate of 2 Hz and aortic diameter of 1.5 cm, $\alpha \approx 13$, which is large compared with 1, consistent with the success of the wave speed theory.

The theory is not so successful in predicting the shape of the pressure wave-form, which varies throughout the circulation as shown in Fig. 3, where the corresponding blood velocity wave-forms are also given. This variation, however, can also be shown to be virtually independent of viscosity, being primarily a consequence of wave reflection, which takes place both at bifurcations and at regions of arterial taper. Wave reflection is responsible for the 'dicrotic notch' in the aortic pressure wave and for the *peaking* of the pressure pulse (McDonald, 1974). The slight *steepening* of the front of the pressure pulse is largely the result of non-linear effects, and is comparable to the steepening of water waves in a channel. The criterion for linear behavior is that the blood velocity u should be small compared with the wave speed c. In the canine aorta the maximum value of u is about 1.2 m s^{-1}, and its average value is about 0.2 m s^{-1} (Caro, Pedley & Seed, 1974). Thus, since the wave speed exceeds 5.0 m s^{-1}, the normal system does behave in an approximately linear way. Non-linear phenomena may, however, be important in certain disease states, such as chronic aortic valve incompetence, when the stroke volume and hence u become very large (Anliker, Rockwell & Ogden, 1971). Note that the fact that the average blood velocity is much smaller than the wave speed means that the blood only travels a few centimetres each beat, while the pulse travels right through the system.

Blood viscosity causes some attenuation of the pulse wave (obscured in Fig. 3 by the peaking), but not as much as the viscoelasticity of the arterial wall (McDonald, 1974). The main importance of viscosity lies in determining the flow response to the propagating pressure gradient. The fact that α (equation (6)) is large indicates that the bulk of the fluid moves in a manner unaffected by viscosity. Nevertheless, viscosity must have some effect, because however small it is, it requires that fluid in contact with a solid wall should adhere to the wall. Thus there must be at least a thin region near the wall in which viscosity is important. This region is called a *boundary layer*, and in the case of oscillatory flow in a tube its thickness is proportional to $(\mu/\rho\omega)^{1/2}$, much less than the radius $\frac{1}{2}d$ because α is large. Womersley (1955) calculated the viscous flow in a straight *rigid* tube subjected to an oscillatory pressure gradient for all values of α. Application of his theory to the arterial pressure gradient wave-form (separated by Fourier analysis into components of different frequencies) predicts a velocity wave-form close to that observed (Fig. 4), while neglect of viscosity leads to gross inaccuracies. This indicates that, while elasticity is important for determining what the pressure gradient is everywhere, it is not important for determining the flow response to that pressure gradient.

The medical interest in the relationship between wall shear stress and atherosclerosis leads us to examine the details of the flow in curves and branches and at the entrance of arteries, where the above straight tube analysis is inappropriate and both theory and experiment become difficult. In these circumstances another dimensionless parameter becomes important. It is the ratio of the convective (not the local) inertia terms to the viscous terms in equation (1), and is represented by the *Reynolds number*:

$$\mathrm{Re} = \rho u d/\mu. \tag{7}$$

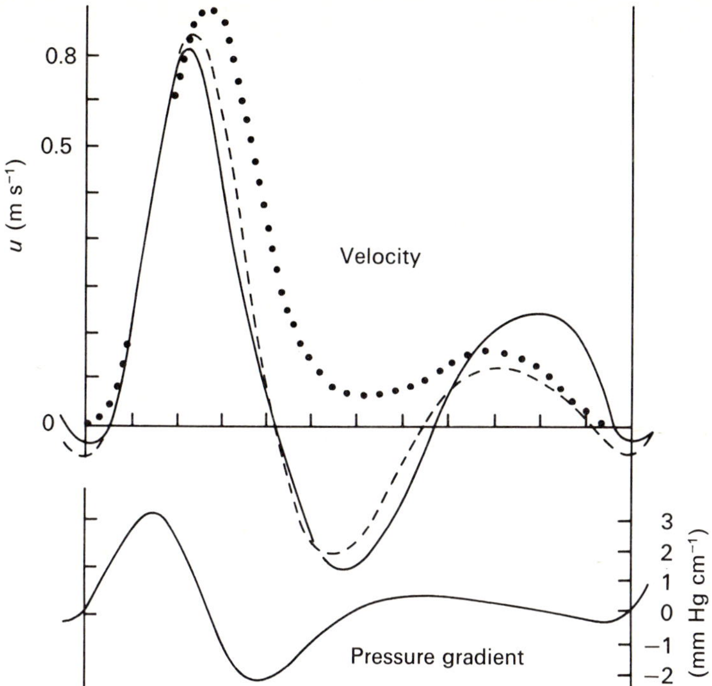

Fig. 4. The velocity wave-form measured in a dog's femoral artery (———), together with that calculated (– – –) from the measured pressure gradient (below) according to the rigid-tube theory of Womersley (1955). Also shown is the wave-form calculated for a non-viscous fluid (• • •). (Modified from McDonald, 1974.)

In all large arteries the mean Reynolds number is very large (i.e. it is about 800 in the ascending aorta of a dog), which again means that viscosity is directly important only in the thin boundary layers which form at the entrance of arteries, on the flow dividers at a branch, and on the outer wall of a curved artery (Fig. 5). However, it is in these regions that the wall shear is determined, and the thinner the boundary layer, the larger the shear. The details of the flow in curves and branches are very complicated, with transverse swirling motions superimposed on the primary through-flow (Schroter & Sudlow, 1969), and sometimes with separation of the flow from the wall, leaving a more or less stagnant region behind.

Another aspect of the flow which may have an important influence on the vessel wall is whether or not the flow is turbulent. [Turbulence is defined as chaotic, random motion; it is the opposite of laminar flow. Note that laminar flows can be very complex (as in Fig. 5), and the term should not be reserved for smooth flow in a long straight pipe, which is called

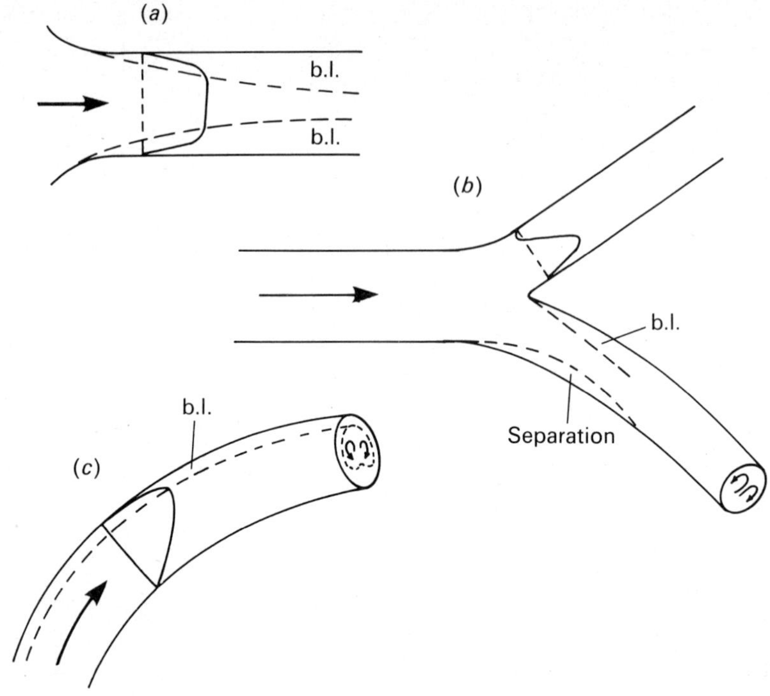

Fig. 5. Details of the flow in tubes with complicated geometries; broken lines indicate boundary layers (b.l.) or regions of separated flow. (a) Entrance flow in a straight tube; (b) flow in a bifurcation; (c) flow in a curved tube. Profiles of longitudinal velocity are shown in each case. Note the secondary motions in (b) and (c).

Poiseuille flow.] A flow can become turbulent for various reasons; usually the more important viscosity is, the less likely the flow is to become turbulent. For example, flow in a long straight rigid pipe becomes turbulent for Reynolds numbers above about 2300 (for water that corresponds to a velocity of less than 10 cm s^{-1} in a 2.5 cm diameter tube). However, flow through a stenosis or constriction can become turbulent at a Reynolds number (based on the diameter and velocity upstream of the constriction) of as little as 200, because the jet which emerges from the constriction is inherently much more unstable than Poiseuille flow. Roach (1972) and her colleagues have shown that post-stenotic dilatation is directly attributable to the effect of turbulence in the downstream jet on the structure of the wall. Cardiovascular murmurs are also usually associated with turbulence (Rushmer, 1970). Thus the presence or absence of turbulence may have important consequences other than increasing the pressure loss, which is unlikely to be important (see below).

Nerem & Seed (1972) have shown that flow in the aorta of a normal dog is sometimes turbulent and sometimes laminar, depending on what drugs have been administered: the flow is always turbulent when the aortic valve is stenosed. Similar observations have been made in man (Stein & Sabbah, 1976). Both experiment and a very crude theory suggest that the criterion for the occurrence of turbulence is that the peak Reynolds number ($\widehat{Re}$) should exceed a critical value given by

$$\widehat{Re}_{crit} = 250\alpha. \tag{8}$$

Normal conscious dogs were found to be very close to critical; and Seed (1972) has found the same to be true in man. Predictions for other mammals (from a scaling point of view) are made below.

Microcirculation

If the mean blood pressure is plotted against distance along a typical systemic pathway (Fig. 6), very little pressure loss is found in the large arteries (confirming that extra pressure losses caused by branching or turbulence are unimportant). There is an enormous loss, however, in the arterioles and, to a lesser extent, the capillaries. This has functional importance because it means that blood flow to different regions can be regulated by muscular control of the calibre (and hence resistance) of the arterioles. It is thus important to understand the dynamics of flow in the microcirculation, which can be regarded as all vessels of diameter less than 100 μm. True capillaries are about 10 μm or less in diameter.

There are two major dynamic differences between flow in the microcirculation and that in large vessels.

(i) α and Re are both less than 1. This means that inertia is unimportant, and the viscous term dominates the right-hand side of equation (1). Thus the propagation of the pressure pulse is not like that of a true wave, in which the elastic forces are balanced by fluid inertia, but is more of a diffusive balance between elasticity and viscous damping forces (Lighthill, 1972). The pulse is rapidly attenuated, but can still be measured in the smallest veins, especially during vasodilatation. The fact that α and Re are small means that the flow is a quasi-steady viscous response to the pressure gradient. However, it is *not* Poiseuille flow, even in fairly long blood vessels, because of the other important difference between the micro- and the macrocirculations.

(ii) Blood is not a homogeneous fluid, but is a suspension of cells, red cells being the most important dynamically. This means that the bulk viscosity of blood varies with shear rate (i.e. blood is *non-Newtonian*) and, more important, blood cannot be regarded even approximately as a homogeneous fluid. The undeformed diameter of a (human) red cell is

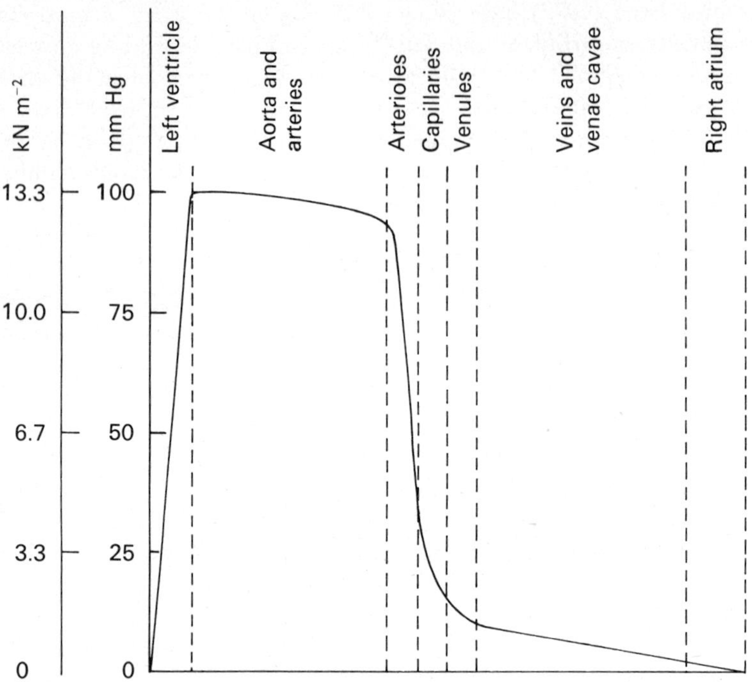

Fig. 6. Distribution of mean pressure along a typical pathway in the systemic circulation. The length scale of the microcirculatory vessels has been expanded for convenience. There is negligible pressure drop along the large arteries and veins, but a very steep pressure gradient in the microcirculation, especially the arterioles.

about 8 μm, comparable with that of most capillaries, so that in capillaries the motion of the cells and the plasma must be considered separately. In arterioles, also, there is a strong interaction between the red cells and the vessel walls, which influences the bulk motion. For example, blood flowing in small tubes exhibits the Fahraeus–Lindqvist effect (Fahraeus & Lindqvist, 1931), which is the name given to the unexpected phenomenon that the apparent viscosity of the blood decreases as the tube diameter decreases, at least until the diameter is close to 8 μm. [Apparent viscosity μ is measured using Poiseuille's law:

$$\Delta p = \frac{128\,\mu l Q}{\pi d^4} \tag{9}$$

where Δp is the pressure drop along length l of the tube, diameter d, and Q is the flow rate through it.] This effect comes about because a very thin region of relatively low cell concentration occurs near the walls (the 'cell-free layer'), so that the apparent viscosity of blood tends towards that of plasma

$(1.8\times10^{-3}$ kg m^{-1} s^{-1}) and away from that of whole blood $(4.0\times10^{-3}$ kg m^{-1} s^{-1}). A complete dynamic explanation of the cell-free layer and the Fahraeus–Lindqvist effect is not available, and most research into arteriolar fluid dynamics has been experimental (Mason & Goldsmith, 1969).

There is more hope of understanding flow in the capillaries, because we need consider only one cell at a time. However, the cells are easily deformable, and are observed to take up a wide range of different shapes as they flow through the capillaries (Skalak & Brånemark, 1969), so analysis is still not easy. A good theoretical understanding has been achieved for an axisymmetric configuaration (Fitz-Gerald, 1969; Skalak & Zarda, personal communication): when the cell fits very tightly, very high pressures are built up in the thin lubricating film of plasma between the cell and the capillary wall, and the flow in the bolus of plasma carried along between the cells has little influence on the cell motion.

I should add that flow in a complete capillary bed is not as simple as that in many equal parallel pathways. Because of the asymmetry of the capillary network, of vasoactive factors, and of the effect of individual red cells or aggregates of them becoming jammed in the capillaries from time to time, a full description of the varying pressure and flow in the network is very complicated (Zweifach, 1974). Note too that special organs require special microcirculatory systems with particular fluid mechanical properties: for example, the kidneys (Barger & Herd, 1973), mammalian pulmonary capillaries (Fung & Sobin, 1969), fish gills (Hughes, 1966), and the retial capillaries used for the thermoregulation of certain marine animals (Schmidt-Nielsen, 1975).

We should remark that the flow in the microcirculation of animals without red cells will be quite different from that of vertebrates, and is likely to be much easier to analyse because the blood (or haemolymph) is homogeneous. Further, the dynamical and diffusive characteristics of the interstitial flow of blood in the tissues of animals with open circulations should also be relatively easy to analyse, because there is a standard semi-empirical law relating the pressure gradient (grad p) and velocity of flow ($\mathbf{u}$) of a Newtonian fluid in a porous medium. This law is known as Darcy's law, and can be written

$$\mathbf{u} = \frac{k}{\mu}\,\mathrm{grad}\,p, \tag{10}$$

where k is a constant called the permeability of the medium (Streeter, 1961); equation (10) is valid as long as the dimensions of the region under study are large compared with the microstructure. The only difficulty lies in obtaining a reliable estimate of k. A similar relationship probably applies to the intervillous sinuses of the mammalian placenta, except that here k may be pressure, and hence time, dependent, because the tissues are elastically deformed by the pulsatile flow (Corrsin, personal communication).

Space and time do not permit a discussion of the fluid mechanics of veins. They are different from arteries in two major respects: (i) they may have valves, which presumably operate in a way similar to that of heart valves, but which have not been studied from the fluid dynamics point of view; (ii) they often operate under conditions where the transmural pressure is close to zero or negative, and tends to collapse the vessel. This can have very interesting fluid mechanical consequences, many of which are incompletely understood (Holt, 1969; Conrad, 1969; Caro *et al.*, 1977).

Scaling considerations

As we have discovered in the field of animal locomotion (Pedley, 1977), a consideration of the scaling or allometry of mechanical factors often leads to new insights into the biology of animals of different sizes. I thought it would be instructive, therefore, to see how the two key dimensionless parameters, mean aortic Reynolds number $\overline{Re}$ and the Womersley parameter α, vary with body size, at least in mammals. I took all biological data from Stahl (1967), who quotes scaling laws for many respiratory and cardiovascular variables. He does not report on aortic diameter, but all volumes and masses (e.g. heart mass) vary in direct proportion to body mass (M), as expected, so it seems reasonable to suppose that aortic diameter (d) varies as $M^{1/3}$. We therefore take

$$d = 0.62M^{0.33}, \tag{11}$$

where d is measured in cm and M in kg. This gives $d = 2.5$ cm for a 70 kg man; more data would be desirable in order to improve the equation. Stahl (1967) also gives

$$f = 4.0M^{-0.25}, \tag{12}$$

for resting mammals, where f is heart-rate, measured in Hz (this gives $f = 1.38$ Hz for a 70 kg man, which is a little high, but the equation represents the best fit to the interspecific data). Combining equations (11) and (12) into equation (6) for α^2 gives

$$\alpha^2 = 57M^{0.42}. \tag{13}$$

For mean Reynolds number we also need average aortic velocity $\bar{u}$, which can be deduced from the cardiac output Q, and aortic diameter d, given above. Stahl gives (again for resting mammals)

$$Q = 3.12M^{0.81}, \tag{14}$$

when Q is measured in cm^3 s^{-1} (although $Q \propto M^{0.75}$ is more commonly accepted), which means that

$$\bar{u} = \frac{4Q}{\pi d^2} = 10.3\, M^{0.15} \tag{15}$$

in cm s^{-1}. This gives only 20 cm s^{-1} for a 70 kg man, which is a little low. Nevertheless we combine (11) and (15) with equation (7) to obtain

$$\overline{\text{Re}} = 160\, M^{0.48}. \tag{16}$$

Note that both $\overline{\text{Re}}$ and α will be greater in exercise.

The smallest mammals have a mass of about 10 g, and for that mass the values of α and $\overline{\text{Re}}$ predicted by equations (13) and (16) are 2.9 and 17.5, respectively. These values are significantly greater than 1, so that inertia still dominates viscosity although the effects of viscosity are no longer negligible. The predictions suggest that the dynamics of the cardiovascular system are essentially the same for all mammals. This is confirmed by a consideration of the speed of propagation of the pulse wave (equation (3) or (5)) which is expected to be approximately invariant with size, because all aortae are made of the same materials (i.e. have the same Young's modulus E) and we would expect the ratio of thickness to diameter to remain constant. The wavelength of the pulse wave (c/f) is therefore predicted to vary as $M^{0.25}$, while the length of the aorta, for example, will vary as $M^{0.33}$. The ratio of wavelength to length therefore increases as body size decreases, but only very slowly. Thus the *Windkessel* model, in which all arteries are lumped together as a single capacitance, is not much better in small mammals than in man. It is thought to be more useful in birds, because their arteries are stiffer (so c is greater), and their heart-rates lower, than those of mammals of the same mass (Calder, 1968).

The above predictions of α and $\overline{\text{Re}}$ for a 10 g mammal raise the question of why there are no mammals with masses much smaller than 10 g (the smallest adult is 3 g). If it has to do with the cardiovascular system (which can only be part of the story), then there could be an intrinsic limitation on the frequency with which heart muscle can contract, independent of the fluid mechanics. It could, however, be that it becomes too hard for the heart to contract at an increasing frequency against an increasingly viscous load (α and $\overline{\text{Re}}$ approaching 1 even in the aorta), so that the energy requirements for operating the circulatory system become unacceptably high. These considerations also suggest that a similar study on humming-birds and small fishes, which have masses well below 10 g, and which (in fish at least) have relatively low heart-rates (decreasing α still further), would be of great interest.

We can use equations (13) and (16), together with equation (8), to predict whether turbulence will normally occur in the aortae of (resting) mammals or not. If we suppose that the peak Reynolds number, $\widehat{\text{Re}}$, is

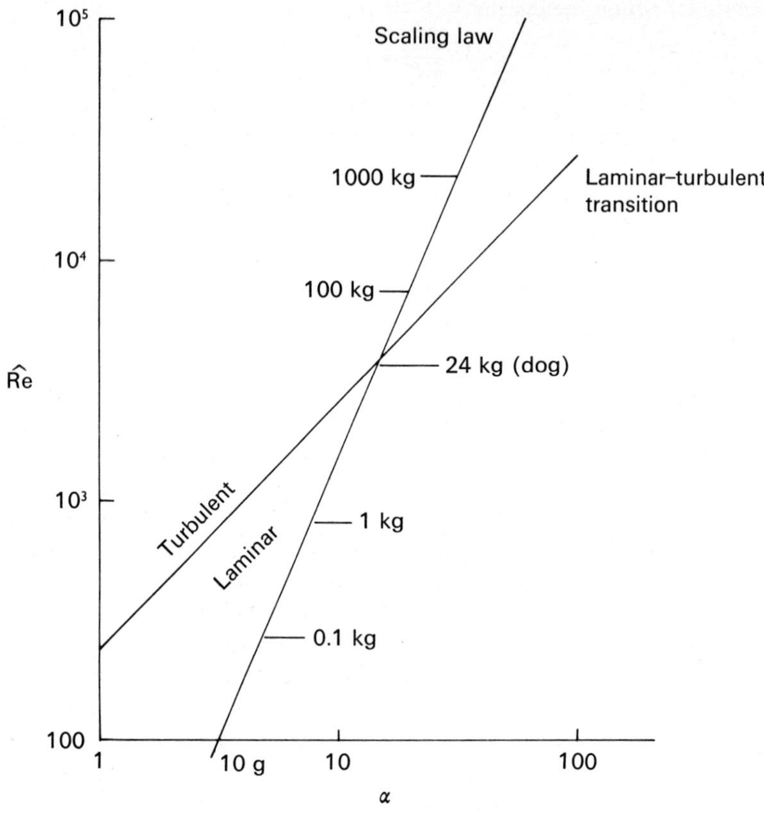

Fig. 7. Log–log plots of peak aortic Reynolds number $\widehat{Re}$ against Womersley parameter α. One line represents the transition from laminar to turbulent flow in the aorta, according to the results obtained (in dogs) by Nerem & Seed (1972) (see equation (8)). The other line represents the allometric equation for all mammals, given by equation (17). Turbulence is predicted to be present in animals larger than dogs, but not in smaller animals.

in all animals about 5 times the mean (as in man and dogs), then combining (13) and (16) we obtain

$$\widehat{Re} = 7.5\alpha^{2.3}. \tag{17}$$

This is plotted (on a log–log scale) in Fig. 7, as is the critical condition (8). This shows very clearly that large animals are expected normally to have turbulence, and small ones are not. The cross-over occurs for a mass of about 24 kg, right in the mass-range of dogs; this is consistent with the observations of Nerem & Seed (1972). Note, however, that (*a*) an individual animal cannot be expected to lie on the scaling curve – there is always so much scatter; (*b*) the uncertainty in the slope of any such line will mean greater uncertainty in the actual values at very large or very small masses; and (*c*) in exercise, the scaling line will be shifted slightly up and to the left, so that an animal which has a laminar flow when resting may have turbulent flow on exercise.

Finally, I would end with a plea. The blood pressures and cardiac outputs of many animals of different sorts are recorded in the literature, but the dimensions of their blood vessels are not, apart from man and a few other mammals. (There are notable exceptions; see Belman (1975), for example.) Without a knowledge of the dimensions, very little fluid mechanical insight can be gained. So I would ask any zoologist who thinks that the fluid mechanics of the circulation may be an interesting field of study, please to measure as many blood vessel dimensions as possible.

Summary

The fluid mechanical principles governing vertebrate circulatory systems are described from a physical point of view, with particular reference to mammals. In large arteries the pressure gradient at any point is determined by the propagation of the pulse wave, which represents a balance between wall elasticity and blood inertia. Blood viscosity is unimportant because the Womersley frequency parameter, α (equation (6)), is large. Viscosity is more important in determining the flow which is generated in response to the pressure gradient, but primarily in thin boundary layers at the entrance of tubes, at branches and in bends, because the Reynolds number, Re (equation (7)), is also large. It is the viscosity and the flow in these boundary layers that determine the shear stress exerted by the blood on the vessel wall, which may have medical implications. The occurrence of turbulence in the aorta is also discussed, and has been found to occur in dogs if the peak aortic Reynolds number $\widehat{\mathrm{Re}}$ exceeds a critical value equal to 250α. The dependence of α and Re on body mass is discussed, and it is found that these quantities are significantly greater than 1 (i.e. inertia dominates viscosity) in all mammals. It is also found that, according to the above result, aortic turbulence is normally to be expected in animals larger than dogs, but not in animals smaller than dogs. Detailed analysis of animals other than mammals is inhibited by the lack of data on vessel dimensions and blood velocities, but a few interesting fluid mechanical problems are pointed out.

This paper was written while I was visiting Professor R. M. Nerem's Physiological Fluid Mechanics Group at Ohio State University, with the support of the National Science Foundation, grant no. ENG 71–02286.

References

Anliker, M., Rockwell, R. L. & Ogden, E. (1971). Non-linear analysis of flow pulses and shock waves in arteries. Parts I and II. *Zeitschrift für angewandte Mathematik und Physik*, **22**, 217–46 and 563–81.

Barger, A. C. & Herd, J. A. (1973). Renal vascular anatomy and distribution of blood flow. In *Handbook of Physiology*, Section 8, *Renal Physiology*, ed. by J. Orloff & R. W. Berliner, Chapter 10, 249–313. Washington: American Physiological Society.

Bellhouse, B. J. (1972). The fluid mechanics of heart valves. In *Cardiovascular Fluid Dynamics*, ed. D. H. Bergel, vol. 1, Chapter 8, pp. 261–85. London: Academic Press.

Belman, B. W. (1975). Some aspects of the circulatory physiology of the spiny lobster *Panulirus interruptus*. *Marine Biology*, **29**, 295–305.

Bergel, D. H. (ed.) (1972). *Cardiovascular Fluid Dynamics*, 2 vols. London: Academic Press.

Burton, A. C. (1965). *Physiology and Biophysics of the Circulation*. Chicago: Year Book Medical Publishers.

Calder, W. A. (1968). Respiratory and heart rates of birds at rest. *Condor*, **70**, 358–65.

Caro, C. G. (1973). Transport of material between blood and wall in arteries. In *Atherogenesis: Initiating Factors*, ed. R. Porter & S. Knight, pp. 127–49. Amsterdam: Elsevier.

Caro, C. G., Pedley, T. J., Schroter, R. C. & Seed, W. A. (1977). *Mechanics of the Circulation*. Oxford: Oxford University Press.

Caro, C. G., Pedley, T. J. & Seed, W. A. (1974). Mechanics of the circulation. In *Cardiovascular Physiology*, ed. by A. C. Guyton, vol. 1, pp. 1–47. (Series 1, Physiology, in the MTP International Review of Science.) London: Butterworths.

Charm, S. E. & Kurland, G. S. (1974). *Blood Flow and Microcirculation*. New York: Wiley.

Conrad, W. A. (1969). Pressure-flow relationships in collapsible tubes. *IEEE Transactions on Bio-Medical Engineering* **BME-16**, 284–95.

Fahraeus, R. & Lindqvist, T. (1931). The viscosity of the blood in narrow capillary tubes. *American Journal of Physiology*, **96**, 562–8.

Fitz-Gerald, J. M. (1969). Implications of a theory of erythrocyte motion in narrow capillaries. *Journal of Applied Physiology*, **27**, 912–18.

Frank, O. (1899). Die Grundform des arteriellen Pulses. Erste Abhandlung: Mathematische Analyse. *Zeitschrift für Biologie*, **37**, 483–526.

Fung, Y. C. & Sobin, S. S. (1969). Theory of sheet flow in lung alveoli. *Journal of Applied Physiology*, **26**, 472–88.

Gordon, M. S. (1968). *Animal Function: Principles and Adaptations*. New York: Macmillan.

Holt, J. P. (1969). Flow through collapsible tubes and through *in situ* veins. *IEEE Transactions on Bio-Medical Engineering*, **BME-16**, 274–83.

Hughes, G. M. (1963). *Comparative Physiology of Vertebrate Respiration*. Cambridge, Mass.: Harvard University Press.

Hughes, G. M. (1966). The dimensions of fish gills in relation to their function. *Journal of Experimental Biology*, **45**, 177–95.

Jones, D. R. & Johansen, K. (1972). The blood vascular system of birds. In *Avian Biology*, vol. 2, ed. D. S. Farner & J. R. King, Chapter 4, pp. 157–285. New York: Academic Press.

Lightfoot, E. N. (1974). *Transport Phenomena and Living Systems*. New York: Wiley.

Lighthill, M. J. (1972). Physiological fluid dynamics: a survey. *Journal of Fluid Mechanics*, **52**, 475–97.

Lighthill, M. J. (1975). *Mathematical Biofluiddynamics*. Philadelphia: SIAM.

McDonald, D. A. (1974). *Blood Flow in Arteries*, 2nd ed. London: E. Arnold.

Mason, S. G. & Goldsmith, H. L. (1969). The flow behaviour of particulate suspensions. In *Circulatory and Respiratory Mass Transport*, CIBA Foundation Symposium, pp. 105–24. London: Churchill.

Nerem, R. M. & Seed, W. A. (1972). An *in vivo* study of the nature of aortic flow disturbances. *Cardiovascular Research*, **6**, 1–14.

Pedley, T. J. (ed.) (1977). *Scale Effects in Animal Locomotion*. London: Academic Press.

Porter, R. & Knight, J. (ed.) (1973). *Atherogenesis: Initiating Factors*. CIBA Foundation Symposium, no. 12 (new series). Amsterdam: Elsevier.

Roach, M. R. (1972). Post-stenotic dilatation in arteries. In *Cardiovascular Fluid Dynamics*, ed. D. H. Bergel, vol. 2, Chapter 13, pp. 111–39. London: Academic Press.

Robb, J. S. (1965). *Comparative Basic Cardiology*. New York: Grune & Stratton.

Rushmer, R. F. (ed.) (1970). *Cardiovascular Dynamics*, 3rd edn. Philadelphia: W. B. Saunders.

Schmidt-Nielsen, K. (1975). *Animal Physiology: Adaptation and Environment*. London: Cambridge University Press.

Schroter, R. C. & Sudlow, M. F. (1969). Flow patterns in models of the human bronchial airways. *Respiration Physiology*, **7**, 341–55.

Seed, W. A. (1972). Hot film anemometry in the human aorta. In *Blood Flow Measurement*, ed. C. Roberts. London: Sector.

Skalak, R. & Brånemark, P. I. (1969). Deformation of red blood cells in capillaries. *Science, Washington*, **164**, 717–19.

Stahl, W. R. (1967). Scaling of respiratory variables in mammals. *Journal of Applied Physiology*, **22**, 453–60.

Stein, P. D. & Sabbah, H. N. (1976). Turbulent flow in the ascending aorta of humans with normal and diseased aortic valves. *Circulation Research*, **39**, 58–65.

Streeter, V. L. (ed.) (1961). *Handbook of Fluid Dynamics*. Chapter 16, pp. 16.1–16.112. New York: McGraw-Hill.

Whitmore, R. L. (1968). *Rheology of the Circulation*. Oxford: Pergamon.

Womersley, J. R. (1955). Method for the calculation of velocity, rate of flow and viscous drag in arteries when the pressure gradient is known. *Journal of Physiology*, **127**, 553–63.

Zweifach, B. W. (1974). Quantitative studies of microcirculatory structure and function, Parts I and II. *Circulation Research*, **34**, 843–57 and 858–66.

22. Convective flows in mammalian lungs

By R. C. SCHROTER

The transport of gases between the internal and external environments of both aquatic and air breathing animals is by diffusion across a membrane. However, in all but the smallest organisms the exchange process occurs in a purpose built organ, the lung or gill, in which the external environment is brought by bulk flow to the gas exchange surface. The geometrical organisation of the passage ways bringing the external medium to this exchange surface and then removing it are vast and well described elsewhere (Schmidt-Nielsen, 1975). In consequence the bulk flow processes involved are many and complicated; this review will consider the bulk flow problems associated with the transport of air in mammalian lungs. However, the physical concepts underlying flow conditions in mammalian lungs are all relevant to other classes of animals.

Perhaps the earliest attempt to describe the bulk flow conditions in human lungs was made by Rohrer in 1915. He made some simple ana-tomical measurements of the airways and applied a simple fluid mechanical approach to predict the pressure–flow relations within them. It was not for many years that the problem of flow in the airways was considered more extensively.

West & Hugh-Jones (1959) and Dekker (1961) have studied the flow of water in models of the human trachea and main bronchi, and have shown that turbulence can occur; they made no detailed observations of the flow. In order to describe flow conditions within the complicated environment of the bronchial tree, it is first necessary to describe flow in simple symmetrical bifurcations. In turn, to describe flows in such a relatively simple geometry we must first consider some basic fluid mechanical pro-cesses. Elsewhere (Pedley, Schroter & Sudlow, 1977) there is a very extended discussion of the fluid mechanical aspects of airflow in the human bronchial tree; in this paper a relatively brief background will be given and the reader is referred to the other publication for a more rigorous description.

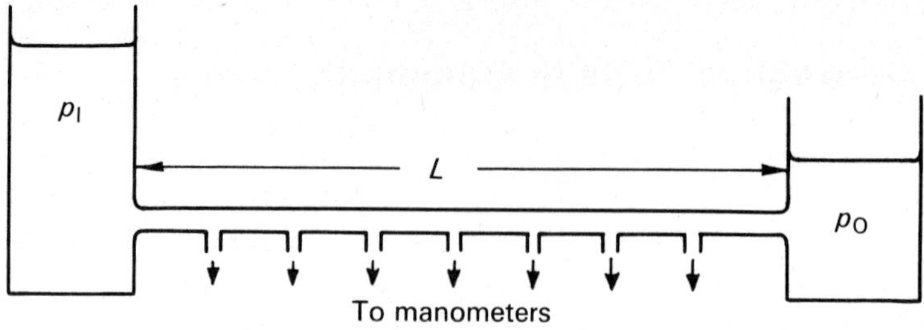

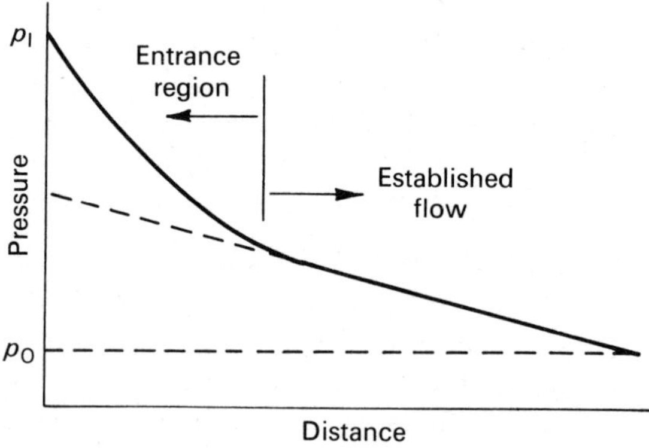

Fig. 1. A schematic view of a long straight tube of length L in which fluid is flowing steadily. The pressure gradient along the tube is initially high in the entrance region and then becomes constant in the established flow region. p_I = inlet pressure, p_O = outlet pressure.

Background mechanics

Steady flow in a straight tube

Let us first consider the steady flow of a fluid within a long straight circular tube. We will assume that the fluid is incompressible; that is, the velocity of flow is always much less than the velocity of sound within the tube. We will also assume that the tube is rigid – any inernal pressure fluctuations cause no changes in tube diameter.

In practice, the first assumption is realistic in virtually all respiratory manoeuvres, except perhaps those of coughing and forced expiration in which airway calibre is so narrowed that extremely high velocities are achieved. The second assumption is also reasonable under most flow conditions, except coughing and forced expiration, because although airways are elastic the internal pressure fluctuations are small compared with those forces maintaining their patency. Airway calibre changes passively with changes in lung volume and hence radial movements of their walls are small compared with the flows through them.

Fig. 1 depicts an imaginary, long, straight, horizontal tube of length L and diameter d units. Its entrance and exit are connected to large reservoirs so that a steady flow with constant inlet and outlet pressures p_I and p_O respectively, can be maintained. At regular intervals along the tube are small lateral pressure tappings enabling us to study the way in which pressure falls along its length. As the inlet–outlet pressure difference is raised the flow rate achieved (Q) is also increased. However, if we look at the distribution of pressure along the length of the tube we see that initially it falls very rapidly with distance, but after a very long distance – which may be as much as 100 diameters – the pressure falls linearly with further distance. The region of the tube in which pressure falls rapidly and non-linearly with distance is called the '*entrance region*' whilst far from the entrance, the flow is '*fully developed*' or '*established*'. In this latter region, the pressure falls linearly with distance – the pressure gradient ($\Delta p/\Delta x$) is constant.

Fully developed flow

The experiment we have described so far may be repeated with tubes of increasing diameter d and it is found that, for a given pressure gradient in the fully developed region, Q increases very rapidly with diameter – in fact as d^4.

If we look at the motion of the fluid within the tube we see that a streak of dye injected into the flow at any radial position is transported downstream as a continuous unbroken filament at the local velocity of the fluid. If the velocity of such streaks at a series of radial positions is measured, it will be seen that the velocity profile is parabolic and axisymmetric. In fact, it was Poiseuille who first performed these experiments and this type of flow is named after him.

In Poiseuille flow

$$\frac{\Delta p}{\Delta x} = 128 \frac{\mu}{\pi} \frac{Q}{d^4}, \tag{1}$$

where μ is the fluid viscosity. It must be stressed that the condition required for Poiseuille flow to occur is that the region is far from bends, constrictions or any other source of disturbance. This criterion is often not met in the bronchial tree.

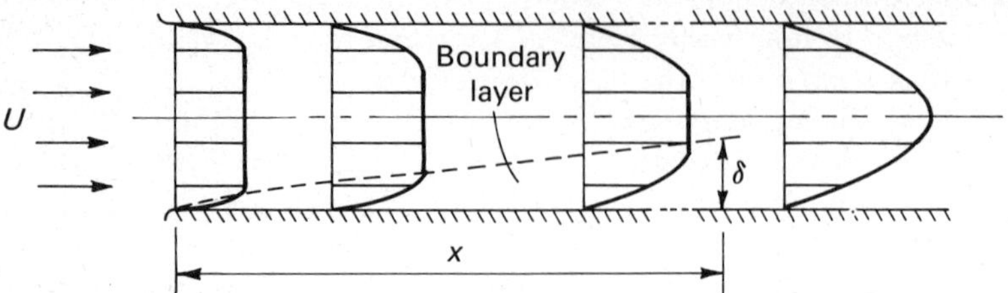

Fig. 2. The changing velocity profile with distance from the entrance of a tube during steady flow. The growth of the boundary layer is also shown.

Entrance flow

If we look at the radial distribution of velocity in the flow in the entrance region of the tube we can see that it changes with distance from the entrance. The fluid entering the tube from the inlet reservoir does so with a uniform or flat velocity profile. On entering the tube, the fluid in contact with the wall is forced to come to rest because of the requirement of 'no slip' on a solid surface. This imposes a velocity gradient within the fluid between the motionless fluid and adjacent elements just within the core of the moving fluid (Fig. 2). As the flow proceeds along the tube, the viscosity of the fluid causes the progressive alteration of the velocity profile as shown in the figure. The initial very high velocity gradient at the wall becomes reduced and progressively more and more of the core fluid is sheared. At the same time, because the peripheral fluid is slowed down by viscosity, the fluid in the core is *accelerated* – otherwise a constant flow could not be maintained. After a considerable distance the whole of the fluid is sheared with the centre line velocity being twice the initial entrance or average velocity. The velocity profile is parabolic and Poiseuille flow has been established, thereafter the velocity profile does not change.

The changes of velocity profile in the entrance region are directly associated with the changes in pressure distribution described above. The initial high velocity gradient at the wall is directly related to the high pressure gradient, as the action of viscosity diminishes the gradient, so the pressure gradient required to maintain the flow is reduced. An excess-pressure driving force is also required to cause the local acceleration of the fluid core. When steady flow conditions are attained no further acceleration of fluid occurs and no further velocity profile adjustments take place, thus the driving pressure gradient becomes constant and independent of position along the tube.

The concept of a boundary layer

As the flow proceeds downstream within the entrance region we have seen that there is an increasing radial distance (δ) from the wall over which the velocity profile is changing and it is within this layer that the viscous effects are dominating the flow. We call this layer the 'boundary layer'. Such layers are formed not only within tubes but whenever a real fluid flows over any surface; thus, a boundary layer is also formed over the surface of a fish swimming in water.

It is possible to compute how quickly the flow within a tube attains fully established conditions – that is, how fast the boundary layer thickness (δ) grows with distance along the tube from the entrance (x):

$$\delta = k\left(\frac{\mu x}{\rho U}\right)^{1/2}$$

where U is the fluid velocity in the core of the tube, ρ is the fluid density and k a numerical constant. At a distance X from the entrance, the boundary layer will reach the tube axis ($\delta = \frac{1}{2}d$) and then:

$$\frac{X}{d} = k'\left(\frac{\rho U d}{\mu}\right)$$

The term ($\rho U d/\mu$) is a dimensionless group called the Reynolds number (Re), whose considerable importance will be discussed in a moment.

Experimental studies have shown that the entrance length X for entry flow in a straight tube is given by:

$$\frac{X}{d} = 0.03 \text{ Re.} \tag{2}$$

Thus, we can see that the entrance length is increased if the flow rate is raised (because raising Q increases the average velocity U), it also increases if the tube diameter is increased. We can obtain an idea of the size of the entry length if we consider flow in a 1 cm diameter tube at a Reynolds number of 1000, X would then be 30 cm or 30 times the tube diameter. If the flow rate were doubled so that Re = 2000, then the entrance length would increase to 60 cm.

In fact, the above equation is only suitable for calculating the entrance length for the range of Re of approximately 10–2500. When Re is very low (less than 1), the inertia forces in the flow are negligibly small, and under the influence of viscosity the flow adjusts very rapidly to established conditions. Then X becomes independent of Re and is approximately $1.5d$ in length. When Re is greater than about 2500 the flow properties change from those we have described so far, streamline flow breaks down and *turbulence* is observed, as described later.

Reynolds number and the idea of scaling

We have just seen that the Reynolds number appears in the boundary layer growth equation and we have commented on the applicability of that equation on the basis of the size of Re. This approach exemplifies two extremely important concepts in fluid mechanics, *similarity* and *scaling*.

In the flow we have been considering we are concerned with the relative importance of inertial and viscous forces operating on the fluid. In fact, we are concerned with this balance in all flow problems. The magnitude or scale of the viscous force will be proportional to the product of the viscosity and the average velocity gradient across the tube $(\mu(U/d))$. Similarly we can scale the inertial force by the kinetic energy per unit volume of the flow (ρU^2). The relative importance of the two may be obtained from their ratio:

$$\mathrm{Re} = \frac{\text{Inertial forces}}{\text{Viscous forces}} = \frac{\rho U d}{\mu}.$$

As already mentioned Re is a dimensionless number and thus its magnitude does not depend upon the units used provided they are self consistent (e.g. the *Système International*).

The principle of similarity allows us to compare flows of different fluids in tubes of various sizes. thus the entrance length expressed as a number of tube diameters (X/d) is a function of the group of variables making up the Re. If we measure the entrance length for water flowing through a narrow tube and compare it with the entrance length for air flowing through a larger tube, the relative lengths will be the same if the Re's are the same in both cases – that is, the flows will be similar. If the Reynolds numbers are different then conditions in the two tubes will be different. Throughout this paper we will be considering flows in terms of Re, for the group is applicable to the characterisation of flows in any situation, not only in tubes. It is always calculated on the basis of the characteristic velocity and dimension of the system – in tube flows this is the average velocity and tube diameter.

Turbulence

When a fluid flows slowly along a tube, all elements of the fluid move along parallel streamlines and their velocities are constant with time. However, as the speed of the flow is increased above a critical value, the steady, smooth motion breaks down and the fluid elements move about in a random fashion. The flow has become *turbulent*. If the time average velocity of the elements is computed, it will be found that the flow has a definite structure and velocity profile.

The pressure gradient for turbulent flow is greater than for Poiseuille flow of the same volumetric rate. This is because the velocity gradient at the wall is steeper in turbulent flow and because the turbulent motions are associated with energy dissipation and hence loss of pressure. In Poiseuille flow the pressure gradient is proportional to Q; in turbulent flow it is proportional to $Q^{7/4}$.

This transition to turbulence occurs when the Reynolds number is in the range 2300 to 2500. When Re is less than 2300 the flow is laminar in both the entrance and established flow regions, between 2300 and 2500 the flow is intermittently turbulent, above 2500 turbulence is continuously present in the established flow region, and as a result the boundary layer growth characteristics are also changed. Just at the entrance the velocity profile is flat, and initially along the tube the profile changes described earlier occur. Further downstream, however, the flow just at the edge of the boundary layer suddenly breaks down and becomes turbulent – a little further on the flows within both the boundary layer and the core break down and also become turbulent. As a result the established velocity profile (which is blunter than in Poiseuille flow) is achieved in a relatively short distance from the entrance. The entrance length in turbulent flow may be estimated from the equation

$$\frac{X}{d} = 0.693 \ \mathrm{Re}^{1/4}. \tag{3}$$

Unsteady flow in a tube

Thus far we have considered only steady motions within tubes but the process of breathing in mammals is essentially an unsteady cyclical flow problem. Thus we must consider under what circumstances we can apply steady flow descriptions to respiration. We may do this by again considering conditions in a long straight tube, considering initially a section far from the entrance. A simple problem that can be conveniently studied is that of the flow resulting from a sinusoidally applied pressure gradient.

If the frequency of the pressure oscillation is very low and the flow rate is slow, then at all times the velocity profile will be parabolic and the flow rate attained will be that expected for the instantaneous applied pressure gradient – we say that the flow responds quasi-steadily. As the frequency of the pressure oscillation is increased, the instantaneous flow conditions will at first still follow the applied gradient; however, at high frequencies, the inertia of the fluid causes modification of the velocity profile.

The fluid in the middle of the tube, moving relatively rapidly will have a greater inertia than that near to the wall thus it will tend to lag behind the changes in the instantaneous applied gradient and the velocity profile in the core will become distorted (Fig. 3). As the oscillation frequency is

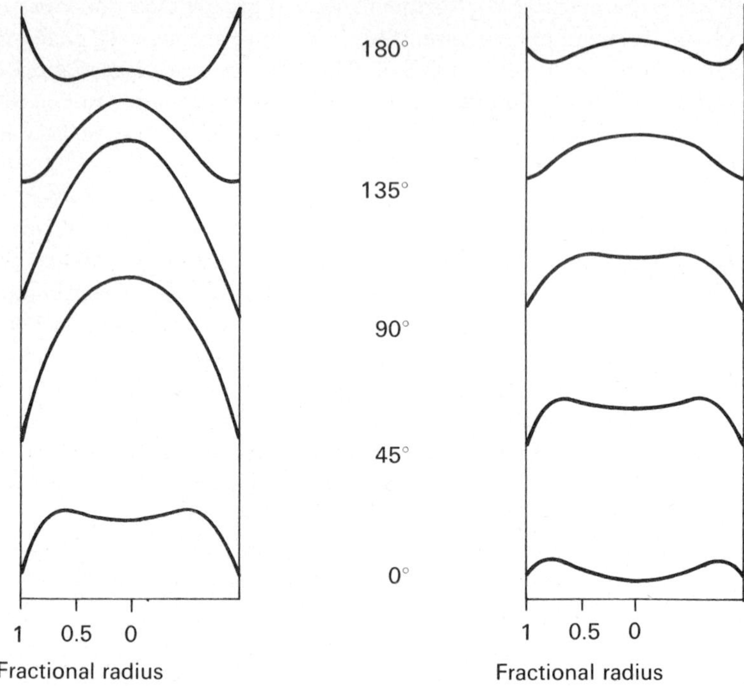

Fig. 3. The effect of an oscillatory applied pressure gradient on the motion of fluid in a tube. In the left-hand panel, the oscillation frequency is sufficient to cause some lag of the core fluid ($\alpha = 3$). At a higher frequency ($\alpha = 5$, right-hand panel) there is considerable distortion.

further raised, the disturbance to the flow increases, the amount by which the core lags increases and the amplitude of the core movement decreases. At very high frequencies there is very little movement of the core, and all motion is constrained to a thin region near the wall. As a result the total flow rate is decreased considerably.

It is possible to determine whether, for a given oscillation frequency, the flow will be significantly disturbed by use of another dimensionless group – the Womersley parameter (α). This group relates the relative importance of inertial and viscous forces within the scale of the period of an oscillation (frequency f)

$$\alpha = \tfrac{1}{2}d\left(\frac{2\pi f}{\nu}\right)^{1/2}, \tag{4}$$

where $\nu = \mu/\rho$.

At low α (less than unity) the flow is quasi-steady and parabolic. As α is increased the flow becomes progressively disturbed. Fig. 4 shows how

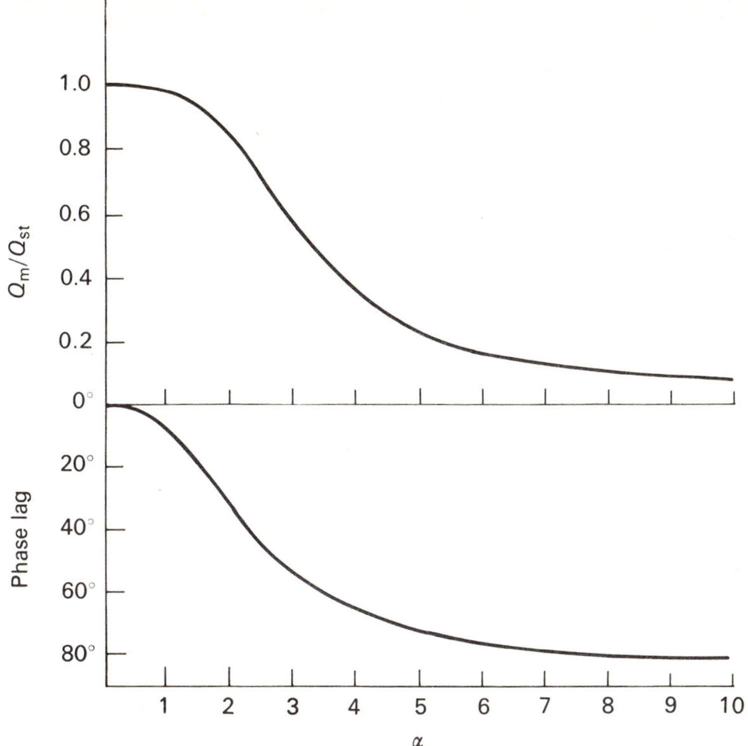

Fig. 4. Graphs showing the progressive loss in maximum flow rate (Q_m) compared with steady flow at peak pressure gradient (Q_{st}) with increasing α and the progressive phase lag.

the phase lag of peak flow rate to pressure gradient increases rapidly as α exceeds unity. Also shown is the maximum flow rate achieved (Q_m) compared with the steady flow (Q_{st}) obtained for the peak applied pressure gradient, this again falls off rapidly as α exceeds unity.

During oscillation of the pressure gradient, conditions nearer to the tube entrance are also affected, though to a lesser degree. We may compute an effective value for α by substituting the boundary layer thickness (δ) in equation 4 in place of the tube radius ($\frac{1}{2}d$); this is because we are concerned with the balance between inertial and viscous forces within the boundary layer:

$$\alpha = \delta \left(\frac{2\pi f}{\nu}\right)^{1/2}. \tag{5}$$

This is the appropriate form of the equation to use when considering unsteady laminar flows within more complicated geometries or around solid bodies. We can see that since δ is less than the tube radius, flow conditions within the entrance tend to have a lower α value than in the established flow region.

Conditions within larger airways may be turbulent during many respiratory manoeuvres and thus we must consider the stability of such flows. A turbulent flow will be less sensitive than Poiseuille flow to an oscillatory pressure gradient and we may obtain an estimate of the value of α from the following equation

$$\alpha = \tfrac{1}{2}\left(\frac{2\pi f d}{0.0075\ U}\right)^{1/2}. \tag{6}$$

The effect of constrictions on tube flow

When the cross-sectional area of a circular tube is decreased locally, the flow conditions within the tube can be markedly altered and we must now briefly look at these effects. Again consider a long straight tube whose cross-sectional area decreases suddenly from A_1 to A_2; within the tube is a fluid flowing in steady laminar motion. If we neglect the effect of viscosity on the fluid, then initially there will be a flat velocity profile across the tube and the magnitude of the velocity will be U_1, as the fluid passes through the section in which the area changes, it will be accelerated and the velocity will increase to U_2 in order to maintain a constant flow rate. By continuity requirements,

$$U_1 A_1 = U_2 A_2.$$

Since we are neglecting the effects of viscosity, there can be no dissipation of energy and Bernouilli's Theorem is applicable:

$$\tfrac{1}{2}\rho U_1^2 + p_1 = \tfrac{1}{2}\rho U_2^2 + p_2, \tag{7}$$

where p_1 and p_2 are the static pressures in the sections up and down stream. Thus we can see that as the flow enters the constriction and is forced to accelerate, the static pressure falls. Conversely, if the flow were into a dilatation, then the velocity would fall and the static pressure would rise locally.

If we were to consider a real fluid passing through such a changing geometry, the same argument would hold true because the amount of viscous energy dissipation over such a short distance would be small.

Of particular interest to us is the situation where flow is into a sudden expansion of the tube. If the flow is other than very slow, the phenomenon of *flow separation* may be observed. When this happens there is no longer simple streamline motion across the whole tube. There is a forward motion in the central core but flow near the walls is sluggish and a recirculation

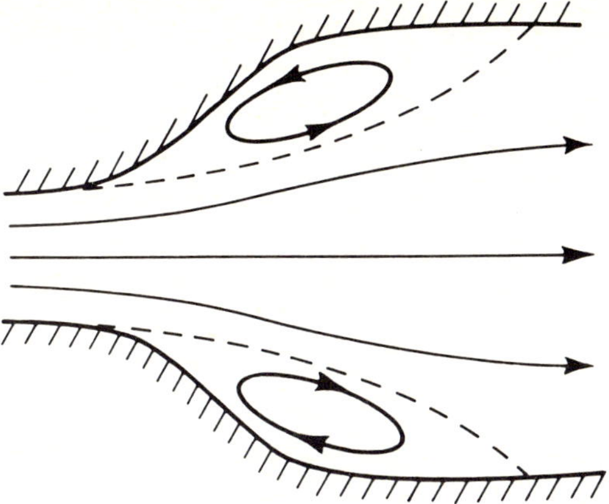

Fig. 5. Schematic view of flow into an expansion region showing boundary layer separation.

of fluid occurs. Fluid trapped within the separation zone is not carried forward with the bulk motion, but may be retained in the vicinity of the expansion for a long time (Fig. 5).

The fluid in the core of the tube emerges virtually as a jet which at low flow rates will be laminar. However, such flows are very unstable and at higher flow rates readily become disturbed and turbulent. If we compute the Reynolds number (based on constricted tube diameter and velocity) of the jet, turbulence occurs when Re exceeds 300–400. Thus, if the flow is exiting into a tube of twice the original diameter the Reynolds number in that tube cannot exceed 150 or the emergent jet will become turbulent. It is possible to prevent separation and hence jet formation if the expansion section has a very gentle rate of increase in area.

Flow in bends

When a fluid flows steadily in laminar motion round a bend, every element of fluid must change its direction of motion. Thus, they must all be given an acceleration inwards towards the centre of the bend; this is supplied by a pressure gradient in that direction. If the flow is originally parabolic, then the faster-moving fluid elements near the tube axis will have greater inertia than those nearer the wall. As a result the fast-moving elements will change direction less rapidly and will be swept out to the outside of the bend; they will be replaced by the slower moving elements near the walls. The result is a transverse circulation or *secondary motion* (Fig. 6). At the

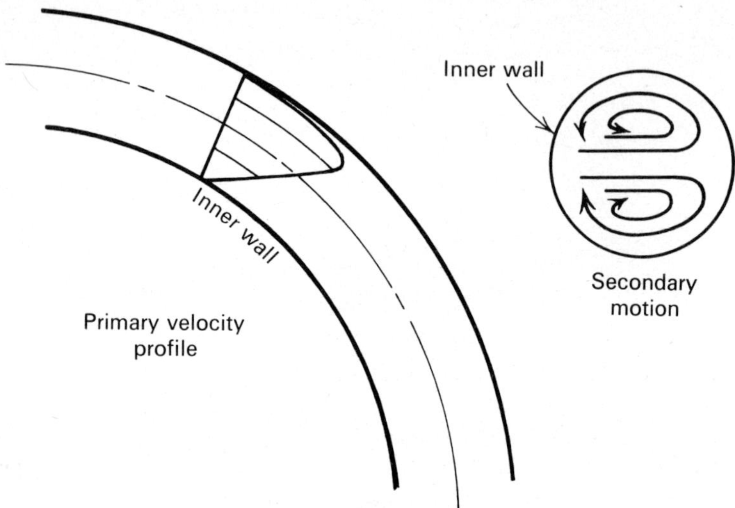

Fig. 6. The effect of curvature in a bend on the primary velocity profile and the direction of the secondary motion.

same time, the primary axial velocity profile will be displaced so that the peak velocity is nearer the outside wall. The higher the curvature of the bend, the greater is the distortion to the velocity profile. At the same time, the pressure driving force required to maintain a given flow is much greater than for a straight tube.

Flow in bifurcations

The mammalian branching tube system of airways presents interesting fluid mechanical problems. Flow conditions are dependent upon the direction of motion, inspiratory or expiratory; they also depend very strongly on the detailed geometry of the bifurcation. We shall discuss here observations of air flow in model symmetrical bifurcations typical of human airways.

Flow visualisation studies have been conducted on laminar inspiratory flows by injecting smoke into the model airway upstream of the junction. Fig. 7 is an end-on view of a daughter tube looking towards the bifurcation, it clearly shows a pair of helical secondary motions in the flow. These motions are in exactly the same sense as in a curved tube (Fig. 6). The magnitude of the secondary motions can be as much as 50% of the axial velocity. If smoke is injected into the outer edge of the bifurcation, then

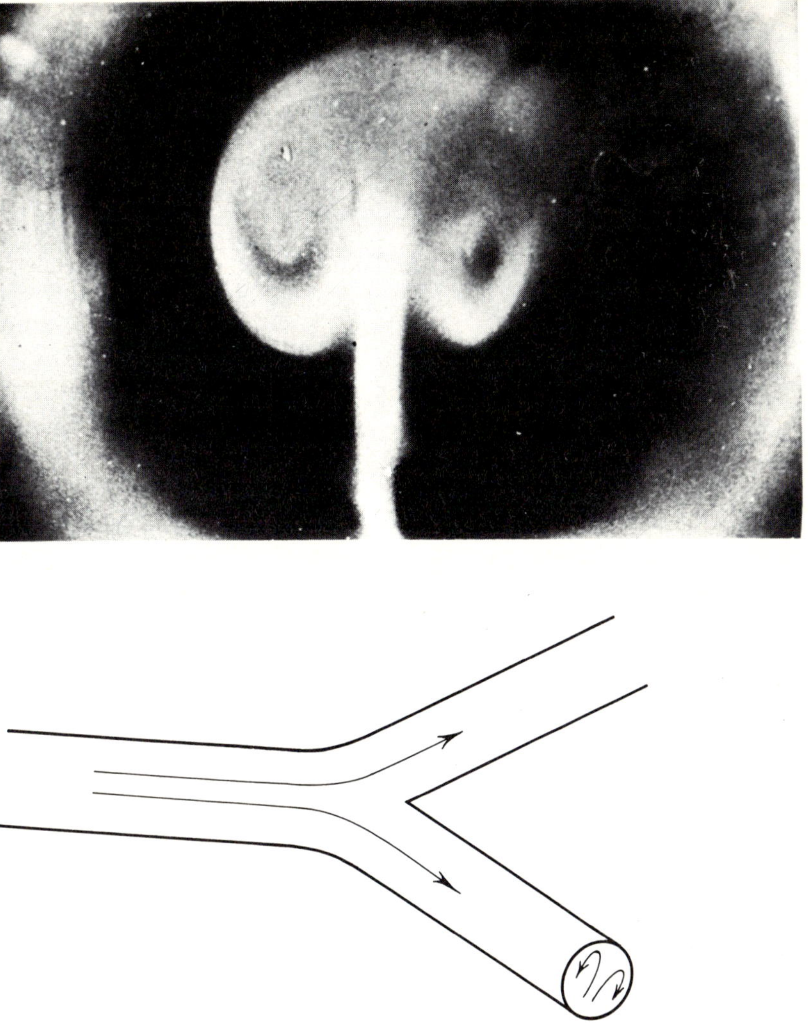

Fig. 7. End-view of a daughter branch on inspiration showing the secondary
motions induced by the bifurcation. The lower panel shows the direction relative
to the junction. (From Schroter & Sudlow, 1969.)

a region of separation can sometimes be seen on the outer edge of the
daughter tube – this happens when the curvature of the outer wall is high
(radius of curvature of wall equal to radius of parent tube) and the rate
of expansion of area in the bifurcation is rapid.

Flow visualisation of expiratory laminar flow shows a somewhat similar

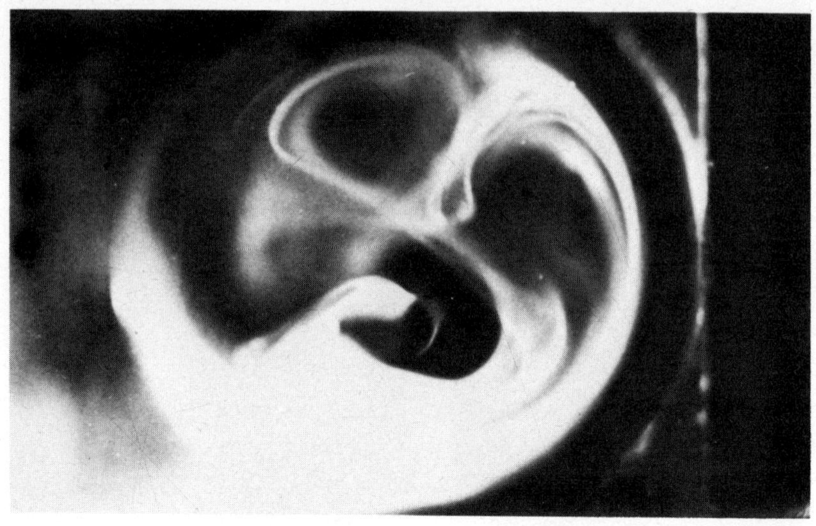

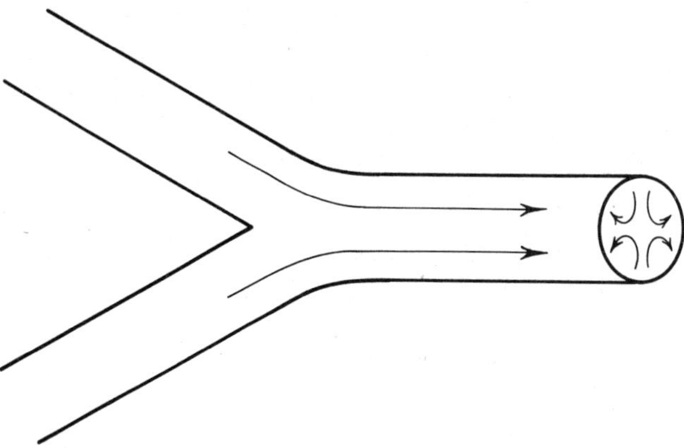

Fig. 8. End-view of parent tube of a single bifurcation during expiration, showing the pair of double helical secondary motions induced by the bifurcation. The lower panel shows the direction relative to the junction. (From Schroter & Sudlow, 1969.)

pattern (Fig. 8) except that two matched pairs of helical motions can be seen. Each pair results from the flow from one daughter tube being turned on passing through the junction.

Studies of the velocity profile within the daughter tubes during inspiration show that the profile which was parabolic on entry becomes severely

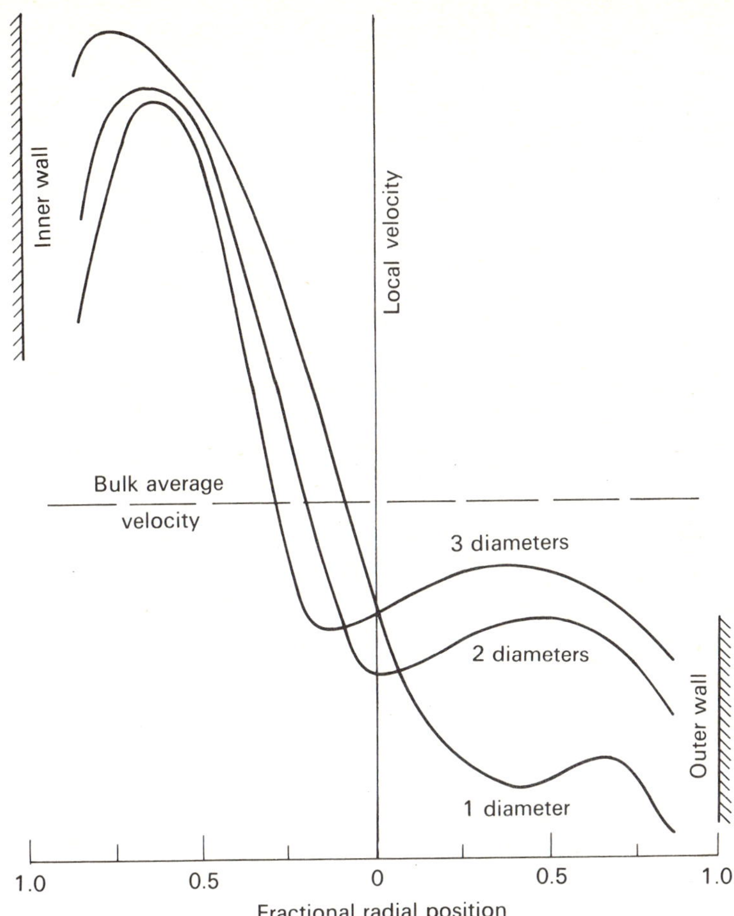

Fig. 9. Velocity profiles measured in a daughter tube in the plane of the bifurcation, the profile entering the junction was parabolic.

distorted with peak velocities near to the flow divider and low velocities on the outer walls (Fig. 9). These disturbances are maintained for a large number of diameters downstream. Expiratory velocity profile measurements show again that the bifurcation severely distorts a parabolic profile into a nearly flat profile.

The flow divider, on inspiration, presents a surface on which a new and hence very thin boundary layer has to grow; on all other surfaces existing boundary layers are modified. Similarly, on expiration the boundary layer characteristics are modified on passage through the junction. As a result the pressure loss across a bifurcation is considerably greater than that in Poiseuille flow. Studies on the flow through a series of model bifurcations show that these processes may all be grossly observed at each bifurcation.

Application to respiratory systems

On the basis of the fluid mechanical descriptions above, it is possible to make a number of points about flow conditions within the airways of the mammalian lung. Since most of our knowledge is based on studies relating to man, discussion will principally relate to the situation in man and thus to other species.

In order to make a detailed analysis of flow in the lungs, it is necessary to have available accurate morphological data. We need to know individual bronchial diameters and lengths as well as bifurcation curvatures; bronchial pathlengths, alveolar sizes and numbers are also important. Such data are largely available for the human lung (Weibel, 1963; Horsfield & Cumming, 1968), but are virtually absent for other species. In man the dichotomously branching network of airways decreases in calibre from the trachea to respiratory bronchioles. Whilst the individual airway calibre decreases peripherally, the total cross-sectional area of airways of each generation increases. Individual airways are short, their lengths being approximately 3.5 times the diameter. Pathlengths from the trachea to respiratory bronchioles are not uniform nor are the number of generations.

During quiet breathing, the Reynolds number in the trachea in man is in the region of 1000 but can rise to more than 10000 on exercise, in the respiratory bronchioles Re falls to order of unity or less. Because the larger airways are so short, we would not expect, on the basis of equations (2) and (3) that established flow would occur within them. In the smallest airways, although local Reynolds numbers are very small, a distance of approximately $1.5d$ is required to achieve established flow and thus profiles will be changing over a significant portion of their length, and we cannot apply Poiseuilles law (equation (1)) to them.

At normal breathing frequencies we can assume that flow in the human bronchial tree is quasi-steady because the value of α in the trachea (based on equation (5)) will be much less than unity and α decreases as the airways become smaller in diameter. In practice, calculations for α in the larger airways, based on equation (5), will lead to an overestimate of the effective value, because even under quiet breathing conditions the flow in the trachea and first few generations of airway will be turbulent; a value of α based on equation (6) would be more accurate. During inspiration, the turbulence within the trachea will originate from the larynx; flow emerging from the laryngeal constriction will do so as a turbulent jet for the reasons explained above. The turbulence will die away fairly rapidly because the airway Re's are considerably less than the critical value of 2300. However, they will persist for two or three generations of bronchi (Owen, 1969).

Thus in man with normal ventilation patterns, it is probably reasonable to assume the flow conditions to be quasi-steady, but considerably more complicated than those of Poiseuille flow.

Pressure–flow relationships in the bronchial tree

It was explained earlier that the pressure–flow relationships in bifurcations are complicated and in many ways similar to conditions in the entrance to a tube. On this basis an attempt has been made to quantify the distribution of pressure drop along the human bronchial tree during inspiration (Pedley *et al.*, 1970*a*, *b*). It was found that the overall pressure drop (Δp_v) from trachea to respiratory bronchiole caused by the action of viscosity could be described by the equation

$$\Delta p_v = k(\rho \mu)^{1/2} Q^{3/2}, \tag{8}$$

where Q is the volumetric flow rate through the trachea. This equation predicts a far higher pressure drop than Poiseuille's Law and the relationship is non-linear; that is, the resistance increases with flow rate. The pressure drop also depends upon both the density and viscosity of the gas mixture breathed. If the pressure drop between trachea and alveolus were to be measured directly, in practice it would be found to be less than that predicted by equation (8), because of kinetic energy changes in the gas as it flows through the lungs. In the trachea velocities are high, but as the flow proceeds towards the periphery they fall to effectively zero within the alveoli. If Bernouille's equation (equation (7)) is applied, it can be seen that as flow velocity falls, the local static pressure rises, thus there is some pressure recovery as we proceed towards the periphery. It is possible to obtain an estimate of the total pressure recovery, since the tracheal velocity profile is virtually flat (Hyatt & Wilcox, 1963). A comparison of the predicted overall pressure drop with experimental observation is shown in Fig. 10, included is the Poiseuille prediction to show how inaccurate it is.

It is also possible, using equation (8), to predict the distribution of pressure along the airways. The pressure falls most rapidly in the large airways, the small airways contributing little to the overall pressure drop. This can be seen in Fig. 11, in which the resistance to flow of each generation of airway is shown. One important implication of the small effect of the peripheral airways on the overall pressure drop is that if, in disease, these airways become severely constricted, then, although the distribution of ventilation is altered, the working of breathing is not greatly increased.

It would appear from Fig. 11 that the highest resistance to flow is in

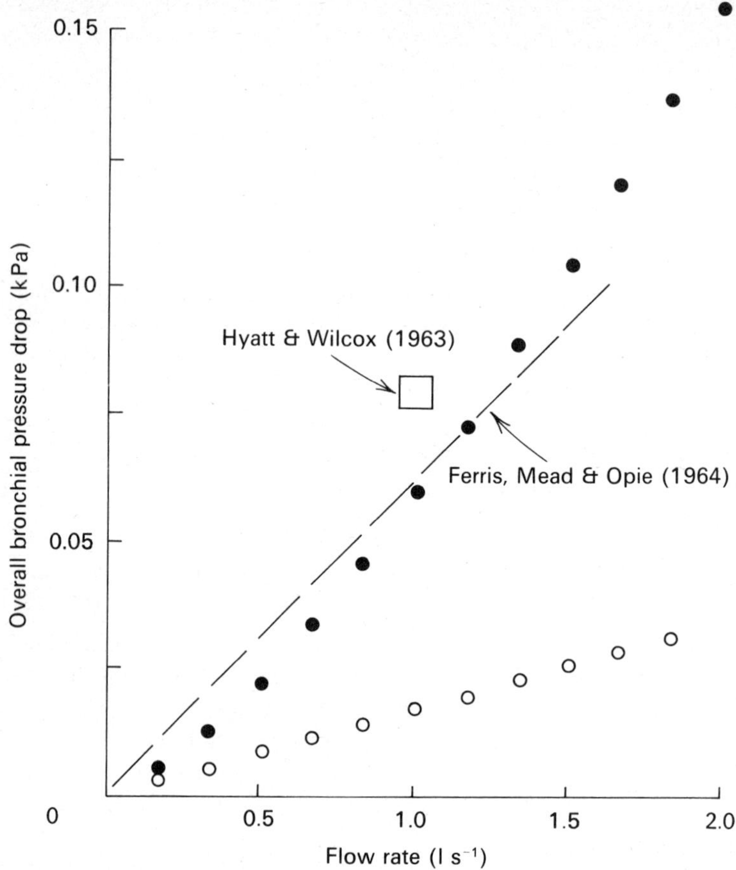

Fig. 10. The variation of overall bronchial pressure drop as a function of flow rate, showing the comparison between predicted (●) and experimentally obtained values. ○ = Poiseuille pressure drop, included for comparison. (From Pedley, Schroter & Sudlow, 1970*b*.)

about the fourth generation of airway; however, this may not be true. The resistance of the largest airways was computed on the assumption that conditions within them were laminar. However, turbulence will be present in the first few airways and the effect will be to raise their resistance to flow; thus the resistance of the largest airways will be underestimated in Fig. 11.

Expiratory pressure–flow relations have not been studied in the same detail. This is mainly because during expiration dynamic compression of airways readily occurs and a reasonable fluid mechanical description of the

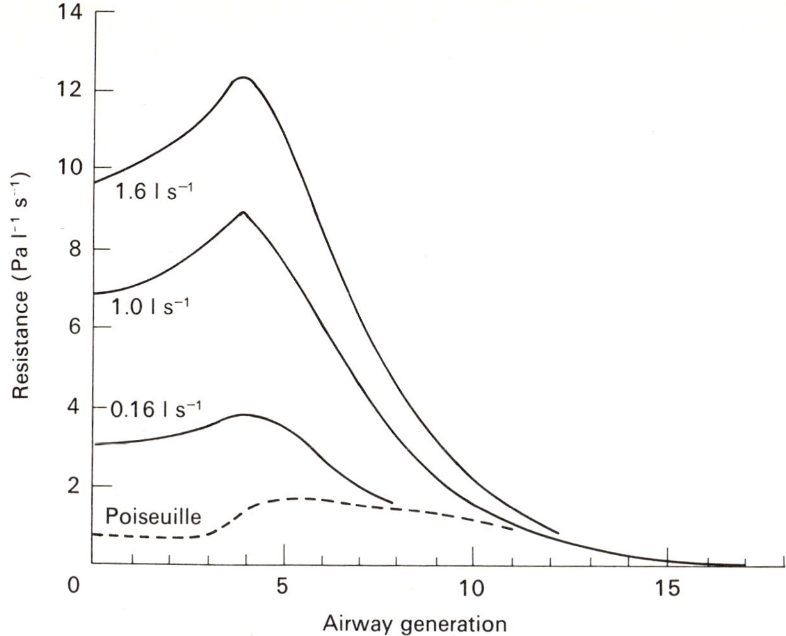

Fig. 11. The variation of resistance along the bronchial tree during inspiration as a function of flow rate. Also shown is the Poiseuille distribution of resistance.

process has not yet been achieved. However, we would expect the viscous pressure drop to be greater than during inspiration because of the flat velocity profiles and the intense secondary motions. The pressure drop is further increased because the flow is being accelerated towards the trachea (equation (7)). These predictions are borne out by observation, at least qualitatively.

There are insufficient morphological data available on the dimensions of the bronchial trees of species other than man to enable a detailed analysis of the pressure flow relations. However, it is reasonable to assume that the bronchial trees are similar – that is, they consist of short tubes of decreasing size but with increasing cross-sectional area towards the periphery.

In consequence the pressure–flow relations in other species will exhibit the same properties as those in man. Thus the quantitative relationship in equation (8) should be applicable, and we would expect the distribution of resistance between large and small airways to be similar to that in man.

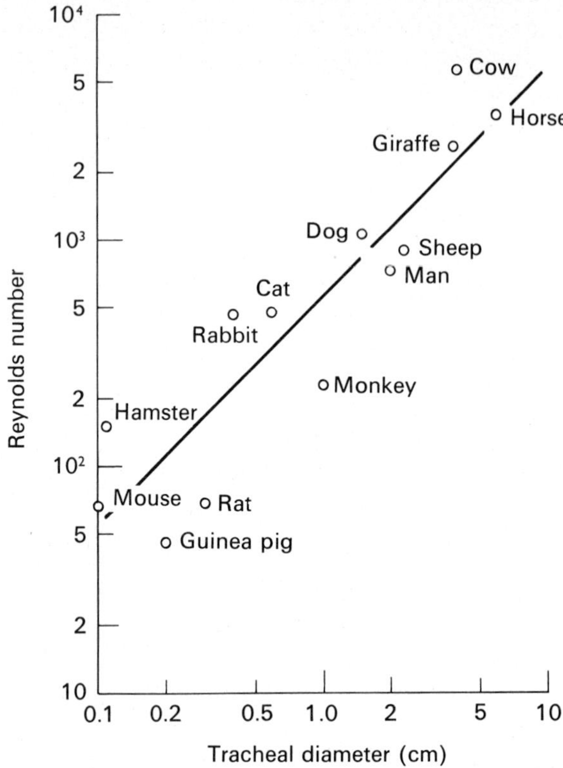

Fig. 12. The variation of tracheal Reynolds number with tracheal diameter (based upon data from Altman & Dittmer, 1971).

Comparison of tracheal flow conditions between species

Data are available for tracheal dimensions (Tenney & Bartlett, 1967) and for tidal volumes, frequency of breathing and minute ventilation at rest in a number of species (Altman & Dittmer, 1971). We may therefore make some interesting comparisons of flow conditions between species.

Fig. 12 shows a comparison of resting tracheal Reynolds number as a function of tracheal diameter. Allowing for the inevitable inaccuracy of the data upon which the calculations are based, it can be seen that there is a roughly first-power relationship between these quantities. The interesting implication of this, as pointed out earlier by Tenney & Bartlett (1967), is that, the velocity of flow in the trachea is virtually independent of animal size; however, the scatter suggests that the velocity in the trachea of the hamster is about four times that of man.

Recalculation of the data for tracheal length and diameter presented by

Tenney & Bartlett indicate that they bear a roughly constant ratio to each other ($L/d = 10$) which is in fact independent of body mass. Thus, in small mammals with low tracheal Re it is possible that flow may be nearly established in the trachea at the distal end. In such animals the value of α is much less than unity even when calculated using equation (4). In larger animals, the resting values of Re are large enough to prevent the boundary layer thickness from approaching the tube radius within the length of the trachea and values of α do not rise above unity. Thus, under most conditions, we can assume quasi-steady flow conditions in virtually all species.

It is interesting to speculate that a possible design criterion for the diameter of the trachea in larger animals relates to the need to keep flow nearly in phase with the applied pressure gradient. Clearly, if α becomes too large the ventilation volume will fall and flow in the trachea will not be in the same direction as flow in the remainder of the lungs. Thus, there would be a maximum minute ventilation related to a limiting value of α. Assuming the appropriate form for α is that given by equation (6), because the tracheal flows may be turbulent, and if the limiting value of α is k, then

$$k = \tfrac{1}{2}\left(\frac{2\pi f d}{0.0075\,U}\right)^{1/2}$$

or
$$U = k'fd.$$

But
$$U = \frac{4Q}{\pi d^2},$$

where $Q =$ volumetric flow rate. Assuming 'square wave' breathing, in which inspiratory and expiratory times are equal, then Q is twice the minute ventilation (V_m). Thus

$$\frac{8V_m}{\pi d^2} = k'fd,$$

and hence
$$V_m \propto fd^3. \tag{9}$$

We may test this hypothesis by comparing predicted and actual minute ventilations, but unfortunately there is little information on maximal values of ventilation and respiratory frequency in many species. However, if it is assumed that maximal values are proportional to resting values of minute ventilation and frequency, then we may test the relationship, as shown in Fig. 13. Allowing for the inevitable innaccuracy in the source data and the gross assumptions of relating resting to maximal values, the agreement appears reasonably good, especially for larger animals. It may be that the criterion is not met for smaller animals because, as already noted, large values of α are never achieved.

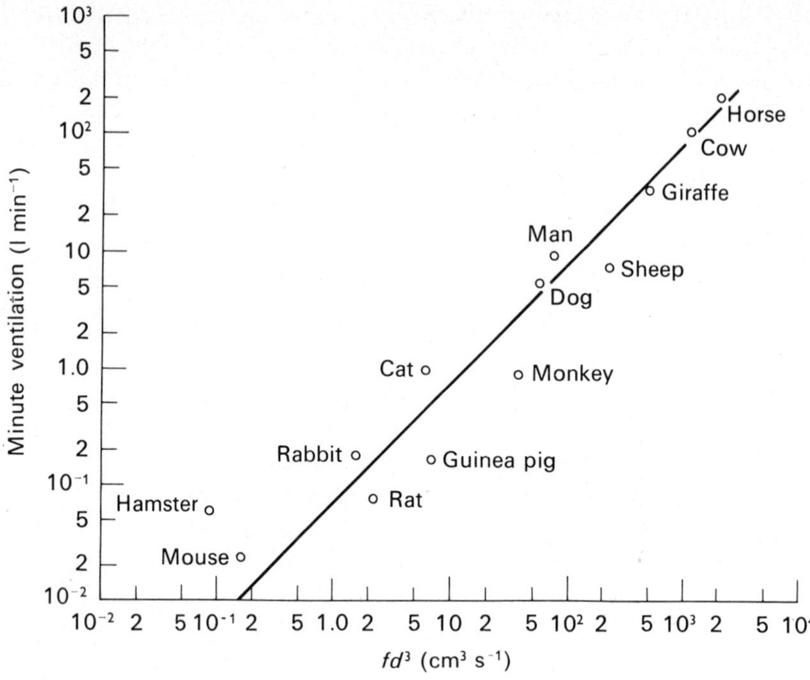

Fig. 13. The variation of minute ventilation with the function fd^3 (based upon data from Altman & Dittmer, 1971).

A second and even more indirect method of testing the hypothesis is to use the published correlations for minute ventilation, respiratory frequency and tracheal diameter as functions of body weight M (Tenney & Bartlett, 1967). Tracheal diameter is proportional to $M^{0.39}$ and respiratory frequency is proportional to $M^{-0.28}$, thus we would expect minute ventilation to be proportional to $M^{0.89}$. The correlation found for minute ventilation as a function of body weight is $V_m \propto M^{0.75}$. Appreciating the inevitable compounding of errors in using body weight correlations in this manner, the comparison is close and gives further support to the idea of the criterion of equation (9).

It would appear that the giraffe may be an interesting animal, in terms of the flow conditions within its trachea. This animal has a remarkable trachea of some 45 diameters in length compared with the normal value of 10 diameters. Thus, under resting conditions, tracheal Re will be about 2600 and a value for α based on equation (6) would be approximately 1.5. However, because the trachea is so long, the velocity profile will have a sufficient distance to develop to much closer to established conditions, and

α calculated on the basis of equation (5) would be approximately 5. Consequently, tracheal flow will be significantly out of phase with applied thoracic pressure swings; any increase in ventilation would inevitably demand a considerable increase in the work of breathing which it may not be capable of. Indeed, it is known that the giraffe does not pant and when forced to run fast can only do so for short bursts after which it has to recover; these two factors would suggest indeed this animal does suffer respiratory limiations.

References

Altman, P. L. & Dittmer, D. S. (1971). *Respiration and Circulation*. Bethesda: Federation of American Societies for Experimental Biology.

Dekker, E. (1961). Transition between Laminar and turbulent flow in human trachea. *Journal of Applied Physiology*, **16**, 1060–4.

Ferris, B. G., Mead, J. & Opie, L. H. (1964). Partitioning of respiratory flow resistance in man. *Journal of Applied Physiology*, **19**, 611–22.

Horsfield, K. & Cumming, G. (1968). Functional consequences of airway morphology. *Journal of Applied Physiology*, **24**, 384–90.

Hyatt, R. E. & Wilcox, R. E. (1963). The pressure–flow relationship of the intrathoracic airways in man. *Journal of Clinical Investigation*, **42**, 29–39.

Owen, P. R. (1969). Turbulent flow and particle deposition in the trachea. In *Circulatory and Respiratory Mass Transport*, CIBA Foundation Symposium, pp. 236–52. London: Churchill.

Pedley, T. J., Schroter, R. C. & Sudlow, M. F. (1970a). Energy losses and pressure drop in models of human airways. *Respiratory Physiology*, **9**, 371–86.

Pedley, T. J., Schroter, R. C. & Sudlow, M. F. (1970b). The prediction of pressure drop and variation of resistance within the human bronchial airways. *Respiration Physiology*, **9**, 387–405.

Pedley, T. J., Schroter, R. C. & Sudlow, M. F. (1977). Gas flow and mixing in the airways. In *Bioengineering Aspects of the Lung*, ed. J. B. West, pp. 163–265. New York: Marcel Dekker Inc.

Rohrer, F. (1915). Der Stromungswiderstand in den menschlichen Atemwegen. *Pflügers Archiv für die gesamte Physiologie*, **162**, 225–355.

Schmidt-Nielsen, K. (1975). *Animal Physiology: Adaptation and Environment*. London: Cambridge University Press.

Schroter, R. C. & Sudlow, M. F. (1969). Flow patterns in models of human bronchial airways. *Respiration Physiology*, **7**, 341–55.

Tenney, S. M. & Bartlett, D. Jr (1967). Comparative quantitative morphology of the mammalian lung: trachea. *Respiration Physiology*, **3**, 130–5.

Weibel, E. R. (1963). *Morphometry of the Human Lung*. Berlin: Springer-Verlag.

West, J. B. & Hugh-Jones, P. (1959). Patterns of gas flow in the upper bronchial tree. *Journal of Applied Physiology*, **14**, 753–9.

23. The aerodynamics of normal hovering flight: three approaches

By C. P. ELLINGTON

The flight of birds and insects may be divided into several convenient categories: gliding, soaring and flapping. Gliding and soaring require little, if any, energy expenditure by the animal, and the aerodynamic treatment is the same as fixed-wing aeroplanes (for an introduction and references see Pennycuick, 1972). In flapping flight the mechanical forces are generated by actively moving wings, necessitating a substantial energy input by the animal, and the aerodynamic analysis is more complicated. One particular mode of flapping flight, hovering, represents both a biological and aerodynamic extreme: the energetic requirement is the highest for any form of animal flight, and the aerodynamic treatment enters a realm where the theoretical framework is possibly stretched beyond its assumptions and adequate experimental investigations are lacking.

Weis-Fogh (1972, 1973) emphasized this importance of hovering flight. Using an analysis which assumes the wings perform as *steady-state aerofoils* during the motions of *normal hovering* he obtained reasonable results for the vast majority of hovering animals investigated. In this paper I shall also concentrate on normal hovering, reviewing the two aerodynamic approaches used to date and presenting a simplified version of a third approach which attempts to synthesize the others.

Normal hovering

Although the size range of animals which habitually hover using aerobic muscle power is enormous, from the small chalcid wasp *Encarsia formosa* (mass 25 μg) to the hummingbird *Patagonia gigas* (mass 20 g), the wing movements of the majority of animals are similar to an idealized pattern called *normal hovering* by Weis-Fogh (1973). In normal hovering (see Fig. 1) the body is oriented almost vertically and the wings beat in a horizontal plane. The kinematic parameters are the angle swept by a wing in the horizontal plane, ϕ, and the wingbeat frequency, n. Distance along the wing length is given by the radius, r, and the positional angle of a wing is γ. The morphological parameters are the wing length, R, and the chord as a function of the radius, $c(r)$. The angular movement of the wing is

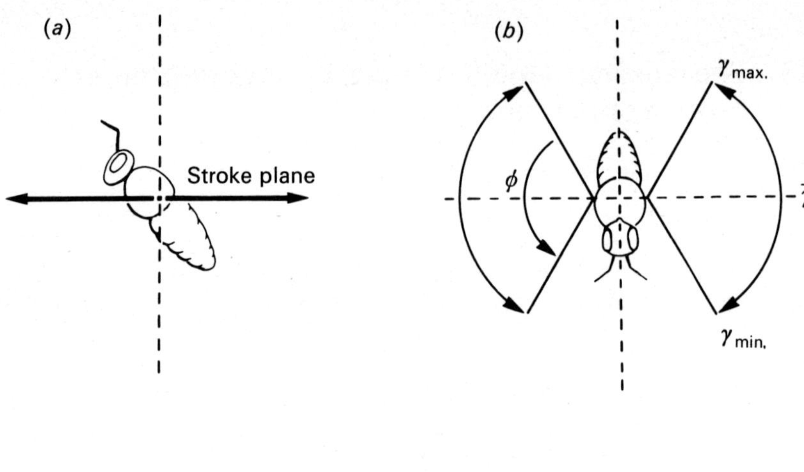

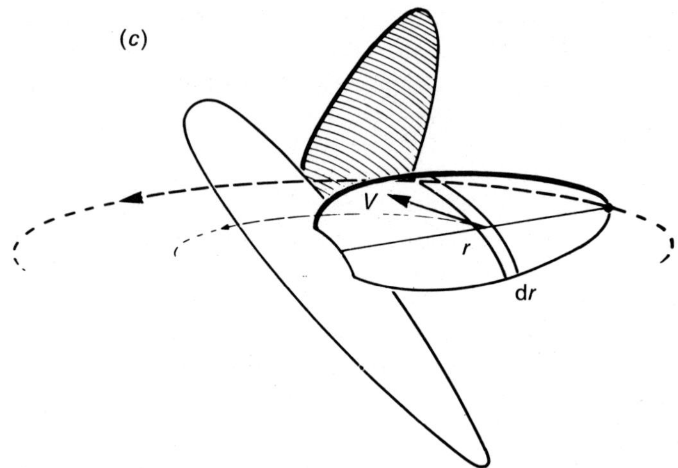

Fig. 1. The simplifications of normal hovering. The horizontal stroke plane seen (*a*) from the side, and (*b*) from above. (*c*) The horizontal velocity of a wing element, d*r*, located at radius, *r* (Weis-Fogh, 1973). For other symbols see text.

assumed to be simple harmonic motion and is given by Weis-Fogh (1973) as

$$\gamma(t) = \bar{\gamma} + \tfrac{1}{2}\phi \sin(2\pi n t),\tag{1}$$

where $\bar{\gamma}$ is the mean angle, and t is time. Thus the velocity V at a point r along the wing axis is

$$V(r, t) = \pi\phi n r \cos(2\pi n t).\tag{2}$$

The assumptions of normal hovering permitted Weis-Fogh (1973) to derive simple analytical expressions which could be applied to a large number of hovering animals. His intention was not a detailed analysis but a semi-quantitative comparative survey.

A main feature of normal hovering is the symmetry of half-strokes, defined as movement of the wings from one extreme of γ to the other. Since aerodynamic interactions of the wings were ignored, the analysis should strictly only be applied to hovering when the wings do not come near each other at the ends of half-strokes. Animals which perform a 'clap and fling' as described by Weis-Fogh (1973) should be omitted from the analysis, as should animals which do not beat their wings in a horizontal plane. This latter group includes the true hover-flies, the Syrphinae (Weis-Fogh, 1973); the dragonflies, Odonata (R. A. Norberg, 1974); and some small passerine birds and bats (U. M. Norberg, 1974). The majority of animals investigated by Weis-Fogh (1973) did perform in a manner similar to normal hovering, and it was found that their wing shape could be approximated by a semi-ellipse. A result of the present analysis is that wing shape is not a strong parameter in the aerodynamic analysis, so that reasons for convergence of many animals on this wing shape may be structural or mechanical.

Aerodynamic approaches to normal hovering

Just as forward flight of animals can be understood by applying the theories of aeroplane wings, it is not surprising that hovering flight of animals, like that of helicopters, is best suited by the theories of airscrews, or propellers. Historically and conceptually the theories of airscrews progressed from the *momentum theory* and *blade element theory* to the *vortex theory*.

The momentum theory is based on the motion imparted to the air during hovering, with the forces acting on the wings (which are the animal's 'propeller blades') being those necessary to generate the motion. The power of this approach lies in its simplicity; the forces on the wing are given, by Newton's laws, as the rate of change of momentum of the air acted upon.

The blade element theory assumes that the forces are generated by the wings acting as aerofoils, and that each element of the wing may be considered independently and described by steady-state aerodynamics. The assumption of steady-state aerodynamics requires that the forces acting on a wing element instantaneously adopt the value which would exist if the element were in a steady flow of the same velocity and attitude.

The vortex theory considers the lift on a wing element to be caused by a circulation of the flow around the wing, which also causes free vortices to be generated into the passing air. The circulation around a wing element can be related to the lift coefficient in the blade element theory, and the trailing vortices are responsible for the air motion used in the momentum theory.

The momentum theory

In the momentum theory the animal is regarded as a black box since only the initial and final states of the air acted upon are considered. Fig. 2 presents a physical picture of the flow field, with the animal at rest with respect to the undisturbed air. The action of the wings sucks air, initially at rest, from above the animal and blows it down vertically. In this picture the wings are similar to an actuator disk, a conceptual device used in the momentum theory of propellers, which was first pointed out by Hoff (1919). The actuator disk ascribes some necessary physical dimensions and properties to the black box. It is infinitely thin with an area equal to that swept by the wings, ϕR^2, and uniformly accelerates air through it by generating a constant pressure difference.

The actuator disk theory shows that the air is accelerated to one-half of its final velocity by the time it reaches the disk, and assumes its final velocity, U_s, in the steady wake below. The air velocity at the disk is called the induced velocity, U_0, since it represents the change in the velocity of the wing relative to the undisturbed air caused by motion of the wing. The volume of air passing through the wing disk per unit time is $\phi R^2 U_0$, and by continuity must equal the volume flow in the wake, $A_s U_s$, where A_s is the cross-sectional area of the wake. The actuator disk theory shows $U_s/U_0 = 2$, so the area of the wake must equal $\phi R^2/2$.

The momentum equation may be applied to the steady wake and states that the thrust, dT, needed to generate a flow of velocity U across an elemental area dA is

$$dT = \rho U^2 \, dA, \tag{3}$$

where ρ is the density of air. This simply equates the force, dT, to the rate of change of momentum of the air, given by the product of the mass flow, $\rho U \, dA$, and velocity, U. Similarly, the kinetic energy per unit time, dP, imparted to the air is $\frac{1}{2} \times$ mass flow $\times$ (velocity)^2,

$$dP = \tfrac{1}{2}\rho U^3 \, dA. \tag{4}$$

Equations (3) and (4) may be integrated over the cross-sectional area of the wake to obtain the total thrust and power. The power is referred to as the induced power, P_i, necessary to generate the induced velocity. The actuator disk uniformly accelerates air through it, so the wake is

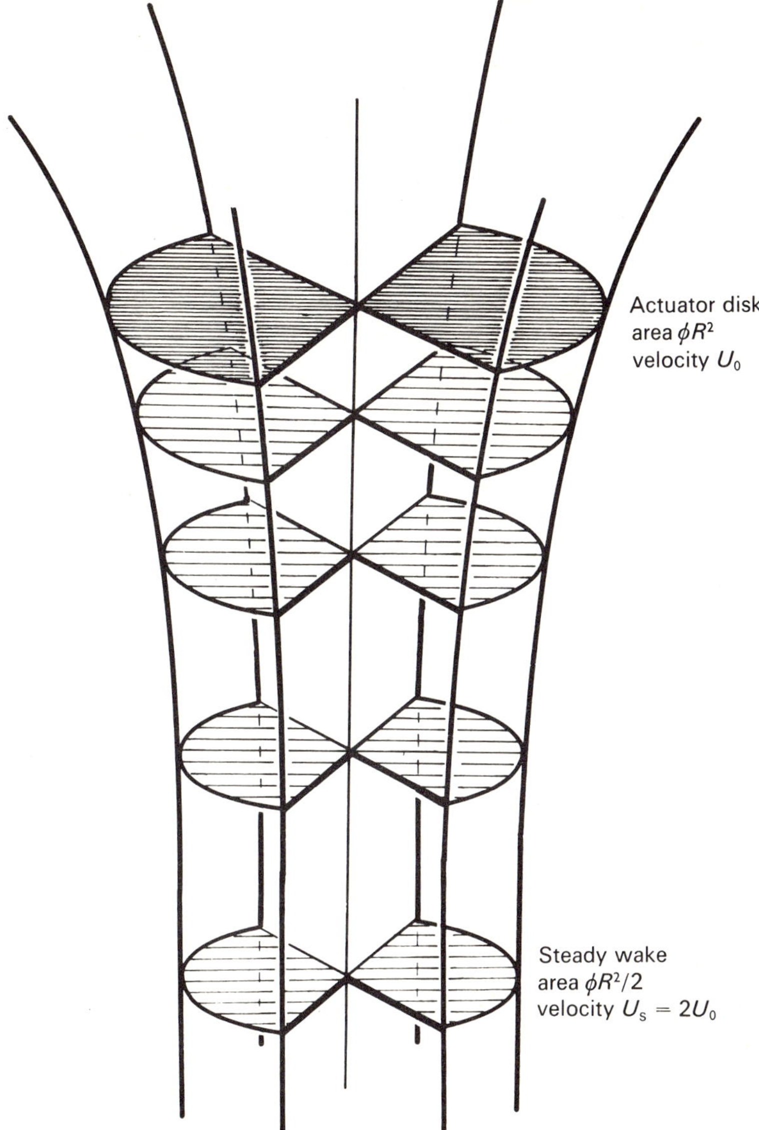

Fig. 2. Velocity field of the actuator disk theory as a contracting jet of air. The cross-section of the flow is defined by the wing disk at the top. Horizontal planes correspond to half-strokes; their spacing proportional to downward velocity. For symbols, see text.

assumed to consist of a jet of area $\phi R^2/2$ moving at a constant velocity U_s:

$$T = \tfrac{1}{2}\rho\phi R^2 \, U_s^2, \tag{5}$$

$$P_i = \tfrac{1}{4}\rho\phi R^2 \, U_s^3. \tag{6}$$

For the animal to hover, the thrust must balance the weight, mg, giving the wake velocity as

$$U_s = \sqrt{(2mg/\rho\phi R^2)}. \tag{7}$$

The actuator disk theory is seen to be a special case of the momentum theory although they are often treated as synonymous in the literature.

For comparative purposes the disk loading, D, may be defined as $mg/\phi R^2$, which is the pressure difference across the actuator disk for a hovering animal. The specific induced power, P_i^*, is often defined as P_i/mg, giving the induced power expenditure per unit body weight. The final form of the actuator disk analysis is then

$$U_s = 2U_0 = \sqrt{(2D/\rho)}, \tag{8}$$

$$P_i^* = U_0. \tag{9}$$

The calculated disk loadings for normal hovering animals range from 0.4 N m^{-2} for the crane-fly *Tipula paludosa* to 11.1 N m^{-2} for the bumble bee *Bombus terrestris*. The relatively high loading for the bumble bee is still well below the average for modern helicopters, 144 N m^{-2} (Nikolsky, 1951). *T. paludosa* and *B. terrestris* represent extremes in the range of disk loadings and consequently specific induced power, 0.4 W N^{-1} and 2.1 W N^{-1}, respectively. The specific induced power for the helicopter is 7.6 W N^{-1}. The high value for the bumble bee is partly reconciled by a larger fraction of flight muscle in the body, but the power output per unit muscle weight must still be higher than the crane-fly.

An advantage of the momentum theory is that it ignores the operating mechanism. This may turn into a disadvantage, however, if energy losses in the mechanism are significant when compared to the kinetic energy of the wake. The frictional drag of wings caused by viscous forces in air represents such a power loss. Weis-Fogh (1973) estimated the specific power needed to overcome these drag forces and found it to be about the same as the specific induced power. Thus the momentum theory may underestimate the power expenditure by roughly a factor of two.

The air velocities just behind the stroke plane have been measured for several tethered insects (Hocking, 1953; Bennett, 1974; Chance, 1974) and a simulated flapping wing (Bennett, 1966). In all cases the wake velocity was inclined to the vertical, indicating that the insects were attempting to fly forward. The measured velocities were averaged over a half-stroke or multiple strokes and the mean used in the momentum equation to calculate thrust. The mean velocity must be used because the flow is unsteady and

PLATE 1

The crane-fly, *Tipula paludosa*, in tethered flight. Air motion is indicated by talcum powder; exposure is $^1/_{60}$ s. Maximum contraction of the wake occurs about 2 cm below the wings.

To face page 332

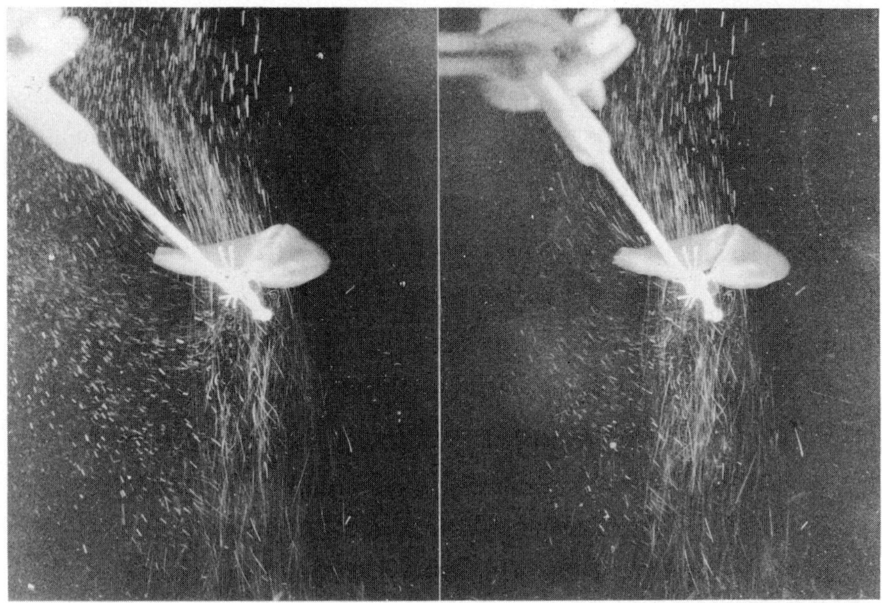

PLATE 2

Stereoscopic pair of photographs, at $^1/_{60}$ s, of *Tipula paludosa* showing flow through the wing disk. Wake velocity near the wake axis is lower than that further along the wings, but the rotational component is relatively greater.

is influenced by the circulation around the wings, which will be described under vortex theory (p. 340). The results are generally satisfactory and support the momentum approach for thrust calculation. Bennett (1966) claims that the thrust generated by his flapping wing was greater than the equivalent steady-state case using a constant rotational velocity. He appears to have calculated the thrust from instantaneous velocity measurements rather than the mean, but this is not clear.

I have been working with G. G. Runnalls in our laboratory on flow visualization techniques applicable to hovering flight. The most successful, and relatively simple, technique is to sift talcum powder through a wire gauze attached to the diaphragm of a loudspeaker driven by an audio-oscillator. The fine cloud of particles drifting down is illuminated by strong side-lighting and may be photographed against a dark background.

The crane-fly, *T. paludosa* was selected for a study of tethered flight. The large coxa may be attached by a drop of wax to a support, leaving the thorax undisturbed, and the distal portion of the legs removed. With its minimal flying abilities the crane-fly performs well and reliably under tethered conditions. Plate 1 shows the flow, photographed at 1/60 s, around a subject. The flow pattern is in general agreement with the actuator disk flow in Fig. 2. Maximum contraction of the wake occurs about 2 cm below the beating wings. The particles are not completely uniform in size so that the streak length is only roughly proportional to velocity, but it is evident that maximum velocity of the wake coincides with maximum contraction. The flow pattern is more irregular than that depicted in Fig. 2, but this is hardly surprising given the simple assumptions of the actuator disk theory. In the vortex theory the discrepancies in the flow patterns will be discussed.

Before leaving the actuator disk theory the assumptions and limitations of it may be summarized.

1. The induced velocity is assumed to be uniform across the actuator disk. This restriction is merely a convenient one since the momentum theory can also be applied to a variable velocity, but there have not been studies on other feasible distributions. Lighthill (1977), however, re-states the argument of propeller theory that the induced power will be a minimum if the induced velocity is constant.

2. The only power input is assumed to be the increase in axial kinetic energy of the wake; any swirling of the wake and the energy involved in this motion is ignored. Again, a more complete momentum theory can account for this, but details of the flow are lacking.

3. The power input to overcome wing drag, especially frictional drag, cannot be calculated by a theory which disregards the aerodynamic mechanism.

4. An infinite number of blades is assumed so that air is uniformly accelerated over the wing disk. Deviations from this flow are most likely to occur at the wing tips and ends of the half-strokes.

The blade element theory

The blade element theory is the most widely used approach in analysing animal flight, the original form being a propeller theory developed by Drzewiecki. The fundamental assumption that steady-state aerodynamics can be applied to flapping animal flight has often been questioned, and adequate experimental work is sorely lacking. Osborne (1951) developed the basic equations applicable to insect flight, but some lift coefficients calculated from his equations were too high to be compatible with what is known of steady-state aerodynamics. In reviewing his approach Weis-Fogh & Jensen (1956) pointed out that some of the experimental data in the literature which Osborne used were incorrect, and that his disregard of lift forces generated by the upstroke limited the analysis to fast forward flight. Jensen (1956) applied the blade element theory to fast forward flight of the locust *Schistocerca gregaria* and found good agreement between theory and experimental measurements. Weis-Fogh (1972) applied the theory to hovering for the hummingbird *Amazilia fimbriata fluviatilis* and the fruit-fly *Drosophila virilis* and obtained reasonable estimates of lift coefficients. As noted earlier, hovering is the most likely case where the theory may not be applicable. In a comparative study Weis-Fogh (1973) used the assumptions of normal hovering to produce simplified expressions of the theory and calculated satisfactory lift coefficients for all animals which performed normal hovering.

If we assume that (i) radial elements of the wings may be considered independently and (ii) their aerodynamic forces are consistent with steady-state measurements, we can calculate the forces exerted by the wings in motion as a series of static pictures. The basic equation of the blade element theory, as equation (3) is of the momentum theory, is

$$dF = \tfrac{1}{2}\rho c V^2 C_F\, dr. \tag{10}$$

The momentum analysis earlier showed that the inertial force associated with a flow of velocity V and an area A is proportional to $\rho A V^2$. For a wing element of chord c and width dr a convenient measure of area exposed to the flow is $c\, dr$, giving the force on a wing element proportional to $\rho c V^2\, dr$. The proportionality constant is $\tfrac{1}{2} C_F$, where C_F is called the force coefficient and the arbitrary factor of $\tfrac{1}{2}$ is introduced to make the form compatible with Bernoulli's equation.

For aerofoils the force is traditionally resolved into a component normal

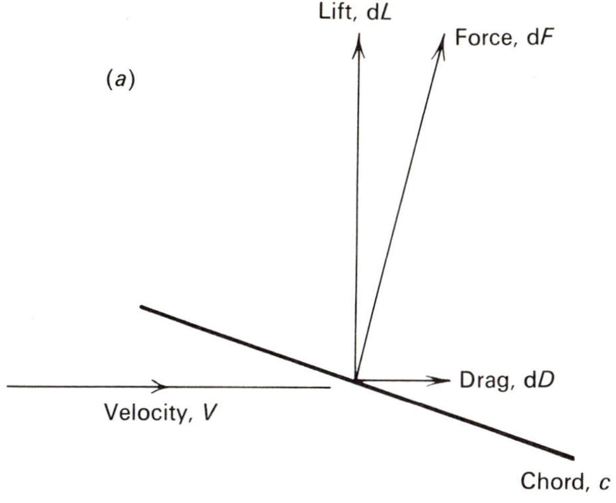

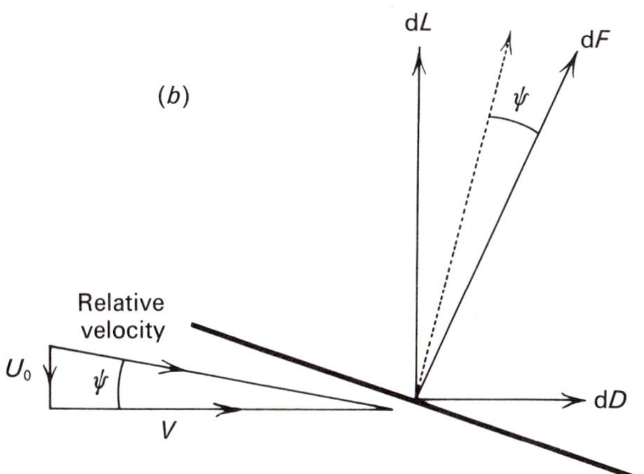

Fig. 3. (a) The horizontal air velocity and resultant force acting on a wing element. The force is resolved into lift and drag. (b) The induced velocity U_0 rotates the resultant velocity and force through an angle ψ. Drag in the stroke plane increases, and the vertical force decreases.

to the flow velocity called the lift, dL, and a component parallel to the flow called the drag, dD (Fig. 3a);

$$dL = \tfrac{1}{2}\rho c V^2 C_L \, dr, \qquad (11)$$

$$dD = \tfrac{1}{2}\rho c V^2 C_D \, dr, \qquad (12)$$

$$C_F = \sqrt{(C_L^2 + C_D^2)}. \qquad (13)$$

Equations (11) and (12) are the heart of the blade element approach. The power expended by the animal in hovering is simply the product of the drag force and velocity,

$$dP = \tfrac{1}{2}\rho c V^3 C_D\, dr. \tag{14}$$

To calculate the total lift during a half-stroke equation (11) is integrated over the wing radius and then averaged over time. Weis-Fogh (1973) approximated $c(r)$ by a semi-ellipse and V by a horizontal sinusoidal oscillation, equation (2). Following Osborne (1951), he assumed that the force coefficients were constant along the radius and in time. This permitted easy manipulation of equation (11) and the result was equated with the animal weight, mg, giving the lift coefficient as

$$C_L = 1.03\ mg/\rho n^2\phi^2 cR^3, \tag{15}$$

where c is the chord at the base of the wing. Thus a lift coefficient is quickly obtained for a minimum amount of information. The lowest lift coefficient Weis-Fogh calculated was 0.5 for several beetles. The highest values for normal hovering were for the hummingbirds, with 2.0 for *Amazilia*. In his more complete analysis, Weis-Fogh (1972) calculated an average lift coefficient of 1.8 for *Amazilia*. A value around 2 is uncomfortably near the accepted upper limit to steady-state lift coefficients, however.

Using C_L as calculated above Weis-Fogh referred to the meagre experimental data on real wings to estimate an appropriate C_D, and then calculated a power expenditure by essentially integrating equation (14) over the radius and averaging the result in time.

By calculating the force coefficients using the horizontal velocity of the wing, the effects of the induced velocity are neglected. The beating wings meet air with a horizontal velocity V, and the actuator disk analysis gives a vertical induced velocity U_0 at the level of the wings. This implies that the velocity of air relative to the wing is inclined at an angle ψ with respect to the horizontal, where ψ is given by

$$\psi = \tan^{-1}(U_0/V). \tag{16}$$

This inclination of the relative velocity tilts the lift vector calculated for a horizontal wind away from the vertical by an angle ψ and effectively increases the resistance force on the wing in its plane of motion (Fig. 3*b*). The work done against this extra force by the wing is equal to the kinetic energy of the induced velocity (Goldstein, 1965). Weis-Fogh calculated an average value of ψ to be around 11°, and added a correction to his power estimates for this increased force.

In hovering the wings rarely beat in a horizontal plane, but usually follow a curve resembling a figure-of-eight. As a consequence of this, the wing

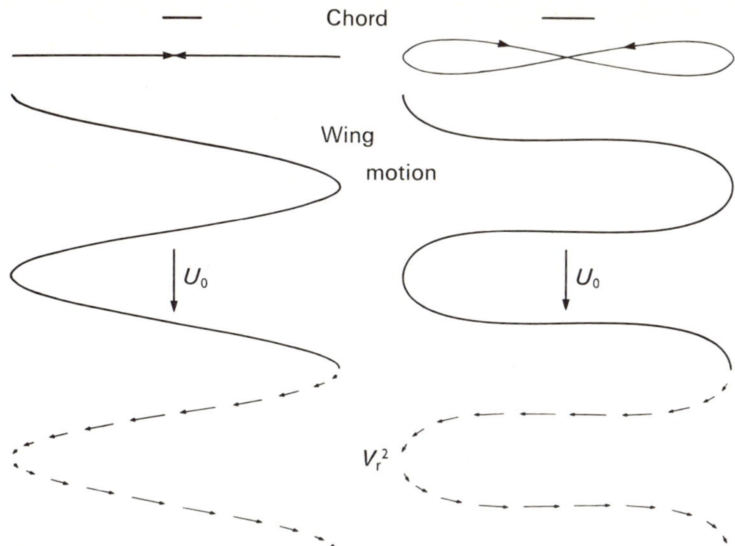

Fig. 4. Calculated wing motions at $0.7R$ for *Amazilia*. On the left is a horizontal sinusoidal oscillation; on the right is a figure-of-eight motion. The solid lines in the middle of the figure show the wing motions relative to the air, obtained by subtracting the induced velocity, U_0, from the wing velocities. The vectors at the bottom of the figure show the square of the relative winds (V_r).

velocity at the middle of a half-stroke is inclined at some angle β with respect to the horizontal. Weis-Fogh (1973) included this tilt of the stroke plane when adding his induced velocity correction and found that the lift coefficient was largely unaffected when the effects of β and ψ were included. This figure-of-eight is present in most hovering animals investigated so far, but has largely been neglected since values of β are only about $10°$. This is approximately the same as the value of ψ, however, and it has a striking effect on the flow pattern.

The top of Fig. 4 shows two wing motions to be considered. Using data on the hummingbird *Amazilia* given by Weis-Fogh (1972), I have calculated a horizontal wing velocity for a radius $0.7R$, on the left. On the right a figure-of-eight is generated by adding a vertical velocity $90°$ out of phase with the horizontal velocity and at twice the frequency. The amplitude of the vertical oscillation is such that the wing velocity at the middle of the half-stroke is inclined at $11°$ to the horizontal. The middle of the figure shows the motions when the induced velocity calculated from the actuator disk theory is subtracted from the wing velocity, a motion described by Hertel (1966) in an analysis of swimming as a *meander*. The square of the relative wind, which is used in the blade element theory, is depicted in the bottom of the figure. The vertical oscillation of the wing acts to bring the relative wind to the horizontal and, for $C_L/C_D = 6$ (Weis-Fogh, 1972), increases the lift by about 5 %. Using the vortex theory, it is shown that this also reduces rotation of the wake and improves the aerodynamic efficiency (Ellington, in preparation).

The analyses by Weis-Fogh (1972, 1973) demonstrate that normal hovering can be understood in light of steady-state aerodynamics. The total aerodynamic power expenditure can be calculated if experimental data on lift and drag coefficients are known. Some assumptions are made in the theory, however, and these have yet to be investigated.

1. R. A. Norberg (1974) points out that although steady-state aerodynamics are a sufficient explanation of hovering flight, it may not be the necessary one. Until adequate experimental work is done this question remains open.

2. Knowledge of the induced velocity field is necessary in calculating the magnitude and direction of the relative wind. The actuator disk theory is used to calculate the induced velocity and this velocity is consequently assumed to be constant over the wing disk.

3. The force coefficients are taken as constants along the radius and in time. The theory will accept variation of them, but as with the induced velocity field there is little knowledge of plausible variations.

4. No aerodynamic interaction of the wings is included, so that the approach cannot be used for a 'clap and fling' form of hovering.

A simplified vortex theory

The vortex theory of airscrews, developed over many years by Betz and by Goldstein, provides the most complete physical and analytical picture of their operation. Using the concept of circulation, it relates the wake velocity to the lift coefficient along the wing, thus combining the two previous theories in a complementary fashion.

To begin with, the actuator disk theory is re-considered. In the wake, an area of flow with velocity U_s is surrounded by still air. The perimeter of the wake may be replaced by a cylindrical sheet representing a surface of discontinuity in the flow field, known as a vortex sheet. The inner surface of the sheet must move with the wake at velocity U_s, while the outer surface is at rest with the still air. The average velocity at which the sheet moves downward is then $\frac{1}{2}U_s$ (Batchelor, 1967).

The sheet does not extend vertically above the wing disk because the flow is continuous before it is acted upon by the wings; vorticity can only be generated by the interaction of the fluid with a solid surface. Therefore,

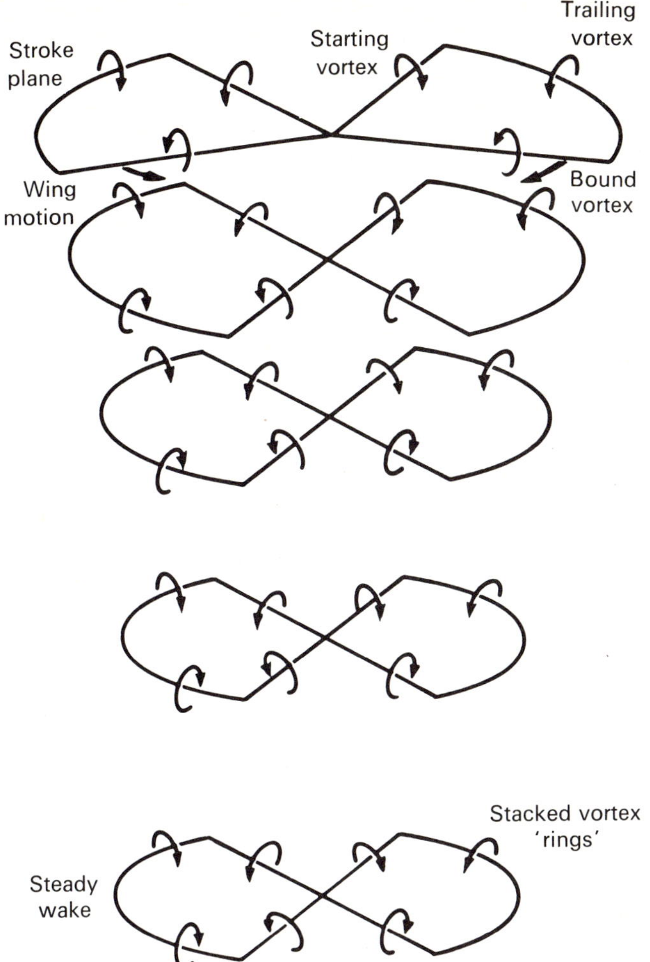

Fig. 5. Representation of the actuator disk flow as a series of stacked vortex 'rings'. At the wing disk (*top*), generation of the rings by starting, trailing and bound vortices is pictured.

the action of the wings during the time of a half-stroke, $1/2n$, must create a length of the vortex sheet given by the product of its average velocity and time, or $U_s/4n$. The strength of a vortex sheet per unit length, Γ', is called its *circulation* and is determined solely by the velocity difference across it, so that $\Gamma' = U_s$ (Batchelor, 1967). The total circulation generated by a half-stroke is the product of Γ' and the length of the sheet,

$$\Gamma_s = \Gamma' \, U_s/4n = U_s^2/4n. \qquad (17)$$

The wake boundary can now be divided into a series of elements which correspond to half-strokes, each with a circulation Γ_s evenly distributed over a length $U_s/4n$. If we consider the circulation to be concentrated at the centre of an element, this permits a view of the wake as a series of stacked vortex rings, albeit distorted ones (Fig. 5). During a half-stroke, the wings must generate a vortex ring of area ϕR^2 and circulation Γ_s. The impulse necessary to generate a ring is $\rho\phi R^2\Gamma_s$ (Batchelor, 1967), and the force needed equals the impulse divided by the period. Therefore, the force required to generate the vortex ring is

$$F = 2n\rho\phi R^2\Gamma_s = \tfrac{1}{2}\rho\phi R^2\, U_s^2, \tag{18}$$

which is the same as equation (5) for the thrust of the actuator disk theory. This type of approach has also been developed recently for application to the forward flight of birds (Rayner, personal communication).

The lift on a wing element can also be described in terms of a circulation. The wing experiences lift because the pressure above it is less than that below, and Bernouilli's equation relates this to a greater air velocity above than below. As the wing moves through the air this velocity difference can be represented by a vortex 'bound' to the wing; the circulation around the wing being a measure of the velocity difference. The lift force is given by aerodynamic theory as

$$dL = \rho\, V\Gamma\, dr, \tag{19}$$

where V is the wing velocity and Γ is the circulation around the wing. Equation (19) may be compared to equation (11) of the blade element theory to yield

$$C_L = 2\Gamma/cV. \tag{20}$$

Finally we note that the total circulation of a fluid initially at rest must always be zero, and that vortex lines must follow a closed path. Thus a wing acquires a bound vortex of strength Γ by generating a 'starting' vortex of strength $-\Gamma$ which remains free in the air. At the ends of the wing the bound vortex is continued by trailing tip vortices back to the starting vortex. The top plane in Fig. 5 shows how the vortex ring of each half-stroke is generated. When the wing stops at the end of a half-stroke the bound vortex is shed into the fluid as a 'stopping' vortex, and the ring is free to move within the wake. Whenever the circulation around the wing changes it must shed an opposite circulation into the wake. Likewise, if the circulation varies along the wing, trailing vortices must be shed whose circulation equals the change in circulation along the wing. In practice, the vorticity generated by the wing will be in a continuous sheet with a variable vorticity distribution rather than a series of discrete vortex lines.

A more complete analysis (Ellington, in preparation) shows that, for a variable distribution of the wake velocity $U_s(r, \gamma)$, equation (17) continues to hold;

$$\Gamma_s(r, \gamma) = U_s^2(r, \gamma)/4n. \tag{21}$$

All that remains is to assume that for a variable induced velocity the area of the wake still contracts evenly by one-half, so that the circulation at radius r in the wake is equal to the circulation at $r\sqrt{2}$ in the stroke plane,

$$\Gamma(r\sqrt{2}, \gamma) = \Gamma_s(r, \gamma) = U_s^2(r, \gamma)/4n. \tag{22}$$

The circulation Γ in the stroke plane is related to the lift coefficient by equation (20).

Using equations (20) or (22) we can now calculate the value of U_s given either a circulation or lift coefficient at the wing. Integrating the momentum equations (3) and (4) over the wake gives the thrust and induced power:

$$T = 2\rho \int_{-\phi/2}^{\phi/2} \int_0^{R/\sqrt{2}} U_s^2(r, \gamma)\, r\, dr\, d\gamma, \tag{23}$$

$$P_i = \rho \int_{-\phi/2}^{\phi/2} \int_0^{R/\sqrt{2}} U_s^3(r, \gamma)\, r\, dr\, d\gamma. \tag{24}$$

In the actuator disk theory U_s is constant, and equation (22) reveals that the circulation around the wing must, therefore, be constant with r and γ. The thrust and induced power calculated from equations (23) and (24) are the same as from equations (5) and (6) from the actuator disk theory. Lighthill (1977) notes that P_i is a minimum for constant U_s, so this value will be called P_i'. Using equation (20) and the normal hovering assumptions for the wing velocity, the lift coefficient must vary as

$$C_L(r, t) \propto 1/r \cos(2\pi nt). \tag{25}$$

Towards the ends of a half-stroke and the base of the wings the necessary C_L becomes far too large for steady-state aerodynamics.

The blade element assumption of constant C_L is investigated next. The circulation around the wing is given by equation (20) as

$$\Gamma(r, t) = \tfrac{1}{2}\pi\phi n C_L rc(r) \cos(2\pi nt). \tag{26}$$

Two cases are considered; a rectangular wing (constant chord), and a semi-elliptical wing. $U_s(r, \gamma)$ is calculated from equation (22) and the thrust is set equal to mg. For the constant chord wing the lift coefficient evaluated by equation (23) is

$$C_L = 0.608 \; mg/\rho cn^2 R^3\phi^2. \tag{27}$$

This is the same equation as given by Weis-Fogh (1973) for a rectangular wing if his shape factor, σ, is given its correct value of $^2/_3$. For the semi-elliptical wing the result is

$$C_L = 1.03 \; mg/\rho c n^2 R^3 \phi^2, \tag{28}$$

which again agrees with Weis-Fogh.

The induced power for the two cases is remarkably similar to P_i'. For the rectangular wing, P_i/P_i' is 1.08, and for the semi-elliptical wing P_i/P_i' is 1.07. The optimum condition of a constant induced velocity is shown not to be a very sensitive one. The shape of the wing, within reasonable bounds, is also not very influential on the induced power.

For the actuator disk theory the induced velocity is constant over the disk; for the blade element theory $(C_L = \text{constant})$ it varies with $\sqrt{(r\,c(r))}$. The radial distribution of induced velocity at the middle of a half-stroke is shown in Fig. 6 for three cases. Plate 2 is a steroscopic pair of photographs in which the radial variation of the wake velocity can be seen over most of the wing for *T. paludosa*. The agreement with the wake velocity profile of the semi-elliptical wing is satisfactory within the limitations of present theory and experiments. The increased wake rotation seen at small values of r represents an energy loss which is considered in the more complete analysis.

If the lift coefficient is a constant, the angular variation of the wake velocity is the same for all wing shapes and is proportional to $\sqrt{\cos(2\pi nt)}$, as shown in Fig. 7.

This simplified vortex theory has proved useful in two ways. By relating the actuator disk and blade element theories we find that they are not consistent; the circulation around the wing must be constant in the actuator disk theory and varies with radius and time in the blade element theory. The induced power for a semi-elliptical wing with constant lift coefficient is only slightly higher than the constant induced velocity assumption, and in general the induced power is fairly insensitive to wing shape and variations in C_L. In the more complete analysis the calculated induced velocity allows a local value of the angle ψ to be defined, and the twist of the wing along its length during a half-stroke is predicted. The limitations of the theory in its present form are as follows:

1. The equations for the circulation and lift coefficient are still those of steady-state aerodynamics.

2. No aerodynamic interaction of the wings is included. In steady-state conditions the interaction can be approximated if the circulation around the wings is given and the wings remain separated by more than half their chord. This promises a value of the theory in giving a first approximation to interactions, at least.

3. The induced velocity is still taken as vertical, but the results are similar to the more detailed treatment.

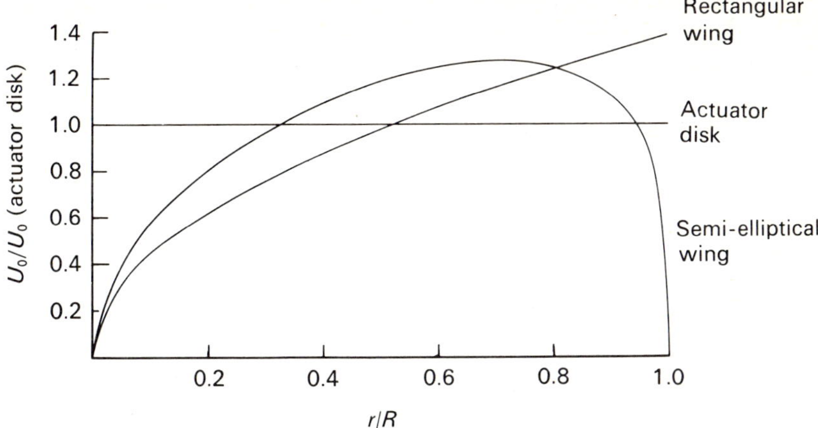

Fig. 6. Radial distribution of induced velocity U_0 at the middle of a half-stroke is calculated by the vortex theory for three cases. The induced velocities are expressed as a fraction of the constant actuator disk value. r = radius; R = wing length.

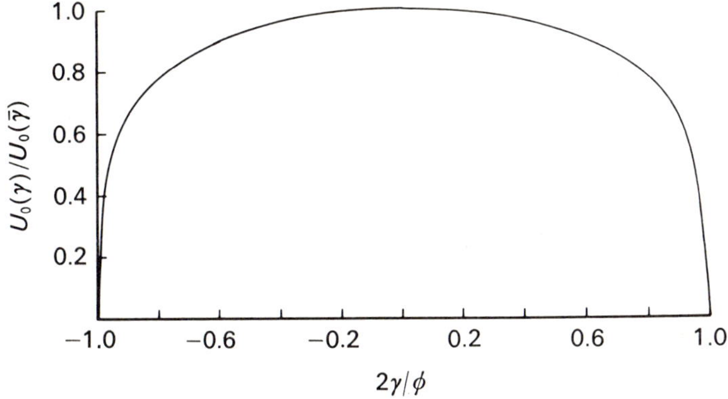

Fig. 7. Induced velocity as a function of positional angle if C_L is constant. Values expressed as a fraction of the induced velocity at $\bar{\gamma}$. For symbols, see text.

4. The expressions for drag on the wing elements, and the associated power requirement, have been omitted in this presentation, since they are the same as the blade element theory.

5. Variation of the lift coefficients can be accommodated by the theory, but again we have little idea what to expect. It may be mentioned that airscrew theory assumes a constant lift coefficient along the blade to be the optimum working condition (Nikolsky, 1951), especially if the lift coefficient corresponds to the maximum lift to drag ratio. This latter consid-

eration would certainly be true, but a hovering animal may be operating near the maximum C_L rather than the maximum lift to drag ratio.

6. The wake velocities are taken as constant over the vertical section $U_s/4n$. If the wake is similar to a series of stacked vortex rings some radial component in the wake velocity must be expected near the wing tips and ends of the half-strokes. This is seen in Plate 1 on the wake boundary to the right, and is equivalent to the tip losses of airscrews. The effects appear small, but they should increase with increased disk loadings.

Concluding remarks

To the biologist, flight is the characteristic mode of locomotion for over three-quarters of the known animal species. The physiological, structural and behavioural implications of this fact cannot be ignored by those who wish to understand how animals work. To the biological readers who have persevered through this aerodynamic analysis in the hope of increasing their understanding, I am grateful. The literature on animal flight is often difficult to understand, even for the specialist in the field. Aerodynamics is not an exact science, as many believe, but full of approximations, simplifications and empirical coefficients. The difficulty lies in the detailed application of theories which are basically simple physical arguments.

The physical constructs and fundamental equations of the three theories have been dealt with in detail. Once they are understood the remainder is usually tedious calculation. The concepts of the vortex theory may be new to many readers, but simplification of the wake into stacked vortex rings may prove useful. The smoke ring, of which a vortex ring is a simplification, is well-known to move perpendicular to its plane for a surprisingly long time. The force necessary to generate it is provided by contracting the mouth, and its circulation is evident by the rotation of the smoke. If smoke rings are blown at a high enough frequency, we have an analogue to Fig. 5. As the frequency increases to the limit, a smooth jet similar to the actuator disk picture is obtainable (at least in theory).

As discussed in the blade element section, satisfactory results are calculated for normal hovering with the assumption of steady-state aerodynamics. Unsteady effects are certainly present to some degree. They are always ignored, if possible, because they are extremely difficult to treat theoretically. To evaluate these effects and the existing theoretical treatments we must turn to experiments on the basic aerodynamic problems.

It is a pleasure to thank Dr K. E. Machin for his encouragement and discussions throughout this study and Mr G. G. Runnalls, whose photographic expertise was invaluable. I also thank Mr D. J. Tyler for building our stereoscopic attachment.

References

Batchelor, G. K. (1967). *An Introduction to Fluid Dynamics*. London: Cambridge University Press.

Bennett, L. (1966). Insect aerodynamics: vertical sustaining force in near-hovering flight. *Science, Washington*, **152**, 1263–6.

Bennett, L. (1974). Insect aerodynamics near hovering. In *Swimming and Flying in Nature*, ed. Y. T. Wu, C. J. Brokaw & C. Brennen, vol. 2, pp. 815–28. New York: Plenum Press.

Chance, M. A. C. (1974). Air flow and the flight of a noctuid moth. In *Swimming and Flying in Nature*, ed. Y. T. Wu, C. J. Brokaw & C. Brennen, vol. 2, pp. 829–43. New York: Plenum Press.

Goldstein, S. (ed.) (1965). *Modern Developments in Fluid Dynamics*, vol. 1. New York: Dover.

Hertel, H. (1966). *Structure, Form and Movement*. New York: Reinhold.

Hocking, B. (1953). The intrinsic range and speed of flight of insects. *Transactions of the Royal Entomological Society, London*, **104**, 223–345.

Hoff, W. (1919). Der Flug der Insekten und der Vögel. *Naturwissenschaften*, **7**, 159–62.

Jensen, M. (1956). Biology and physics of locust flight. III. The aerodynamics of locust flight. *Philosophical Transactions of the Royal Society*, B, **239**, 511–52.

Lighthill, M. J. (1977). Introduction to the scaling of aerial locomotion. In *Scale Effects in Animal Locomotion*, ed. T. J. Pedley. London: Academic Press.

Nikolsky, A. A. (1951). *Helicopter Analysis*. New York: Wiley.

Norberg, R. A. (1974). Hovering flight of the dragonfly *Aeschna juncea* L., kinematics and aerodynamics. In *Swimming and Flying in Nature*, ed. Y. T. Wu, C. J. Brokaw & C. Brennen, vol. 2, pp. 763–81. New York: Plenum Press.

Norberg, U. M. (1974). Hovering flight in the pied flycatcher (*Ficedula hypoleuca*). In *Swimming and Flying in Nature*, ed. T. Y. Wu, C. J. Brokaw & C. Brennen, vol. 2, pp. 869–81. New York: Plenum Press.

Osborne, M. F. M. (1951). Aerodynamics of flapping flight with application to insects. *Journal of experimental Biology*, **28**, 221–45.

Pennycuick, C. J. (1972). *Animal Flight*, Studies in Biology, No. 33. London: E. Arnold.

Weis-Fogh, T. (1972). Energetics of hovering flight in humming-birds and Drosophila. *Journal of Experimental Biology*, **56**, 79–104.

Weis-Fogh, T. (1973). Quick estimates of flight fitness in hovering animals, including novel mechanisms for lift production. *Journal of Experimental Biology*, **59**, 169–230.

Weis-Fogh, T. & Jensen, M. (1956). Biology and physics of locust flight. I. Basic principles in insect flight. A critical review. *Philosophical Transactions of the Royal Society, London*, B, **239**, 415–58.

Index

Where an author is mentioned in the text of his own chapter, no page entry is included